Towards Sustainable
Road Transport

Towards Sustainable Road Transport

Ronald M. Dell

Patrick T. Moseley

David A. J. Rand

AMSTERDAM • BOSTON • HEIDELBERG • LONDON
NEW YORK • OXFORD • PARIS • SAN DIEGO
SAN FRANCISCO • SINGAPORE • SYDNEY • TOKYO

Academic Press is an imprint of Elsevier

Academic Press is an imprint of Elsevier
The Boulevard, Langford Lane, Kidlington, Oxford OX5 1GB, UK
225 Wyman Street, Waltham, MA 02451, USA

First edition 2014

British Library Cataloguing in Publication Data
A catalogue record for this book is available from the British Library

Library of Congress Cataloging-in-Publication Data
A catalog record for this book is availabe from the Library of Congress

ISBN: 978-0-12-404616-0

For information on all Academic Press publications
visit our web site at books.elsevier.com

Printed and bound in United States of America

14 15 16 17 18   10 9 8 7 6 5 4 3 2 1

Working together
to grow libraries in
developing countries

www.elsevier.com • www.bookaid.org

The Open Road

They were strolling along the high road easily….. when far behind them they heard a faint warning hum, like the drone of a distant bee. Glancing back, they saw a small cloud of dust, with a dark centre of energy, advancing on them at incredible speed, while from out of the dust, a faint 'Poop-poop!' wailed like an uneasy animal in pain. Hardly regarding it, they turned to resume their conversation, when in an instant (as it seemed) the peaceful scene was changed, and with a blast of wind and a whirl of sound that made them jump for the nearest ditch, it was on them!........

……..'Glorious, stirring sight!' murmured Toad…..'The poetry of motion! The *real* way to travel! The *only* way to travel! Here to-day – in next week to-morrow! Villages skipped, towns and cities jumped – always somebody else's horizon! O bliss! O poop-poop! O my! O my!'…..

….'O what a flowery track lies spread before me, henceforth! What dust-clouds shall spring up behind me as I speed on my reckless way! What carts I shall fling carelessly into the ditch in the wake of my magnificent onset!'….

….The following evening the Mole, who had risen late and taken things very easy all day, was sitting on the bank fishing, when the Rat, who had been looking up his friends and gossiping, came strolling along to find him. 'Heard the news?' he said. 'There's nothing else being talked about, all along the river bank. Toad went up to Town by an early train this morning. And he has ordered a large and very expensive motor-car.'

Kenneth Grahame,
The Wind in the Willows, 1908

Contents

For historic Milestones in transport visit the companion website:
http://booksite.elsevier.com/9780124046160

Preface

'Sustainable development' was defined in 1987 by the World Commission on Environment and Development (the Brundtland Commission) as being development that:

'meets the needs of current generations without compromising the ability of future generations to meet their own needs'.

Circumstances have changed markedly since this straightforward definition was proposed. For instance:

- global population has expanded greatly, thereby exacerbating demands on the Earth's resources for food, water and minerals
- the natural environment has been seriously damaged by mankind's activities
- the threat posed by climate change (commonly referred to as 'global warming') has been recognized.

Consequently, the scope of 'sustainable development' has been amplified and extended.

Whereas significant progress has indeed been made in the re-use of metals and various other materials, world reserves of oil and natural gas are being exploited at an ever-increasing rate. In addition, there remains serious concern over the harmful effects on global ecology caused by major oil spills, the dispersion of fragmented plastics in the oceans, and deforestation. The likely consequences of this widespread degradation are now widely recognized but, to date, little progress has been made towards a successful remediation. There is conflict between the long-term needs of the environment and the short-term financial interests of land-owners, farmers, and the manufacturing industry. The identification of an acceptable route to a sustainable future has proved exceedingly difficult for politicians, particularly in an era of restricted economic growth, fiscal deficits, and burgeoning unemployment.

The present concept of 'sustainability' now encompasses the following societal aspirations:

- universal access to clean water and sanitation
- freedom from hunger and a nutritious diet for all
- eradication or control of diseases
- reduction of child mortality and improvements in maternal health
- elimination of poverty and a more equitable distribution of wealth
- a basic standard of education for all, with the opportunity to progress to more advanced levels according to ability
- racial and gender equality
- democracy and the freedom to participate in decision-making
- human rights and the universal rule of law, order and justice
- population control, without which most of the foregoing aims may prove unachievable.

At the beginning of the 21st century, many of the items in the above formidable list were adopted by all 193 member states of the United Nations in a set of eight Millennium Development Goals, which are to be accomplished by 2015. The statement of intent is clear but the target timescale represents a serious challenge.

This book focuses attention on road transport — a key aspect of human activity among the many that require attention if sustainable development on a global scale is to be achieved. The work examines the prospects for its evolution from a system that consumes irreplaceable resources and degrades the environment, towards one with a *modus operandi* that will be both supportable and benign.

The accommodation of conventional road vehicles powered by internal combustion engines (ICEs) has reached a 'tipping point'. As a result of the predicted expansion of the human population from its current level of seven billion to perhaps nine billion by the second half of the present century, coupled with the aspirations of people in developing countries to match the levels of personal transport that are enjoyed by those in developed nations, a huge growth in the numbers of such vehicles is expected. If automotive power continues to depend on the ICE, the surge in road transport will obviously lead to increased congestion on the roads with consequent intensified air pollution, frustration for motorists, and greater fuel consumption. The attendant release of carbon dioxide will also be exacerbated. There is no doubt that the concentration of this greenhouse gas in the atmosphere has risen steadily since the onset of the Industrial Revolution in the late 18th century. The precise impact of this accumulation on global climate compared with the effects of natural phenomena, such as fluctuations in insolation levels, cloud formations and emissions from volcanoes, is still the subject of much debate. Nevertheless, moderation of climate change is perhaps the most important long-term challenge that confronts the world today and, as long as uncertainty over its origins prevails, it is prudent to work towards limiting the emissions of carbon dioxide. The possible consequences of not doing so are simply too serious to ignore.

A further consideration relating to the long-term prospects for road transport is that reserves of readily-recoverable petroleum are finite and are being depleted rapidly. For instance, already more than half of the economically-viable oil in the North Sea has been extracted and consumed —1999 was the year of peak production. The search for new oil fields is now moving into deeper waters offshore where prospecting and production are more difficult and this will inevitably give rise to higher prices at the pump.

Many other petroleum-like fuels (oil sands and heavy oils, for example) can be refined to produce petrol and diesel, and other fossil fuels (coals, natural gas) can also be transformed from their native state to liquid fuels. Unfortunately, all these alternative energy sources invoke a heavy penalty in terms of environmental degradation and most require complex processing that results in increased cost. Bio-fuels, too, involve the expenditure of considerable amounts of energy in the harvesting and processing of crops over vast tracts of prime agricultural land, so that it is difficult to justify the claim that such fuels are 'carbon neutral' in terms of greenhouse gas emissions.

The petroleum and automotive industries have gone to some lengths to reduce the impact of the ICE on the environment. Over the past few decades, oil companies have made major advances in refining technology and now curtail the sulfur content of motor fuels to extremely low levels as in 'ultra-low sulfur diesel', for example. Meanwhile, the automotive components industry has perfected catalytic converters for use with petrol engines to meet ever-tighter environmental legislation that has been introduced to control the release of hydrocarbons, carbon monoxide, and nitrogen oxides. Ceramic filters have been developed to trap the fine particles of carbon in diesel exhaust emissions that are a potential cause of lung disease.

Automotive manufacturers face the conflicting demands of customers for vehicles with ever-improved performance, safety and comfort — but without any appreciable increase in cost. This dilemma has been resolved to a large degree through advances in design and the economies-of-scale that result from mass production. Although the outcomes of such developments are welcome to motorists, many are peripheral to the issue of sustainability. Rather, it is the refinement in engine and power-train technology that is most relevant. The combined use of direct fuel-injection and turbo-chargers to allow the use of smaller engines with no decrease in power appears to have been particularly effective in reducing both emissions and fuel consumption. The 'stop–start' principle, which eliminates fuel usage whenever the vehicle is stationary, is particularly beneficial in instances of heavy traffic.

It has long been recognized that to provide rapid acceleration and good hill-climbing performance, the power capability of traditional engines is grossly over-designed in comparison with that required when a vehicle is cruising on level ground. Thus, considerable fuel economy may be achieved by fitting a smaller engine that is adequate for cruising, together with an auxiliary power source. This is the basis of the 'hybrid vehicle'. There are many possible versions of such a vehicle, but most in use today depend on electric motors to provide the necessary boost. The power-trains may have either series or parallel configurations; variants of these with progressively increasing levels of 'power-assist' are also available. Hybrid technology introduces complex issues of power flow and control that are managed by on-board computers, advanced types of power electronics, and suitable storage batteries. Regenerative braking, whereby some of the energy dissipated in braking is recovered in the form of electricity and fed back to the battery, is also helping to improve fuel efficiency. The progressive electrification of the power-train is on-going and further advances are inevitable.

In principle, the ultimate stage in the process of replacing the ICE power source in vehicles by an electric equivalent is the move to a battery electric vehicle (BEV). Given that the driving range is inferior and battery recharging takes much longer than refuelling a petrol tank, the uptake of BEVs is likely to be restricted to urban/suburban use, e.g. for public and commercial services (particularly those operating over fixed routes) and for commuting to work. The 'plug-in' hybrid electric vehicle (PHEV) addresses the problem of range limitation — the design draws most of its energy from mains electricity but has an ICE for use during long journeys. Whereas electrically-propelled vehicles are to be welcomed on the grounds of reduced pollution and conservation of petroleum supplies, they only contribute to true sustainability if the mains electricity on which they run stems from non-fossil sources, i.e. from nuclear power or from renewable forms of energy.

Fuel cell vehicles (FCVs), which operate on hydrogen, represent an alternative approach to 'carbon-free' transport. Notably, FCVs have effectively no range limitation, emit no local pollution, and emit carbon emissions only to the extent that fossil fuels are employed in obtaining the hydrogen; conventionally, hydrogen is produced predominantly by the steam-reforming of natural gas, which does involve the release of carbon dioxide. Unfortunately, the large-scale introduction of FCVs poses major problems. Aside from challenges associated with the on-board storage of sufficient hydrogen and from the high costs incurred in fabricating fuel cells — both of which might reasonably be expected to be resolved in due course — there is the over-riding challenge of providing a national/international infrastructure for the manufacture, transmission, distribution and dispensing of hydrogen in bulk quantities. This is primarily a question of economics, although there are certainly some technical aspects to be addressed. Notwithstanding these obstacles, several automotive companies see FCVs as a prospective step towards sustainability and are planning for their introduction within the foreseeable future.

Until relatively recently, road transport was a straightforward matter. All vehicles were equipped with comparatively simple ICEs, fuel was widely available at moderate prices, and there were few restrictions on pollutant emissions. The situation has now changed: hybrid vehicles are more complex and more costly to manufacture than their predecessors and designs are still evolving; the price of fuel has escalated; strict pollutant emission limits have been imposed; and, finally, the spectre of greenhouse gases must be confronted. Major shifts are occurring in the standard-of-living of previously undeveloped nations and millions more people around the world will expect to own their own transport. Globalization has led to ever-increasing volumes of freight transport that necessitate the construction of more durable roads. At the same time, however, restricted economic growth is imposing serious constraints in many developed countries.

Looking ahead a few decades, there is likely to be: a wider range of vehicle drive-trains; much innovation based on electronics and computer control such as 'drive-by-wire' to replace mechanical operation; and, ultimately, the driverless car. Such advances would be taking place within a rapidly evolving social framework and a changing sense of political and environmental priorities. It should be clearly understood, however, that none of the improved technologies will be adopted in sufficiently large numbers to address concerns over global climate change unless they are affordable. The management of world economics is thus expected to play a pivotal role.

Predicting the future with confidence is very difficult, as witnessed by all the unanticipated developments of the past century — most notably, the explosive growth in electronics and communications. Nevertheless, it is important to strive for a clear, forward-looking view of those factors that may threaten the environment and the security of the planet, as well as the quality of life of its inhabitants. In the area of road transport, as with many other areas, it is important to review all of the present technical developments that might contribute to a sustainable future. This is the prime objective of the present book.

Ronald M. Dell
Patrick T. Moseley
David A. J. Rand

Please also visit the companion website for additional material, including historic Milestones at: http://booksite.elsevier.com/9780124046160

Biographical Notes

The authors are senior research chemists who have spent their entire professional careers working in the energy field.

 **Ronald Dell PhD DSc CChem FRSC** graduated from the University of Bristol. He lived for several years in the USA, where he worked as a research chemist, first in academia and then in the petroleum industry. On returning to Britain, Ron joined the UK Atomic Energy Research Establishment at Harwell in 1959. During a tenure of 35 years, he investigated the fundamental chemistry of materials used in nuclear power and managed projects in the field of applied electrochemistry, especially electrochemical power sources. Since retiring in the mid-1990s, he has interested himself in the developing world energy scene and has co-authored with David Rand several books on batteries, on clean energy, and on hydrogen energy.

 **Patrick Moseley PhD DSc** graduated from the University of Durham, England. He worked for 23 years at the UK Atomic Energy Research Establishment at Harwell, where he brought a background of crystal structure and materials chemistry to the study of lead–acid and other types of battery, thus supplementing the traditional electrochemical emphasis of the subject, and to the study of sensor materials. From 1995, Pat was Manager of Electrochemistry at the International Lead Zinc Research Organization in North Carolina, USA, and Program Manager of the Advanced Lead–Acid Battery Consortium. In 2005, he became President of the Consortium. He is also a director of Atmospheric Sensors Ltd. Pat has been an Editor of the *Journal of Power Sources* since 1989 and, together with David Rand, was a Co-editor of the *Encyclopaedia of Electrochemical Power Sources* published by Elsevier in 2009. In 2008, he was awarded the Gaston Planté Medal by the Bulgarian Academy of Sciences.

 **David Rand AM PhD ScD FTSE** was educated as an electrochemist at the University of Cambridge. Shortly after graduating in 1969, he emigrated to Australia and has spent his research career working at the government's CSIRO laboratories in Melbourne. In the late 1970s, David established the CSIRO Battery Research Group and remained its leader until 2003. As a Chief Research Scientist, he was CSIRO's scientific advisor on hydrogen and renewable energy until his retirement in 2008. He has served as the Vice-President of the Australian Association for Hydrogen Energy. David has been the Asia–Pacific Editor of the *Journal of Power Sources* since 1983, and the Chief Energy Scientist of the World Solar Challenge since 1987. He was elected a Fellow of the Australian Academy of Technological Sciences and Engineering in 1998, and received the Order of Australia in 2013 for service to science and technological development in the field of energy storage.

Acknowledgements

The authors have consulted many sources of information including books, reports, reviews, and web sites. As the present volume is a broad-brush overview, it is not seen fit to quote detailed references. Nevertheless, our indebtedness to the work of others must be acknowledged. Similarly, we are grateful to the holders of copyright who have kindly consented to the use of their illustrations.

The assistance of Dr Jacquie Berry, Professor Neville Jackson, Dame Professor Julia King and Sir Robert Watson is gratefully acknowledged.

The authors would like to express their thanks also to their wives: Sylvia Dell, Heather Moseley and Gwen Rand, for their patient support and encouragement during the writing of this book.

Acronyms, Initialisms, Symbols and Units used in this book

Acronyms and Initialisms

4WD	four-wheel drive
ABS	anti-lock braking system
a.c. (or AC)	alternating current
AFC	alkaline fuel cell
AGM	absorptive glass-mat (battery separator)
ALABC	Advanced Lead–Acid Battery Consortium
AMT	Automated Manual Transmission
APU	auxiliary power unit
AWD	all-wheel drive
BEV	battery electric vehicle
B-ISG	belt-driven integrated starter–generator
BMS	battery-management system
BRIC	Brazil, Russia, India, China (group of nations at similar stages of development)
BRT	bus rapid transit
CAFE	corporate average fuel economy
CAN	controller area network
CAPP	Canadian Association of Petroleum Producers
CB	conduction band
CC	climate change
CCGT	combined-cycle gas turbine
CCS	carbon capture and storage
CHIC	Clean Hydrogen in European Cities
CHP	combined heat and power
CI	compression-ignition (diesel engine or vehicle)
C-ISG	crankshaft-mounted integrated starter–generator
CNG	compressed natural gas
CR	compression ratio (of engine)
CTL	coal-to-liquids
CSIRO	Commonwealth Scientific and Industrial Research Organisation (Australia)
CUTE	Clean Urban Transport for Europe
CV	commercial vehicle
CVT	continuously variable transmission
d.c. (or DC)	direct current
DCA	dynamic charge-acceptance
DCT	dual-clutch transmission (*or* double-clutch transmission)
DME	dimethyl ether

DMFC	direct methanol fuel cell
DoD	depth-of-discharge
DPF	diesel particulate filter
DSSC	dye-sensitized solar cell
EBD	electronic brake force distribution
ECTOS	Ecological City Transport System
ECU	electronic (or engine) control unit
EFB	enhanced flooded battery
EMS	engine-management system
EPA	Environmental Protection Agency (USA)
E-REV	extended-range electric vehicle
ESC	electronic stability control
ETP	Energy Technology Perspectives (IEA)
EU	European Union
EV	electric vehicle
E2W	electric two-wheeled cycle
FC	fuel cell
FCEB	fuel cell electric bus
FCHV	fuel cell hybrid vehicle
FCV	fuel cell vehicle
FEHRL	Forum of European Highway Research Laboratories
FTA	Federal Transportation Agency (USA)
GDP	gross domestic product
GE	General Electric
GHG	greenhouse gas
GM	General Motors
GNI	gross national income
GPS	global positioning system
GRT	group rapid transit
GTL	gas-to-liquids
GVW	gross vehicle weight
HC	hydrocarbons
HCCI	homogeneous charge compression ignition
HEV	hybrid electric vehicle
HF	high frequency
HGV	heavy goods vehicle
HHV	higher heating value
HRPSoC	high-rate partial-state-of-charge
ICE	internal combustion engine
ICEV	internal combustion-engined vehicle
IEA	International Energy Agency
IGBT	insulated-gate bipolar transistors
IGCC	integrated gasification combined-cycle

IPCC	the Intergovernmental Panel on Climate Change
ISG	integrated starter–generator
LAB	lead–acid battery
LDV	light-duty vehicle (cars and vans/light trucks)
LED	light-emitting diode
LFP	lithium iron phosphate (material for positive electrode of lithium-ion battery)
LH$_2$	liquid hydrogen
LHV	lower heating value
LIB	lithium-ion battery
LNG	liquid natural gas
LPG	liquid petroleum gas
MCFC	molten carbonate fuel cell
MEA	membrane electrode assembly
MOSFET	metal-oxide-semiconductor field-effect transistor
MPV	multi-purpose vehicle
NCA	lithium nickel cobalt aluminium oxide (material for positive electrode of lithium-ion battery)
NEDC	New European Drive Cycle
Ni-MH	nickel-metal-hydride(battery)
NMC	lithium nickel manganese cobalt (material for positive electrode of lithium-ion battery)
NO$_x$	nitrogen oxides
NREL	National Renewable Energy Laboratory (USA)
NVH	noise, vibration and hardness
OECD	Organisation for Economic Co-operation and Development
OICA	Organisation Internationale des Constructeurs d'Automobiles
OPEC	Organization of the Petroleum Exporting Countries
PAFC	phosphoric acid fuel cell
PEMFC	proton exchange membrane fuel cell (*or* polymer electrolyte membrane fuel cell)
PHEV	plug-in hybrid electric vehicle
PM	particulate matter
PNGV	Partnership for a New Generation of Vehicles
PRT	personal rapid transit
PSA	pressure swing adsorption
PSoC	partial-state-of-charge
PST	power-sharing transmission
PSV	public service vehicle
PTFE	polytetrafluoroethylene
PV	photovoltaic
RAPS	remote-area power supplies
RDS	Radio Data System
RFC	regenerative fuel cell
RON	research octane number

SAE	Society of Automotive Engineers
SEI	solid | electrolyte interface
SI	spark ignition
SLI	starting–lighting–ignition
SoC	state-of-charge (of battery)
SoF	state of function
SOFC	solid oxide fuel cell
SPEFC	solid polymer electrolyte fuel cell (alternative term for PEMFC)
SSV	stop–start vehicle
SUV	sports utility vehicle
tdc	top dead centre
UHV	ultra-high-voltage
ULSD	ultra-low-sulfur diesel
UN	United Nations
USB	universal serial bus
VB	valence band
VRLA	valve-regulated lead–acid battery
WEO	World Energy Outlook (IEA)
WGS	water-gas shift (reaction)
WHO	World Health Organization
WSC	World Solar Challenge

Symbols and units

atm	1 standard atmosphere pressure (= 101.325 kPa)
A	ampere
Ah	ampere-hour
bar	unit of pressure (= 0.1 MPa)
bbl	barrel of oil (approx 159 L)
bhp	brake horsepower (= 745.7 W)
cc	cubic centimetre (unit of capacity equivalent to 1 mL)
cm	centimetre
cwt	hundredweight (= 112 lb = 51 kg)
C	capacitance of a capacitor or circuit (units: Farads)
C_1	one-hour rate for charging/discharging a battery
$C°$	degree Celsius
d	pre-decimal British penny coin (240 *d* = £1)
d	day
dm	decimetre

η	electrode overpotential (V)
η_+	overpotential at a positive electrode (V)
η_-	overpotential at a negative electrode (V)
e^-	electron
eV	electron volt
E	electrode potential (V)
E°	standard electrode potential
E_g	band-gap energy (eV)
ft	foot (linear measurement = 305 mm)
ft lbf	unit of torque (1 foot pound force = 1.356 Nm)
F	Faraday (1 Faraday = 96 458 coulombs per mole)
g	gram
gall (UK)	1 gallon (UK) = 4.546 L
gall (USA)	1 gallon (US) = 3.785 L
G	Gibbs free energy (J mol^{-1})
GJ	gigajoule
Gt	gigatonne (= 10^9 tonnes)
Gtoe	gigatonne oil equivalent
GW	gigawatt
GWe	gigawatt electrical output
GWh	gigawatt-hour
ΔG	change in Gibbs free energy of a chemical reaction (J mol^{-1})
ΔG°	standard change in free energy (J mol^{-1})
h	hour
$h\nu$	energy of a photon
h^+	electron hole (in conduction band of a semi-conductor)
hp	horsepower
ΔH	change in enthalpy of a chemical reaction (J mol^{-1})
in	inch (linear measurement = 25.4 mm)
I	current (A)
J	joule (unit of energy 1 J = 1 Ws)
k	kilo (10^3)
km	1 kilometre = 0.6214 miles
km h^{-1}	unit of speed (*or* kmph)
kPa	kilopascal
kW	kilowatt
kW$_e$	kilowatt electrical output
kWh	kilowatt-hour
K	kelvin (a measure of absolute temperature)
lb	pound weight (= 0.45 kg)
L	litre (*or* lit)

m	metre
mile	1 mile (statute) = 1.609 km
min	minute
mol	mole (mass of 6.023×10^{23} elementary units of a substance)
mpg	miles per gallon (UK or US gallon)
mph	miles per hour (unit of speed)
ms	millisecond
M	mega (10^6)
MJ	megajoule
MPa	megapascal (unit of pressure ~ 10 atm)
Mt	megatonne
Mtoe	million tonnes of oil equivalent
MW	megawatt
MW$_e$	megawatt electrical output
MWh	megawatt-hour
ν	frequency of electromagnetic radiation
n	number of electrons involved in a chemical reaction
nm	nanometre (10^{-9} m)
N	newton (unit of force 1 N = 1 kg m s^{-2})
N m	newton metre (unit of torque)
Nm^3	normal metre cubed (1 m^3 of gas at normal temperature and pressure)
ppm	parts per million
ppmv	parts per million by volume
psi	pounds per square inch (pressure) (1 psi = 6.895 kPa)
Pa	pascal (unit of pressure = 1 N m^{-2})
Q	theoretical capacity (of a cell, Ah)
rpm	revolutions per minute
rps	revolutions per second
s	second
S	entropy (J K^{-1} mol^{-1})
Δ*S*	change in entropy of a chemical reaction (J K^{-1} mol^{-1})
t	tonne (*or* te) (= 1000 kg)
toe	tonnes of oil equivalent
T	temperature
TM	trademark
TWh	terawatt-hour (10^{12} Wh)
vol.%	percentage by volume (in a gas mixture)
V	voltage
*V*º	standard cell voltage
Vr	reversible ('open circuit') voltage

wt.%	percentage by weight
W	watt (= 1 J s^{-1})
W$_e$	watt, electrical power
Wh	watt-hour
x	variable in stoichiometry

Fuel consumption

The fundamental unit, used in Europe and much of the world, is expressed as:
litres of fuel consumed per 100 km travelled ('lit/100 km').
In Britain, it is common to speak of 'miles per imperial gallon' (mpg).
To convert mpg to lit/100 km take the reciprocal and multiply by 282.5.
For example: 30 mpg = 1/30 gal/mile = 4.546/30 lit/mile = 4.546/30 × 0.6214 × 100 = 9.42 lit/100 km.
In the USA, it is common to speak of 'miles per US gallon'.
To convert to lit/100 km take the reciprocal and multiply by 235.2.

The Evolution of Unsustainable Road Transport

1.1 Bicycles and beyond

Until the early 19th century, transport had altered little for 2000 years. Then, quite suddenly, the whole situation changed with the discovery and development of the steam engine and the advent of railways. Not only were speeds far in excess of anything previously thought possible, but also long-distance travel was brought within the means of ordinary citizens. Nevertheless, the vast majority of people were still tied to areas within walking distance of their homes. The development of the bicycle alleviated this restriction – the machine could roughly treble the distance that workers could travel and therefore enabled them to seek jobs further afield. Bicycles, which were usually bought by instalments, also proved perfect for day trips to the countryside at weekends.

The earliest bicycles had no pedals or transmission and were driven by the rider's feet pushing along the ground. The first verifiable claim to the invention of such a machine is attributed to Baron Karl von Drais in Germany (1817); an advertisement for his so-called *draisienne* is shown in Figure 1.1(a). Dennis Johnson, a coach maker of Long Acre in London, introduced this mode of transport into England in 1818 where it became known as the *hobby horse* or *dandy horse*; see Figure 1.1(b); the contrivance reached America in 1821. A year later, a modified version to suit ladies was produced – probably by Johnson, who exhibited it at his riding school. The machine, which weighed about 66 lb (30 kg), had a wooden dropped frame that somewhat resembled the present-day design of the ladies' bicycle. The saddle was supported on an iron pillar fixed to the lower part of the frame. There is very little evidence that the ladies of the early 19th century took up the pastime of riding on two wheels, although Johnson's advertisements assured them that such activity could be enjoyed without loss of decorum!

It has been said that the Scotsman Kirkpatrick Macmillan produced, in 1839, a 'powered' hobby-horse propelled by pedals at the front that worked backwards and forwards to turn the rear wheel via connecting rods. The claim was made, however, by one of his relatives some 50 years after the alleged event and none of MacMillan's supposed bicycles have survived. A counter suggestion that another Scotsman, Gavin Dalzell, built a rear-drive machine in 1845 is equally without strong foundation. The true breakthrough in the quest for a 'mechanical horse' has therefore been attributed to two events that took place in France.

In 1862, a baby carriage maker in Nancy called Pierre Lallement mounted a pedal on the front wheel of his bicycle; see Figure 1.1(c). He did not, however, pursue mass production of this prototype. About the same time and independently of Lallement, Pierre Michaux together with his son Ernest developed a design that had pedals attached to the hub of a larger front wheel; see Figure 1.1(d). They called their machines *velocipèdes* and within four years were manufacturing 400 a year. By 1869, bicycle racing was established on the roads of France. Both in England and in the USA, the *velocipède* earned the name of *boneshaker* because of its rigid frame and iron-banded wheels that resulted in a 'bone-shaking' experience for riders. The burgeoning interest in the USA had been initiated by Pierre Lallement, who had left Paris in July 1865, settled in Connecticut and patented the velocipede in 1866; see Figure 1.1(e).

FIGURE 1.1

Evolution of the bicycle: (a) the *draisienne,* 1817; (b) Johnson's *hobby horse* specification, 1818, printed in 1857 by Eyre and Spottiswoode; (c) Ernest Lallement riding his invention; (d) Ernest Michaux on a velocipede, 1868; (e) Pierre Lallement's original patent for the first pedal-driven bicycle, US Patent No. 59,915, 20 November 1866; (f) Starley's *Ordinary* bicycle and origin of the *penny-farthing* nickname; (g) couple on quadricycle, 1886; (h) ladies *Rover Safety* bicycle, 1889; (i) John Boyd Dunlop (1840–1921)

(All images sourced from Wikipedia and available under the Creative Commons License)

The *boneshaker* configuration not only required riders to twist their bodies when steering but also, because of the inherently low gearing, yielded relatively low speeds for the pedalling invested. The latter problem was ameliorated by greatly increasing the size of the front drive wheel. In 1870, James Kemp Starley of Coventry in England designed the *Ordinary*, better known now by its derisive *penny-farthing* nickname; see Figure 1.1(f). By 1874, the bicycle incorporated Starley's enduring innovation of suspension wheels with wire spokes set at a tangent to the hub to resist the winding stresses of pedalling. It was a highly efficient machine that was capable of speeds of up to 20 mph (32 km h^{-1}), but there were the obvious difficulties of mounting and dismounting, as well as the likelihood of severe injuries emanating from the falls taken by riders on hitting an obstacle. The size of the large wheel was chosen to suit the inside leg measurement of the rider, rather like a pair of trousers, and varied from 1 to 1.5 m in diameter. As women, too, wanted to take to wheels, various types of tricycle and quadricycle became popular; see Figure 1.1(g). One of the added attractions of these models, for the young at least, was that they were for the most part 'made for two'. In 1877, Starley invented the differential gear for his *Coventry Salvo* tricycle, which was the first fully successful example of a chain-driven machine.

The use of the *Ordinary* only by the relatively long-legged and audacious triggered a search for less hazardous solutions to the gearing problem. The answer was to use a chain to connect the pedals with the drive wheel, an arrangement that was adopted in the first 'safety' bicycles, such as Harry J. Lawson's *Bicyclette* of 1879, and the highly successful *Kangaroo* of 1883 that had a front wheel of more reasonable size and chain-drives on either side of the fork.

In 1885, John K. Starley, nephew of James (*v.s.*), devised the configuration that effectively stabilized bicycle design – the *Rover Safety* bicycle; see Figure 1.1(h). This was a low-profile design with more or less equal-sized wheels and chain drive to the rear wheel. The first frames were curved, but the archetypal diamond-shaped frame was quickly established, as were sprung saddles. The geometry allowed the rider to adjust the height and position of both the seat and the handlebars. The machines, which were virtually modern bicycles, were immensely successful. They opened cycling up to a broad section of the public, since it really was possible for anyone to ride them.

The remaining key innovation of the 19th century was the pneumatic tyre, which was re-invented in 1888 by John Dunlop, a Scottish veterinary surgeon; see Figure 1.1(i). Though Robert William Thomson had patented the idea in 1845, the vulcanized rubber tyres produced by Dunlop were the first to enter the market; he formed the Dunlop Rubber Company in 1889. When such tyres were applied to bicycles in the early 1890s, they soon ousted the solid rubber versions that had been fitted to wheels from about 1868. Not only did they greatly increase the comfort of the bicycle, they also significantly increased its speed. Although there were later developments in gears, which made riding easier in hilly country, and in braking (whether the back-pedal type or the shoe type invented by Sir Harold Bowden), bicycle design was now fundamentally established.

In all parts of the world, as standards of living and individual wealth have risen, so also have the aspirations for the preferred mode of private personal transport. Communities that have been content with pedal power, as long as affordability has been the deciding factor, inevitably seek for something better as soon as their income allows. During the 20th century, the populations of developed countries 'graduated' from horse riding, horse-drawn carriages and bicycles to motorcars. This change marked a paradigm shift from a system that made virtually no use of hydrocarbon fuels, and thus produced only tiny amounts of carbon dioxide, to an alternative that began to consume the global resources of oil and to bring about the sharpest change in the composition of the planet's atmosphere in centuries. Although bicycles are still used in niche circumstances, such as in city centres and for leisure/exercise activities, private personal transport is now ruled by the motorcar in Western nations and the populations of other parts of the world have launched plans to hasten their vehicle fleets to the same level of development. In China, for example, the first step has been to advance from pedal-driven bicycles to e-bikes (see Section 5.3.3.3, Chapter 5), but now the further move to widespread car ownership is gathering pace. In the remainder of this chapter, the evolution and dominance of unsustainable forms of road transport (i.e. those that consume hydrocarbon fuels and produce undesirable emissions) are charted.

1.2 Steam takes to the road
1.2.1 Early pioneers: 1765–1840

As the bicycle had become a convenient and less-expensive alternative to the horse for the individual rider, it is scarcely surprising that the industrial revolution should spawn thoughts about the replacement of the horse as the motive power for carts and carriages also. The only viable alternative power source at the time was steam.

The first practical steam engine was developed in England in the early 18th century by Thomas Newcomen (1663–1729). This was a static 'atmospheric' engine, so called because it engaged the weight of the atmosphere to produce the working action. The machine was both massive in size and very slow in operation, with one piston stroke every minute or two. Between strokes, the cylinder that housed the piston had to be externally cooled, to condense the steam, and then reheated. The Newcomen engine was used predominantly to pump water out of mines. Many years later, James Watt (1736–1819), who was unimpressed by the slow speed of operation and poor efficiency of the 'atmospheric' engine, devised an improved version which had a separate condenser so that the cylinder could remain hot the whole time and thus the engine would be far less wasteful in its use of heat and fuel. Watt was therefore not the inventor of the steam engine but of the separate condenser, which was a spectacularly successful modification and was patented in January 1769. James Watt is, however, recognized as the 'father' of the steam engine, thanks to the many new features he introduced, such as the use of a flywheel and a 'governor' to control and maintain a constant speed. The start of the Industrial Revolution can be largely credited to his work and he has been honoured for posterity by naming the fundamental unit of power the 'watt'.

Early in the 19th century, steam engines were deployed for the propulsion of ships, railway locomotives and traction engines. Practical steam locomotives were introduced in 1814 when George Stephenson's *Blucher* became the first to be fitted with flanged wheels and to run on a track laid with smooth cast-iron rails. By the middle of the 19th century, railway tracks were being laid across Britain and many other countries so that travel by steam train became ever more popular and thereby gave rise to progressively lower fares.

The steam carriage was introduced by Richard Trevithick, a British mining engineer. In 1801, Trevithick completed his first full-sized 'road locomotive' in partnership with another engineer, Andrew Vivian, who financed the work; an illustration of the vehicle is given in Figure 1.2(a). The finished engine weighed more than 1.5 t and had four wheels on a square-frame chassis and a small passenger platform. It worked at a steam pressure of 414 kN m^{-2} (60 psi). A single cylinder was set vertically into the top of one end of the boiler, with the furnace and chimney at the other end. In effect, the vehicle was little more than a boiler and engine on wheels that belched steam from its

(a)

(b)

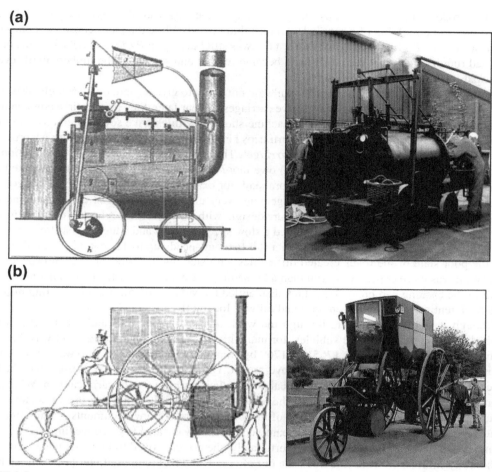

FIGURE 1.2

(a) Illustration by Trevithick's son Francis of the *Puffing Devil* and a replica built by the Trevithick Society, (b) artist's impression and replica of Trevithick's *London Steam Carriage*

(Images sourced from Wikispaces and Wikipedia and available under the Creative Commons License)

chimney; not surprisingly, Trevithick named it the *Puffing Devil*. On Christmas Eve 1801, Trevithick together with Vivian and about six other men climbed aboard and set out uphill at 6.4 km h^{-1} towards Camborne Beacon in Cornwall. They did not reach the top of the hill and were said to have covered only between 0.8 and 2.4 km in total. Nevertheless, this was the first time that people had travelled under mechanical power. The next day, *Puffing Devil* was used to visit Vivian's family, who lived nearby at Crane Manor – a round trip of some 3 km. This was the longest journey that the engine completed, although few details of the expedition have survived. The locomotive's third and final journey took place on 28 December. With Trevithick tending the engine and Vivian steering, they hit a rut in the road and *Puffing Devil* overturned. The machine was heaved upright by passers-by and dragged into an outbuilding of Knapp's Hotel on Tehidy Road. While the party consoled themselves over the accident with food and drink, either a fire was still burning in *Puffing Devil*'s furnace, or the boiler had run dry, such that the metal parts became red hot and set fire to the timbers until 'everything capable of burning was consumed'.

Despite the setback, Trevithick and Vivian were not about to give up and, on 26 March 1802, they patented their ideas for steam engines to drive carriages. A year later, they built their second vehicle, which looked more like a carriage than a machine shop. Trevithick gave several demonstrations of his invention in London by undertaking return trips between Holborn and Paddington; accordingly, he called the vehicle '*The London Steam Carriage*'. There is no exact record of its construction for the well-known drawings (see Figure 1.2(b)) owe more to the artists' imaginations than to their eye for engineering detail. It is difficult to comprehend, for example, how the passengers ever got into the vehicle body, nor are the steering arrangements very clear. It can be seen that the engine was based on Trevithick's established high-pressure design, with the horizontal cylinder recessed into the boiler. An arrangement of spur gears provided a slow speed for hills and a higher gear for use on the level. Although the steam carriage ran, with a full load, at speeds up to 12 mph (19 km h^{-1}) on the level, the poor state of the roads created difficulties with steering and gave an uncomfortable ride. Moreover, it was more expensive to run than a horse-drawn carriage and it frightened the horses. As a result of the ensuing bad press, Trevithick dismantled the vehicle to power a hoop rolling mill and turned his attention to stationary engines and railway locomotives.

Although it would appear that, through the work of Newcomen, Watt and Trevithick, Britain effectively pioneered steam road vehicles, for many historians the vital starting point was the wagons constructed in Paris between 1765 and 1769 by Nicholas Joseph Cugnot, who was a Swiss military engineer in the employ of the French government. The wagons were, as far as is known, the first full-scale vehicles to be moved by mechanical power; their purpose was to carry cannon. With great ingenuity and daring, Cugnot abandoned the atmospheric principle and made use of steam under pressure. The most credible reconstruction of his work – there are few documents – maintains that around 1765 he had already produced a steam vehicle that was capable of conveying four passengers at a speed of 3 mph (4.8 km h^{-1}). Unfortunately, the boiler proved to be inadequate, such that after some 15 minutes of travel the truck ran out of breath and had to pause to build up a fresh head of steam.

Notwithstanding the above setback, the French government ordered a second vehicle, which was built in 1769. The three-wheel design was made entirely of wood, had mechanical transmission, and was capable of drawing a load of 4 or 5 t. The single front wheel was used for both driving and steering. The engine was mounted at the front and consisted of a double-walled boiler (the space in between containing the fuel) from which steam passed through a copper tube into the two vertical brass cylinders. The intake and exhaust of steam were controlled by hand-operated valves. An ingenious system of connecting rods and cogwheels was used to transmit the alternating motion of two pistons to the front wheel. Shoe brakes were fitted at the front only. The spoked wheels were of the artillery type and were iron-rimmed. Between 20 November 1770 and 2 June 1771, the wagon was tested on several occasions near Vincennes. During one of these tests, it crashed into a wall and was severely damaged – this was hardly surprising since the tiny tiller had to turn the mass of the boiler and the engine as well as the front wheel. The trials were abandoned because a change of government policy cut off the supply of money needed for further development. Following the Revolution, the pension assigned to Cugnot by King Louis XV in 1772 was withdrawn and he died in Brussels on 10 October 1804 in complete poverty. Cugnot's second machine is still preserved in the Conservatoire des Arts et Métiers in Paris, although the boiler is not the original and part of the mechanism has been lost; see Figure 1.3.

With the practicality of steam power proved, Great Britain became the home of the steam engine. By 1820, there were at least 5000 such engines at work in the nation's mills and factories, in addition to the hundreds of pumping engines, which allowed mines to be driven ever deeper. By contrast, there were no more than 200 engines in France and fewer than 100 in Prussia. It is not surprising, therefore, that between 1820 and 1840 the steam road vehicle forged ahead and England was again the centre of creative activity. Progress was helped by outside factors, such as the invention of the laminated spring by Obadiah Elliot in 1805 and the improvements in road construction by Telford and McAdam (see Section 1.6.1). These developments, along with the use of artillery wheels with iron tyres, resulted in spring-mounted steam passenger vehicles that were lighter, more comfortable, and able to move faster. Though the use of high-pressure steam kept the engines themselves compact, it was more difficult to restrict the size of the boilers, and many shapes and forms were tried.

One of the vehicles that did most to influence the development of new designs, yet achieved no practical results itself, was the steam carriage designed by Julius Griffith in 1821. It was built in the Birmingham workshop of Joseph Bramah and had a change-speed gear system to provide greater engine flexibility. A condenser was used to reconvert the exhaust steam into water and thereby lengthen the intervals between refuelling stops – even 80 years later, most steamers let their steam go to waste. Unfortunately, the boiler proved too small to generate sufficient steam for propulsion. Nevertheless, Bramah's workshop was a meeting place for many of the leading engineers of the day and Griffith's design served as a benchmark for their powers of invention.

An outstanding early experimenter was a former surgeon from Cornwall named Goldsworthy Gurney. He built a 3.75-t carriage that featured an early form of servo-assisted steering – the driver's tiller acted on two small outrigger wheels that supplied the leverage to turn the main forecarriage to which they were linked. Developing 14 hp, its twin-cylinder engine had a voracious appetite for fuel and consumed about 10 gallons (38 litres) of water and 20 lb (9 kg) of coke for every mile covered.

FIGURE 1.3

Joseph Cugnot's *fardier à vapeur* of 1769 as preserved at the Conservatoire des Arts et Métiers, Paris

(Image sourced from Wikipedia and available under the Creative Commons License)

Nonetheless, it was with this vehicle in 1829 that Gurney undertook the first long-distance motor trip at a sustained speed in history, from London to the fashionable spa city of Bath; an illustration of the vehicle is given in Figure 1.4(a). Despite colliding with the Bristol Mail Coach near Reading and later having to be escorted into Bath under guard after being attacked by a Luddite mob outside Melksham, the coach completed the return journey of 210 miles (338 km) at an average speed of 14 mph (23 km h^{-1}). The driving mechanism was poorly protected from road dirt and, after having broken down, was awkward to repair. Though the carriage was intended for a regular London-to-Bath service, it only made the journey once.

(a)

(b)

FIGURE 1.4

(a) An 1827 illustration of the Goldsworthy Gurney steam carriage; (b) steam coaches invented and built by Walter Hancock in the 1830s

(Part (a) image sourced from Wikipedia and available under the Creative Commons License. Part (b) reproduced with permission from Quarto Publishing Ltd.)

Gurney was also working on a smaller, lighter type of carriage to act as a drag or 'steam horse' to pull vehicles that were originally intended for horse draught. In 1830, three of these drags were bought by Sir Charles Dance, who used them to operate a passenger service between Cheltenham and Glouces-ter. During four months of operation, the service made 396 journeys and carried nearly 3000 passengers before opposition from local carriage owners, who rolled large stones in the path of the steamers (caus-ing one of them to fracture its crank axle), plus the introduction of prohibitive turnpike tolls caused its withdrawal. Subsequent attempts to operate these drags by Ward in Glasgow in 1831, again by Dance in London in 1833, and by Gurney himself between Plymouth and Devonport as late as 1837 all proved completely fruitless. In those days, many ordinary citizens were opposed to the introduction of the new-fangled 'steamers' on all sorts of grounds: their weight was said to damage roads and bridges, they made pungent smoke and horses were scared of them. Consequently, after having spent £30 000 of his personal wealth in trying to establish steam traction, Gurney returned to other projects – notably, to improving the heating, lighting and plumbing of the House of Commons, work for which he was knighted in 1863.

Meanwhile, the efforts of Walter Hancock of Stratford in East London proved to be more successful. Hancock realized that the problems with contemporary steamers lay mainly in poor boiler design. In 1827, he devised an ingenious form of boiler made up of a number of flat chambers that were 50 mm wide and arranged side by side. Into the 20-mm spaces between the chambers projected bosses that touched one another and thus formed abutments to give extra heating surface. The pressure was 690 kN m^{-2} (100 psi) and, since the boiler was only 0.7 m square and 1 m high, it was eminently suitable for its task. Indeed, it was found to be exceptionally efficient; the fuel consumption was half that of Gurney's engine for a similar power output. In 1830, Hancock introduced the *Infant*, a 'toast-rack'-bodied car-riage (i.e. forward-facing seats), which in February 1831 was used to inaugurate the first self-propelled passenger service in London. In 1832, came the larger vehicle *Era*, which was followed the next year by *Autopsy*. These three vehicles are illustrated in Figure 1.4(b).

In 1836, four Hancock carriages were put into regular duty in London, and a chain of service stations was established. Over 20 weeks, 12761 passengers were carried on 721 round trips from Moor-gate, over a total mileage of 4200, at an operating cost of 2*d* a mile. This was a most convincing dem-onstration of the potential of steam traction. Hancock's crowning achievement was *Automaton*, a 20-seat vehicle that was capable of nearly 34 mph (55 km h^{-1}) when fully laden. On 24 October 1836, it was driven from London to Epping, on the most uneven main road out of London.

Two factors worked against the further development of the steam road carriage, namely, the expanding network of railways and opposition from the vast industry that had evolved around the horse. The antagonism came not only from stagecoach proprietors, but also from thousands of driv-ers and grooms as well as farmers, hay and fodder merchants, and landowners. These interests were particularly well represented on the Turnpike Trusts, which administered the highways. Their strat-egy was simple but lethal: tolls on mechanical vehicles were increased to such an extent that it became uneconomic to run such transport. Railway engines ran on smooth, level rails and therefore faced none of the problems experienced by steam carriages on uneven, badly-maintained roads; moreover, rail travel promised a quicker return on capital. Consequently, financial interests and the best engineering talents were drawn into the railways, which grew and prospered between 1830 and 1860, whilst the road-steamer branch of transport withered away. There was a brief revival in the late 1850s, and this time Parliament came to the assistance of the railway companies, which now had a near monopoly, by passing the 'The Locomotive Act' in 1865. The legislation limited the speed of mechanical vehicles to 2 mph in towns and 4 mph in open country, i.e. 3.2 and 6.4 km h^{-1}, respectively! Further, every 'road locomotive' was to have three persons in attendance, one of whom must walk 60 yards ahead of the machine and carry a red flag by day and a red lantern by night. This was the notorious 'Red Flag Act', which effectively put a brake on the progress of road transport in Great Britain for 40 years.

1.2.2 Steam traction engines

The traction engine was a logical development of the portable steam engine (the 'Cornish boiler') that had been built by Trevithick in 1812. This was a derivative of his high-pressure engine, in which boiler and engine were combined in one unit and the exhaust turned up the chimney to create a draught on the fire; the revised configuration about doubled the efficiency of his previous models (*v.i.*). In the following years, portable engines came to be produced in every manufacturing nation in the world. Their design adopted the same basic formula of an engine on top of a horizontal locomotive-type boiler, usually with cylinders arranged over the firebox where the heat was greatest (the so-called 'overtype' arrangement; see Section 1.2.4).

The increasing use of the portable engine caused many active inventors in the 1840s to question the absurdity of dragging a potential prime mover about the countryside by a team of horses. Although Trevithick had made his engine self-moving with his *Puffing Devil* locomotive-cum-carriage in 1801, the new attempts to make portable steam engines become mobile bore no relation to his use of connecting rods and cranks between the crosshead and the wheels. Instead, rotary motion of the crankshaft was transmitted to the wheels by pitch chain and sprockets. Thus was born the concept of the steam traction engine, a machine that, with very little refinement, continued in production for almost a century. Not surprisingly, Britain was the major centre of its early development.

The first move in the true direction of the traction engine was the self-propelled threshing machine built by Ransomes, Simms and Head of Ipswich and demonstrated during the Royal Show at Bristol in 1842. This had a vertical boiler, a pitch chain for coupling the wheels to the engine, and a horse-steered forecarriage. There followed an active period of experimentation during which engines took many different forms. It is not clear who was the first person to apply chain drive to a portable engine in the style that is recognizable today, but Thomas Aveling of Rochester was a notable pioneer. In 1859, he fitted a long driving chain between the crankshaft and the rear axle of a Clayton & Shuttleworth portable, and a year later produced the first of his engines designed for traction with a single cylinder arranged over the firebox and driving a forward-mounted crankshaft. In 1862, Aveling moved the cylinder to the front of the boiler with the crankshaft over the firebox and carrying a large flywheel just behind the chimney. This rearrangement simplified the drive and became the standard arrangement.

A succession of other improvements in design followed. For instance, the rudimentary footplate and separate tender evolved into a man stand with tank and bunkers of the true traction engine form. Steering, too, progressed from the straightforward horse-steering to steerage via a ship's wheel and chains operated by a steersman placed on a platform before the smokebox, and finally to the long-lived chain and bobbin steering that was worked from the man stand. The Aveling engine of 1870 brought together the essential components of the typical traction engine, namely: (i) steam-jacketed cylinders at the leading end of the locomotive-type boiler which formed the chassis; (ii) crankshaft and countershaft in a crankbox integral with the firebox side plates; (iii) all-gear drive and steering from the man stand, under which was placed the water tank and upon which the coal was carried in a bunker behind the driver.

Three refinements remained to be accomplished – the use of differential gears, compounding and road springs. The differential did not become general until the end of the 1890s mainly because, until the advent of the alleviating clause of the Locomotives on Highways Act of 1896 (see Section 1.2.3), it was practically impossible to use a traction engine for anything outside agriculture or parochial haulage, where the removal of the driving pin from one wheel at really sharp turns was not a crippling disadvantage. On a soft-surfaced road, a little wheel slip to compensate for the absence of a differential took place easily, though it was less readily tolerated on properly surfaced town roads. In 1879, Richard Garrett & Sons exhibited the first compound portable during the Royal Show at Derby. Compounding in traction engines – the use of two or more cylinders of increasing size to permit greater steam expansion and efficiency – was introduced by Fowler in 1881. It was effectually brought into use, however, by the remarkable performance displayed by Edwin Foden's compound traction engine at the trials organized by the Royal Agricultural Society at Newcastle in 1887. As with the differential, the springing of engines acquired commercial significance only after the 1896 Act had made road haulage possible on a serious scale. No weather protection was provided until late in the 19th century, when full- or

FIGURE 1.5

British steam engines: (a) Ransomes, Sims and Jefferies general-purpose engine, 1918; (b) Foden compound traction engine, 1899; (c) Fowler ploughing engine, 1916; (d) Aveling and Porter steamroller, 1929

(Parts a and c sourced from Wikia web sites, parts b and d sourced from Wikipedia; all available under the Creative Commons License)

half-length awnings were fitted if the customer so required. Examples of preserved Ransomes and Foden traction engines are shown in Figure 1.5(a) and (b), respectively.

By the late 1870s, British traction engines had developed into fine, rugged machines with the minimum of 'frills'. Although speed on the road was limited by the Red Flag Act of 1865 (*v.s.*), under the influence of Thomas Aveling a special variant, the steamroller, was introduced. The use of such machines brought about a dramatic improvement in road surfaces. From 1870 until the end of construction in the late 1940s, the general form of the steamroller was a two-speed, unsprung, traction engine, but mounted on rollers instead of wheels. The front axle, held in a massive fork, carried a two-part roller, and the rear wheels had thick, smooth treads so that they also acted as rollers. When the vehicle moved in a straight line, the overlap between the front and rear rollers was sufficient to produce a smooth, rolled track equal in width to the distance across the hind wheels. This basic design is still used, though today's version is diesel powered. A preserved Fowler engine and an Aveling and Porter steamroller are presented in Figure 1.5(c) and (d), respectively.

As the years passed, the policy of rugged simplicity and realistic pricing proved so successful that the firm of Aveling & Porter was able to concentrate primarily on steamrollers. In all, more than 8000 were made in Rochester and the brass rampant horse of Kent that each machine displayed on its steering head came to be recognized throughout the world. Most of the producers of traction engines in other countries added steamrollers to their ranges as time went on, as there was a large market for them. Designs were very close to those for traction engines, and several manufacturers fitted differentials. Some steamrollers were given compound cylinders for greater economy; others were converted from traction engines, especially after 1920 when the use of these engines in agriculture was declining.

Though the traction engine is thought of as essentially a 19th-century machine, and it is true that the problems of its design were largely worked out before 1900, the numbers built up to that year were comparatively small. At the turn of the century, Fowler had built about 9000 engines, which were mainly designed for ploughing and with a heavy emphasis on exports, Burrell about 1800 and Aveling roughly the same number. Overall, the total number of traction engines produced in Britain up to the end of the century – discounting ploughers and rollers – was about 12 000, a good proportion of which were sent overseas. An upsurge in the numbers of traction engines was brought about in part by the recovery of British agriculture from the depression that had lasted nearly 20 years from 1879, but it was given impetus by the provisions of the Locomotives on Highways Act of 1896. The latter also created interest in the possible useful employment of light steam tractors, which were simply very small traction engines. Tractors reached the peak of their popularity about the end of World War I, with numbers tapering off during the 1920s until the impost of heavy taxation put an end to all manufacturing activity in about 1932. The building of steam traction engines continued, but on a reduced scale. Ironically, it is generally accepted that the last traction engine of all was built by Ransomes in 1942. Thus the British company that had demonstrated the first traction steam engine in 1842 had also written the final word, after a century of continuous production.

Traction engines became widely used in North America as well as in Europe, where large-scale production was confined to France and Germany. Although the early developments of American, British and European engines proceeded in parallel, from the 1860s onwards their courses began to diverge. The ultimate solution adopted by most engine builders everywhere embodied the same overall principles (the locomotive boiler acting as the main frame and carrying an overtype horizontal engine that transmitted its motion to the rear wheels through a gear train). Nevertheless, many of the actual details of the American versions were remarkably different. For instance, Case and Avery insisted on putting the cylinders and motion underneath the boiler; the company produced some fine, modern 'undertype' machines which, because of the uncluttered boiler, had more of the railway than the road in their outline. The machines were also made in very large sizes so that they were equipped to burn wood or straw and thereby make use of local fuel resources. In this respect, it needs to be remembered that the use of steam road haulage in America was so slight as to be practically negligible and that the work of traction engines fell broadly into the two operations of threshing and ploughing.

1.2.3 Steam cars

Although legal repression and public hostility interrupted mainstream development of the steam carriage in Britain, a few pioneers continued to build such vehicles for their own use. In the late 1850s, Thomas Rickett produced a few three-wheeled two-seaters, one of which was sold in 1858 to the Marquis of Stafford, and another to the Earl of Caithness. J.W. Boulton built several carriages between 1848 and 1860, and another early venture was the *Cornubia*, built by the Tangye brothers in 1868 and subsequently exported to India. Two companies, Yarrow & Hilditch and Garrett & Marshall, displayed private steam vehicles at the 1862 Great Exhibition in London.

In 1855, inventor Richard Dudgeon astounded New Yorkers by driving from his home to his place of business in a steam carriage. After losing this vehicle in a fire, Dudgeon constructed a second steamer in 1866. This resembled a liaison between a dwarf traction engine and two park benches, but it was able to carry 12 passengers at 14 mph (22.5 km h^{-1}).

Among the outstanding steam-powered creations were those designed by Amédee Bollée, a bell founder of Le Mans, France, who at the age of 26 in 1871 produced his remarkable *l'Obéissante* ('Obedient'); see Figure 1.6(a). Years ahead of its time, it featured independent front suspension with geometrically perfect steering and an iron chassis frame, on which was mounted a metal-panelled 12-seater body. Each rear wheel was driven by a separate V-twin engine; when *l'Obéissante* turned a corner, the

(a)   **(b)**

FIGURE 1.6

(a) Amédée Bollée's steam vehicle, *l'Obéissante*, carried 12 passengers at a speed of 15 km h^{-1} (9.3 mph);
(b) Léon Serpollet at the wheel of his steam car, *Oeuf de Pâques*, in 1902

(Part (a) image sourced from Wikipedia and available under the Creative Commons License. Part (b) image by kind permission of:
Mr Bricolage Lillebonne)

steam supply to the engine on the inside of the curve was reduced to give the effect of a differential. There was also a separate two-speed gear incorporated in each power-train. After a couple of years of trials and modifications, Bollée judged that his creation was ready to be shown to a wider audience. Accordingly, on 9 October 1875, he drove it to Paris, where it became a sensation. His next steam-car, *la Mancelles* (the Girl from Le Mans) of 1878, was equally advanced in concept. With the engine at the front under a bonnet and driving the rear wheels by a shaft and side chains, it set the pattern for the layout of motorcars for years to come. Again, the front suspension was independent, but this time it incorporated parallel leaf springs. Although the price was reasonable at 12 000 francs, there were few sales. Eventually, Bollée's experiments ran him deeply into debt and he redirected his activities to bell founding.

The peak of steam design was attained by Léon Serpollet, whose vehicles were light, quiet and manoeuvrable. In 1891, he obtained permission to drive through 'all the streets of Paris equally', which was in effect the very first licence for a motor vehicle. His major step forward in steam technology was the invention of the multi-tube flash boiler in 1896, the principle of which may be likened to dropping water on a red-hot iron. A small stream of water is pumped through a coil of tubing which is subjected to an intense heat by burners. Therefore, immediately on entry into the tube, the water is 'flashed' and instantly converted into high-pressure superheated steam. This technology resulted in a lighter and more compact boiler with adequate steaming capacity that took less time to raise from a cold start, typically within 5 minutes. Moreover, the steam was only generated on demand, and explosion of the boiler was impossible as it contained so little steam at any given time. The only major precedent was the 'Patent Steam Chamber' invented by American John McCurdy in 1824. Serpollet used paraffin to heat his boilers, instead of wood or coke, and this departure gave not only greater power, but also less pollution.

With backing from Frank Gardner, a fabulously wealthy American who had built his own petrol-engined cars in Paris between 1898 and 1900, Serpollet was to spearhead steam car production on the European continent. His factory produced and sold 200 steam vehicles during 1900; clients included such eminent people as the Shah of Persia and the Prince of Wales, later King Edward VII. Serpollet cars were also making their mark in the great Continental city-to-city races and thereby established a reputation for reliability and hill-climbing performance, if not for winning speeds. In 1900, however, one of the cars became the first vehicle to break the 100 km h^{-1} (62 mph) barrier. Subsequently, on 13 April 1902, Serpollet's ovoid steam car *Oeuf de Pâques* ('Easter Egg') took the world land speed record by reaching 120.80 km h^{-1} (75.06 mph) over the flying kilometre on the Promenade des Anglais at Nice; see Figure 1.6(b). The18-hp model, introduced in 1904, was designed more in line with existing convention, i.e. the boiler remained at the back of the chassis while the four-cylinder engine was placed longitudinally under a bonnet behind a large circular condenser with the appearance of a petrol-engined car radiator. Sadly, Serpollet died of consumption at the early age of 44 in February 1907 and the company died with him. As a consequence, enthusiasm in Europe for the steam car faded away.

British steams cars were even less successful. The above-mentioned Red Flag Act brought the mechanical transport of the day more or less to a grinding halt. The subsequent Highways and Locomotives (Amendment) Act of 1878 merely confused issues by laying down pettifogging regulations that related to tyre widths, weights and permissible hours of travelling, together with a curious little rule whereby a mechanical vehicle 'should consume its own smoke' – possibly the first environmental legislation! The Locomotives on Highways Act of 1896 lifted the restriction on speed of mechanically-propelled vehicles to 14 mph (22.5 km h^{-1}). By 1896, however, Britain had been left out of the steam car race and most of the machines used on its roads were imported. The new legislation did, however, give birth to the annual London to Brighton Veteran Car Run, which is the oldest motoring event in the world. The inaugural competition, named The Emancipation Run, was held on 14 November 1896. Nevertheless, very few steam cars were produced in Britain. The Turner-Miesse, superficially similar to the Serpollet, first appeared in 1902 and was produced in ever-decreasing numbers until 1913. Other companies had very short lives and built only a handful of cars.

Whereas France led the way in the design of steams cars, it was the USA that gained pre-eminence in their production – remarkably, through the relatively crude product of two identical twins, Francis E. and Freeland O. Stanley, from Newton, Massachusetts. The Stanleys had a natural flair for invention. In turn, they mass-produced violins, invented a home generator for illuminating gas, fabricated some of the first X-ray equipment, and invented the dry photographic plate which netted them a fortune on selling the patent rights to the Eastman Kodak Company. In 1896, by way of a hobby, they began to build a steam car, although they 'knew but little about steam engines and less about boilers'. The brothers took over a bankrupt cycle works next to their dry-plate factory and equipped it for vehicle production. In 1899, they sold their design to the Locomobile Company for US$250 000 and undertook to abstain from making cars for two years. The Locomobile steamer, shown in Figure 1.7(a), was mass-produced on a scale then unique – 4000 were turned out in under three years, at a time when there were only some 4000 other cars running in the USA. A huge depot was set up in South Kensington, London, where 'Locomobiles were bought like eggs', and the model enjoyed a brief vogue in Britain. But though steam could be raised 'in less time than it takes to harness a horse and carriage', the vehicle design was too frail to be a continued success.

The Locomobile Company dropped steam in favour of petrol in 1903 and the irrepressible Stanleys bought back their own manufacturing rights. The brothers produced an improved design, which had its twin-cylinder engine geared directly to the back axle in a layout that would not change during the remainder of the marque's long life; two examples of their later models are shown in Figure 1.7(b), (c). In 1905, standard Stanley engines with increased boiler capacity were fitted in low, streamlined bodies – probably the first car bodies to be developed by wind-tunnel tests – for attempts on speed records. A Stanley steamer broke the land-speed record by achieving 127.7 mph (205.5 km h^{-1}) on Ormond Beach, Florida, in January 1906. Despite a healthy production of 500 cars in 1917, the brothers decided to retire and sold the company in May 1918. Unfortunately, F.E. Stanley was killed in a driving accident a few months later; both he and his brother were fast, reckless drivers. In the face of overwhelming competition from internal-combustion-engined vehicles (*v.i.*), the Stanley Company closed down in 1927.

The White steamer was the main rival of the Stanley during the first decade of the 20th century. The car was built by Rollin H. White, the son of a well-known sewing machine manufacturer in Cleveland, Ohio. The first designs, manufactured in 1900, were conventional steam buggies on similar lines to the Locomobile, but in 1903 a new model appeared; see Figure 1.7(d). In general form, this resembled a normal petrol-driven car with a front-mounted twin-cylinder compound engine, a condenser in the 'radiator' position to save frequent stops for water and, unique among its contemporaries, a 'semi-flash' steam generator. Unlike the pure flash boiler, the White unit had reserve capacity which could be called on in an emergency. This feature made the car especially competitive in short-distance speed events. Certainly, victories on the racetrack contributed to the impressive sales figures achieved by White; 1500 vehicles were produced in 1906 and many were exported to different parts of the world. In 1909, president-elect William Howard Taft selected a 40 hp (30 kW), seven-seat *White Model M* tourer as the first official car of the president of the United States. Nevertheless, even such distinguished patronage proved insufficient to keep sales at an acceptable level. After 1910, the steamer was supplanted by a petrol car, and by 1918 the White Motor Company had abandoned the manufacture of private cars entirely in favour of trucks.

(a)

(b)

(c)

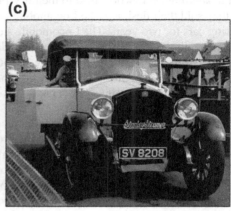

(d)

FIGURE 1.7

(a) F.O. Stanley and his wife, Flora, driving an 1899 Stanley-Locomobile steamer; (b) 1912 Stanley *Runabout*, the boiler is in front and its double-acting twin-cylinder engine is fully enclosed underneath the car; (c) 1920s Stanley, engine and transmission followed the *Runabout*. (d) 1903 White *Type-C*

(Images sourced form Wikipedia and available under the Creative Commons License)

During the 1920s, numerous makers appeared briefly on the scene but mostly produced only one or two cars. The sole vehicle to enjoy any lasting success was the *Doble*, built between 1924 and 1932 in Emeryville, California, by Abner Doble, who had already designed the flash-boilered *Doble-Detroit* of 1918–1922. Doble aimed at making his steamers as automatic as possible. The burner was ignited automatically and, thanks to a venturi booster blower, could raise a full head of steam in under a minute. Enthusiasts claimed it would run on anything combustible, from coal dust to petrol. The paradox was that in making the steamer as easy to operate as possible, Doble had to invoke a highly complicated mechanism and the amount of handwork involved in its fabrication meant that his cars were expensive. Henry Ford could manufacture a car that anyone could drive, and sell it for US$290; see Section 1.4.2), whereas Abner Doble was unable to build his vehicle for under US$8000. Ford eventually sold over 15n million *Model Ts*; Doble's output was just 42 cars. During the 1930s, the steam car ceased to be commercially viable and consequently became the preserve of enthusiasts, most of whom spent a great deal of money to produce vehicles which were less satisfactory than the petrol cars that they were intended to supplant.

1.2.4 **Steam vans and lorries**

Though steam wagons were built and operated in other countries, Britain led the way and persevered the longest with the advancement of such vehicles for light and heavy haulage. In England, Thornycroft anticipated a relaxation of the law relating to motor vehicles and began development work on a steam van; the 30-cwt prototype was built in 1896. A three-ton wagon soon followed and both steamers performed well in the commercial vehicle trials of the day; they carried off many awards. Given this success, other manufacturers soon followed Thornycroft's example; notably, the Lancashire Steam Motor Company (the beginning of British Leyland), Clarkson and Capel, and the Liquid Fuel Engineering Company (strangely abbreviated to LIFU).

The British steam wagon, as it was termed, fell into the two principal categories mentioned above, namely: (i) the 'overtype', which followed traction engine practice with a horizontal and short locomotive-type boiler in front of the driver, and cylinders and motion on top; (ii) the 'undertype' with a vertical boiler in front of, or behind, the driver and a horizontal engine slung underneath the chassis. There were a few departures from the norm, notably the vehicle produced by the Yorkshire Patent Steam Wagon Company that had an undertype engine with a T-shaped boiler arranged transversely across the front of the vehicle, see Figure 1.8(a), and the Fowler that had a vertical boiler and cylinders in V formation at the rear of the cab. Although the Yorkshire design improved the weight distribution and made better use of the limited space, it required the use of a more complicated boiler.

Wagons with vertical boilers tended to be poor steamers and by 1906 most manufacturers were employing the horizontal locomotive-type boiler – a practice that Mann of Leeds was the first to introduce in 1901; see Figure 1.8(b). The final and most successful form of overtype had a boiler shell forming an extension of the chassis frame, like a mini-traction engine but with cab and seats in place of a tender. It was produced by various firms, of whom the largest was Edwin Foden of Sandbach in Cheshire; see Figure 1.8(c). Every one of these manufacturers was an established builder of traction engines and so was well able to build practical steam vehicles. With the sole exception of the Sentinel Waggon Company of Shrewsbury, whose excellent water-tube boiler and dynamic designs were to outlast all the rest, the vertical boiler companies faded from the scene. The Sentinel steam waggon (the company always spelt the name with two 'g's) capitalized on the one major inherent drawback of the overtype; namely, the length of the latter's boiler and cab greatly impaired the carting capacity unless an excessively long wheelbase was used.

FIGURE 1.8

Steam lorries and trucks: (a) *Denby Maiden*, Wagon No. 117, Yorkshire Patent Steam Wagon Company, 1905; (b) *Uncle Walt*, 5 Ton Type No.881–TF 1598, Mann Patent Steam Wagon & Cart Co.Ltd, 1914; (c) *Britannia*, C Type 6 Ton, Edwin Foden Sons & Co. Ltd, 1926 – famous for its 'Grand Global Tour' of 1968 to 1972. By kind permission of The Steam Museum, Ramsgate, UK; (d) *Type S4*, Wagon No. WV 4705, Sentinel Waggon Works Ltd, 1934

(Parts a and b sourced from Wikia websites, part d sourced from Wikipedia and all available under the Creative Commons License)

  Progress was relatively rapid and by the end of 1903 a fleet of 39 steam wagons were in service in Britain. Among these were the first Leyland vehicles, which were operated by the Road Carrying Company of Liverpool; remarkably, three others were exported to Ceylon (Sri Lanka) to serve as mail vans. Even though Leyland was to continue to produce steam lorries for some years to come, the fate of this technology was sealed from the moment that the company introduced its first internal-combustion-engined wagon, nicknamed *The Pig*, in 1904. Although this vehicle proved to be a failure, the further development of petrol lorries by the company continued unabated.

  In the 1920s, as the competition from petrol became more intense, the builders of locomotive-boiler lorries also gradually stopped production and Sentinel was left to carry on alone. Having amazed the world in 1923 with its twin-cylinder undertype *Super Sentinel* design, the firm made another breakthrough in 1930 with the introduction of a massive 15-t eight-wheeler, the *DG8*, which had four steered wheels in front and four driven wheels at the rear. The vehicle was more than 9 m (30 ft) in length and

FIGURE 1.9

Sentinel steam vehicle in service as a bus in Whitby, North Yorkshire

(Image sourced from Wikipedia and available under the Creative Commons License)

carried 0.76 t (15 cwt) of coal and 1045 litres (230 gallons) of water. Three years later came the ultimate Sentinel, the *S Model*, with a four-cylinder single-acting engine and a four-, six- or eight-wheel form designated as *S4, S6* or *S8;* see Figure 1.8(d). Unlike earlier Sentinels, the *S Model* was mounted on pneumatic tyres and a steam-driven tyre pump was a clever feature of the design. This was a fine vehicle – only a thin plume of steam above the small chimney showed that it was a steamer at all! The water-tube boiler was behind the driver, who had a comfortable cab (at least in cold weather) and a rotating firegrate enabled automatic stoking. There was no gear changing, so that the *S4* was a source of great astonishment to other lorry drivers as it accelerated silently from traffic lights. Some *S4*s remained in service until well after World War II, and a handful came back into operation during the 1956 Suez crisis. In the 1930s and 1940s, several German firms produced heavy steam vehicles and steam railcars with engines that followed closely the Sentinel design. It is a testament to the robustness of the Sentinel engineering that one of their vehicles, a *DG6P* (double-geared, 6-wheeler, pneumatic) originally manufactured in 1931, has been converted to function as a bus, in which form it has been carrying passengers during the summer months in Whitby, North Yorkshire, for several years; see Figure 1.9.

A few other makers built undertypes, the most famous of which, and the greatest of Sentinel's competitors, was Foden's splendid *Speed Six* and *Speed Twelve* range. The former model, named after E.R. Foden's beloved Bentley, appeared in 1929. It had a 90-bhp engine, pneumatic tyres and a reputed maximum speed of 60 mph (96.6 km h^{-1}), or three times the existing legal speed limit. Nevertheless, only a couple of years later, the company turned to the diesel engine and abandoned steam for good.

A major reason why the steam wagon had survived so long in Britain was that coal was less heavily taxed than petrol and hence steamers were far cheaper to run. In an effort to equalize the situation, the 1933 Road Traffic Act introduced a differential system of taxation for commercial vehicles that trebled the annual licence fee for steamers. Duties on diesel-engined vehicles were lighter, and on petrol commercials were the lightest of all. The Act made the running costs of the three types of vehicle approximately the same, but the operators did not see it that way, as there was suddenly no financial inducement to tolerate the disadvantages of steam wagons, such as the poor forward vision, the hot cabs and the inadequate brakes.

1.2.5 Steam buses, trams and cable cars

Although Thomas Hancock and Sir Charles Dance had proved the practicability of public steam omnibus services in the 1830s, the revival of the mechanically-powered vehicles at the end of the century saw relatively few attempts at operating steam-powered buses. Test trials took place in France in 1897 and these featured vehicles built by De Dion–Bouton. An event in Britain that was to have a far-reaching effect on the nation's motor industry was the formation of the Lancashire Steam Motor Company in 1896. This small enterprise built a number of steam buses. The initial model, which had seats for 18 people and a top speed of 8 mph (12.9 km h^{-1}), was delivered to the Dundee Motor Omnibus Company in May 1899; see Figure 1.10(a). The firm was a success and in 1907 it became Leyland Motors Ltd, probably the first of the pioneering manufacturers to concentrate purely on commercial vehicles and the foundation of what became the British Leyland empire.

A handful of other British manufacturers constructed steam buses but with little success and it was therefore left to Thomas Clarkson to demonstrate the viability of steam for public transport. During the period 1902–1908, his Chelmsford works produced over 80 buses; an example is shown in Figure 1.10(b). Eventually, however, the various bus operators closed ranks against the steam bus, with most of them standardizing on the successful Milnes-Daimler petrol chassis. In retaliation, Clarkson established his own bus company, the National Steam Car Company, on 2 November 1909. By the beginning of 1914, he had 173 buses running in London, with a combined total of 11 others in Harrogate and Nottingham. The vehicles soldiered on through World War I, but when the question of their renewal arose the Company chose to introduce petrol vehicles; the last National steam bus ran in London on the 18 November 1919.

An overseas rival to the Clarkson had appeared in 1907 – the Darracq–Serpollet from France. Each of these buses was reckoned to use up 28 gallons (127 litres) of paraffin per day in heating the water in its boiler; see Figure 1.10(c). By 1912, the Metropolitan Steam Omnibus Company was operating 100 of these vehicles, one of which was used to start an express service from London to Maidstone. In the Isle of Wight, the Ryde Seaview & District Motor Service ran their Darracq–Serpollet until 1923; see Figure 1.10(d). This is widely regarded to be the last of the regular steam bus services to operate in Britain.

By contrast, efforts by the American company Brooks Steam Motors in Buffalo to produce steam city buses between 1927 and 1929 proved to be spectacularly unsuccessful, even when Thomas Clarkson's son was engaged as the chief engineer. During the late 1930s, the German Henschel company constructed a number of steam buses and trucks to the design of the American Abner Doble, on the premise that use of the steam engine would make Germany independent of imported petrol. The venture failed to attract interest and only a small number of vehicles were produced.

Two apparently insuperable obstacles prevented steam from becoming a serious contender in the search to find suitable engines for commercial passenger vehicles, namely:

(i)   the inherent inefficiency of steam plant in response to the constantly fluctuating demands of stopping and starting, traffic, and gradients;
(ii)   finding enough drivers of a sufficiently high calibre to handle the complicated system of controls for heat, water and engine operations; a difficulty that was compounded in the summer when the heat from the under-bonnet boiler could become unbearable.

Either obstacle would perhaps have been susceptible to further investment and research, but taken together they were overpowering and the brief age of the steam bus drew to a close.

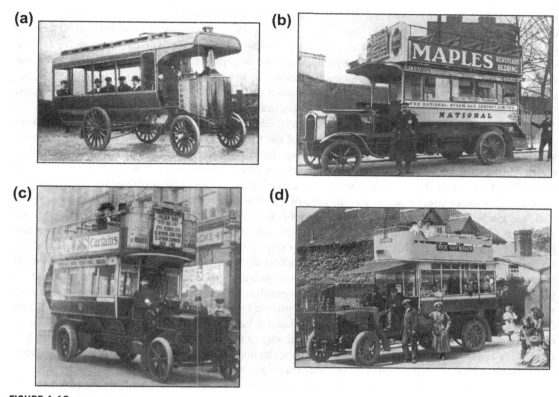

FIGURE 1.10

Steam buses: (a) Lancashire Steam Motor Company, 1899; (b) Thomas Clarkson of Chelmsford, 1905; (c) Darracq–Serpollet, 1912; (d) Darracq–Serpollet, 1907

(Parts c and d Courtesy of: Bounty Books, a division of Octopus Publishing Group Ltd. 2007)

If the steam bus was short-lived, then the steam tram was positively ephemeral! This alternative form of public transport was even more hedged around by restrictions than the traction engine. To placate public opinion, locomotive designers had to provide machines that were silent, had no exposed parts which might threaten the safety of pedestrians and draught horses, and would not degrade or endanger wooden structures by smoke and sparks.

The desire to come up with a cleaner form of city transport led to a number of experiments. One idea was to design a traction engine that was capable of running along the horse-tram network, without being expensive to operate or requiring complex auxiliary equipment. The first practical solution of this kind was the compressed-air tram. A Polish-born Frenchman, Louis Mékarski, developed a reliable form of air engine, in which the twin problems of unwanted condensation in the cylinders and excessive cooling caused by the expansion of compressed air were overcome by mixing steam heated to 165 °C with the air. The air tanks were replenished at termini by means of a compressor. The vehicles were quiet and smokeless, apart from a slight hissing and chuffing sound, and passengers liked to travel in them. Since the trams were not heavy, they were able to run along the existing horse-tram lines without expensive reconstruction being needed. Nevertheless, it was only in France that a regular service of any duration was established. Mékarski trams were put to work in Nantes in 1879 and some 80 were in operation in 1900; see Figure 1.11(a), (b). These operations encouraged expansion of the technology to the Paris area; some 208 vehicles were employed and the majority of these lasted until 1913–1914. Air trams eventually fell out of favour principally because of two defects – the necessity to recharge frequently and the unreliability of their air brakes. Similar experiments with various engines that operated with superheated steam, hot caustic soda solution or ammonia gas all proved to be unsuccessful.

A more practical substitute for horsepower was cable traction. The proposal to haul vehicles by an endless rope wound around pulleys at each end of the track, with the pulleys driven by steam power, was not a new notion. It had first appeared during the 18th century in Britain, i.e. before the use of steam locomotives, and had been applied in the hauling of rail-guided wagons on steep inclines in coal mines. The fact that a cable traction system was revived in the USA can be attributed to the regular grid pattern of American street plans and to the hills of San Francisco in particular. The successful inventor was Andrew S. Hallidie, who had inherited from his father a factory that produced steel cables for the funiculars of the Sierra Nevada silver mines. Hallidie saw the problems caused by the horse trams, and he noted that they were unable to run up the steep slopes of some quarters of San Francisco. He came up with a relatively simple solution; namely, an endless steel cable laid in a narrow groove between the tramlines and driven at a constant speed by a central steam engine. Streetcars (the American name for trams) could be attached to or detached quickly from the cable by means of a special gripping device that was mounted on board and dropped into the groove by the driver, or 'gripman'. (In fact, William Joseph Curtis had patented a similar friction coupling back in 1838.) On 1 August 1873, Hallidie opened the world's first cable-operated street tramway in Clay Street, a hill with a 1-in-5 gradient in places; see Figure 1.11(c). Despite early difficulties, the line proved effective and safe so that, by 1880, San Francisco had 112 miles (180 km) of cable tramways which were run by different companies.

The advantages of cable cars were self-evident: they were clean and quiet, and compared with horse trams they moved at twice the speed, were cheaper to run and could carry much heavier loads. It was also possible to attach various numbers of cars to the cable as required. A further advantageous feature was that the cars travelling downhill when attached to the cable assisted the traction of other cars – a simple and effective method of energy regeneration. The major disadvantage was the capital cost of the cable installation and power house – US$100 000 a mile – and there was always the threat of a break in the cable. Nonetheless, cable tramways were installed during the 1880s on many high-volume city routes, especially in the USA, where by 1894 about 5000 vehicles were operating over 620 miles (1000 km) of track. Networks were also built in large cities in England, Scotland, France, Portugal, Australia and New Zealand. The first in Europe was London's Highgate Hill line, opened in 1884. The last cable system of any size was opened in Melbourne, Australia, in 1885. At its peak, it employed over 1200 cars and trailers and had 47 miles (75 km) of double track; see Figure 1.11(d). The last car operated on 26 October 1940. Sydney was also serviced by cable cars from 1884 to 1905.

By the end of the 19th century, however, the nemesis of steam-powered road vehicles was already in view. External-combustion engines suffered from several disadvantages in comparison with internal-combustion engines (ICEs), for example:

- the boiler of a steam engine represents a rather massive burden that is not present with an ICE;
- either type of engine must carry the necessary fuel on board but steam engines additionally must also convey a large volume of water;

FIGURE 1.11

(a) Mékarski tram in Nantes being recharged with compressed air, circa 1880; (b) preserved Nantes tram; (c) first cable line running on Clay Street hill, San Francisco, 1873; (d) cable tram on the St Kilda Line in Melbourne, 1905

(Images sourced from Wikipedia and available under the Creative Commons License)

- to operate efficiently, a steamer requires a larger radiator than an ICE;
- a steamer takes more time to start from cold due to the need to raise steam.

Although the steamer has good torque characteristics and thus needs no complex transmission system, the disadvantages listed above proved decisive and today steam-powered vehicles are operated only by a few enthusiasts.

1.3 The age of electricity

1.3.1 Birth of the electrical power industry

In 1820, the Danish scientist Hans Oersted noted that electricity flowing along a wire deflected the needle of a magnetic compass held nearby. This was not entirely surprising because a number of scientists had speculated that there was some connection between electricity and magnetism. What was unexpected, however, was the way the effect depended on the position of the compass relative to the wire. The forces then known – gravitation, electricity and magnetism – all acted along a line joining the two given objects together. By contrast, Oersted's experiment revealed that the compass needle was not deflected toward or away from the wire, but at right angles to the compass–wire direction. This observation suggested that the magnetic force produced by an electric current acted not in a straight line, but in circles around the wire. Consequently, it occurred to the British electrochemist Michael Faraday (1791–1867) (see Figure 1.12(a)) that such an effect might be used to obtain continuous motion. In December 1821, he designed two sets of apparatus that demonstrated such motion was indeed possible: in one, a bar magnet rotated in mercury round a wire carrying an electric current, while in the other the suspended wire moved round a magnet held in a pool of mercury; see Figure 1.12(b). The importance of these experiments was quickly recognized by the scientific community; Faraday had in fact invented the electric motor by converting electricity into mechanical motion.

Faraday next deduced that, if a flow of electricity produced magnetism, then magnetism should be capable of generating an electric current. Remarkably, it was not until 1831 that he had the occasion to look properly for this inverse effect. He constructed a simple apparatus that consisted basically of an iron ring of about 15 cm across; see Figure 1.12(c). Around one side of the ring were wound a number of turns of wire that were insulated from the ring and connected to a battery. On the opposite side of the ring, wire was wound in a similar way and extended some distance to run near a compass needle. The idea was that current in the first coil of wire would produce magnetic effects that would be guided around the ring and through the second coil of wire. If this magnetism was converted to electricity in the second coil, it would be observed via a deflection of the compass needle. Faraday found that whereas passage of a steady current did not affect the compass, the needle jumped every time the battery was connected, or disconnected. This explained why the conversion of magnetism to electricity had not been observed previously: current was only 'induced' in the second coil when the amount of magnetism was changing, not when it was steady.

Faraday's ring was actually the first electrical transformer. The electrical pressure, i.e. the voltage, in the second coil of wire depended not only on the power of the battery, but also on the relative number of turns of wire in the first and second coils. Thus high-voltage electricity could be converted into low-voltage, or vice versa, simply by changing the number of turns in each coil. The discovery of electromagnetic induction did not, however, demonstrate the production of electricity from a permanent source of magnetism, such as a bar magnet. During 1831, Faraday solved this problem in a further series of electrical experiments. His simplest apparatus consisted of a coil of wire into which a bar magnet could be pushed. Every time the magnet was pushed in, or pulled out, a current flowed in the wire. At a rather more complex level, Faraday showed that a metal disc rotated between the ends of a horseshoe magnet produced a continuous flow of electric current between the centre and edge of the disc; see Figure 1.12(d). He had now invented the dynamo – a method of generating a continuous supply of electricity from motion, such as that provided by a steam engine.

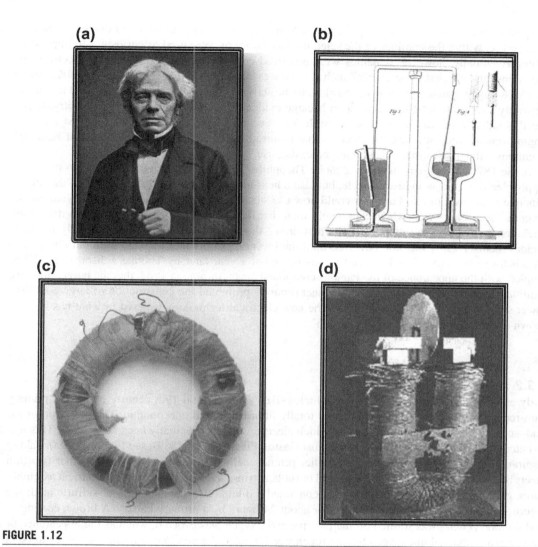

FIGURE 1.12

(a) Michael Faraday; (b) published version of 'electromagnetic rotations' apparatus made by Joseph Newman to Faraday's instructions; (c) the first transformer, 29 August 1831; (d) the first dynamo, November 1831

(Parts a, b and d reproduced by permission of The Royal Society of Chemistry. Part c Courtesy of: Bounty Books, a division of Octopus Publishing Group Ltd.)

By making the first electric motor, transformer and dynamo, Michael Faraday provided the three key components required to create an electrical industry – an exceptional achievement from a man who was born in poverty, malnourished in childhood and rudimentarily educated, and who failed to acquire a knowledge of mathematics beyond simple arithmetic!

It took some time for the large-scale generation of electricity to become viable. The reasons were economic, rather than technical. Steam power was already well established, and electricity could not compete for many of the existing applications. Thus, the first electric motors and dynamos remained more or less within the confines of the scientific laboratory or were used as curiosities to impress the public. Eventually, however, dynamos were produced commercially, in particular for electroplating operations, and the first large unit for such service was commissioned at a Birmingham, UK, plant in the 1840s. Early dynamos were installed close to the activities that they served, but the need to transmit electricity over quite long distances soon became evident. The question was how the electricity produced could best be transmitted elsewhere. It soon became clear that the voltage required for wide-ranging transmission was not the same as that required for domestic use. So the third of Faraday's inventions – the transformer – came into increasing use.

In the 1870s, a Belgian engineer, Zenobe Theophile Gramme, built not only a dynamo that could supply electricity on an industrial scale, but also a near-identical machine that operated on the reverse principle to act as a motor. Factories could now use steam or water power to turn huge dynamos and generate electrical power, which they could then distribute to smaller electric motors to drive their machines, rather than the complex belts and drives that were needed to connect directly to steam engines themselves. It was not, however, until the invention of the incandescent light bulb by Joseph Wilson Swan in England in 1878 and its improvement a year later by Thomas Edison in the USA, coupled with the appearance of the Parsons compound steam turbine in 1884, that the burgeoning and additional demand for electricity by the general public prompted the construction of large coal-fired power stations and distribution systems. The new electrical technology ushered in what has become known as the 'Second Industrial Revolution'.

1.3.2 Battery road vehicles
1.3.2.1 Cars
Early experiments with battery electric vehicles (BEVs) in the mid-19th century employed primary (non-rechargeable) batteries, which proved totally impractical and uneconomic. The invention of the lead–acid battery provided the means by which electric vehicles eventually became feasible. This was invented in 1859 by the French electrochemist Gaston Planté (1834–1889); see Figure 1.13(a). Taking inspiration from the popular design of earlier primary cells, Planté loosely rolled together two thin sheets of lead separated by a sheet of flannel or rubber strips, and then inserted the cylindrical assembly into a glass jar of the same geometry that contained a dilute (10 wt.%) solution of sulfuric acid; see Figure 1.13(b). The plates were charged for about 24 hours by a primary battery. A brown coating of lead dioxide (PbO_2) was formed on the positive and oxygen was evolved, whereas the appearance of the negative remained unchanged but hydrogen was liberated. Gaston Planté found the device to be 'a secondary element of great power'. On 26 March 1860, he demonstrated a battery of nine such cells to members of the French Academy of Sciences; see Figure 1.13(c). He wrote later that 'by passing through this apparatus the current from five small Bunsen cells, we obtained a very bright spark when the two terminal wires of the battery were brought into contact for an instant'. The device proved to be the first practical rechargeable ('secondary') battery and it rapidly found widespread application for the storage of electricity; an early battery of Planté cells is shown in Figure 1.13(d).

Towards the end of the 19th century, two alkaline rechargeable batteries also appeared, namely the nickel–cadmium and nickel–iron systems, which were invented, respectively, by Ernst Waldemar Jungner in Sweden (1899) and Thomas Edison in the USA (1901). Both designs employed an electrolyte of potassium hydroxide and a positive electrode of nickel oxide. Due to high material costs,

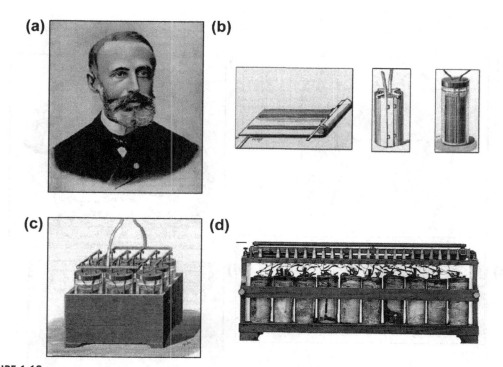

FIGURE 1.13

(a) Raymond Louis Gaston Planté (1834–1889); Gaston Planté's illustrations of (b) his seminal design of the lead–acid cell and (c) a battery of nine such cells; (d) an early battery of Planté cells

however, these two batteries failed to enjoy the same degree of market success as lead–acid for electric vehicle applications.

The construction of the world's first full-size electric vehicle is generally attributed to Gustave Trouvé of France, who, in November 1881, displayed his electric tricycle at the International Exhibition of Electricity in Paris; the battery consisted of six lead–acid cells. Unfortunately, no drawing of the tricycle appears to have survived. A few weeks later, Englishmen William Ayrton and John Perry built an electric tricycle with two large wheels at the rear, with the right one driven, and a small wheel up front; see Figure 1.14(a). The vehicle was fitted with electric lights and was propelled by a pack of 10 lead–acid cells that provided 0.5 hp. The speed was changed by switching the batteries on and off, one after another. The tricycle had a range of between 10 and 25 miles (16–40 km) and a maximum speed of 9 mph (14 km h^{-1}) as determined by the prevailing terrain.

The electric vehicle is a very simple machine: it can be said to have only five basic components and eight moving parts, four of which are wheels. In essence, the system consists of the battery, a controller, the motor, the transmission, and the vehicle chassis and body, and it was for this reason that the technology attracted much interest in the early days of powered transport. The first commercial battery vehicle is thought to be that produced in the early 1880s by the Paris Omnibus Company. Later, in 1894, H. Krieger introduced electric hansom cabs in Paris with great success. They had two electric motors

FIGURE 1.14

Early battery vehicles: (a) electric tricycle, William Ayrton and John Perry, 1881; (b) battery bus, Electric Motive Power Company, 1894; (c) *La Jamais Contente*, Camille Jenatzy, 1899; (d) passenger wagon, William Morrison, 1891; (e) Pope-Waverley electric runabout, 1905; (f) Thomas Edison (left) with a Bailey Electric after a 1000-mile endurance run using his new battery, which was recharged nightly; *c.* 1910 over the front axle, and each drove one of the front wheels. In case one motor broke down, the other was capable of propelling the vehicle on its own.

In January 1889, the Ward Electrical Car Company was granted permission by the London Metropolitan Police to carry out experimental runs with an electric battery bus. It earned some marks for silence and the lack of odours associated with hot oil or steam and the later familiar stench of incomplete combustion of vapourized petroleum spirit, but none for speed since it appeared to be capable of only 7 mph (11 km h^{-1}). In 1891, Walter C. Bersey began to operate an electric battery double-decker, with seating for 26 passengers, between Victoria and Charing Cross. Another such experimental service was conducted in Liverpool three years later by the Electric Motive Power Company; see Figure 1.14(b). This resembled an orthodox 26-seat horse bus with garden seats on the top deck and the battery stowed between the wheels. ('Garden seats' were wooden slatted seats that accommodated two persons and were transversely mounted to face the direction of travel; they were preferred to the alternative longitudinal arrangement, which was less comfortable and frequently led to squabbles between people sitting back to back.) Battery-powered taxicabs were built by Bersey in 1897 for the London Electric Cab Co. Ltd and operated for two years. He followed this up by constructing a lorry for the Post Office. In this lorry, three months before the repeal of the Red Flag Act, he exceeded the 2 mph (3.2 km h^{-1}) speed limit and was prosecuted. Earlier, in 1894, Bersey had built a luxurious phaeton, in which Edward, Prince of Wales, took his first drive in a car. The London Electrobus Company, which was formed in 1906, was popular amongst the public for their quiet and clean buses, although the batteries produced acid fumes which were less well received. At the peak of its success the company had 21 buses in operation, but was forced to close in 1910 following accusations of fraud. The vehicles were sold to the Brighton, Hove and Preston United Omnibus Company and continued in service for another six years.

During the same period, great rivalry developed between Count Gaston de Chasseloup-Laubat, who drove electric cars designed by Charles Jeantaud, a vehicle manufacturer in Paris, and Camille Jenatzy, who was a racing driver and built his own electric cars in Belgium. They broke one world speed-record after another. In the end, it was Jenatzy who triumphed, by attaining a speed of 105.8 km h^{-1} (65.8 mph) in a car with the proud name of *Jamais Contente* ('Never Satisfied'); see Figure 1.14(c). For the year 1899, it was a very respectable speed. Electric record-breaking was to end in 1902, however, when Walter Baker crashed while driving a streamlined car powered by a 40-cell battery and killed two people on Staten Island, USA.

Whereas the construction of the first battery-driven car in America is attributed to Fred Kimball in 1888, the first truly successful vehicle was a six-passenger wagon built by William Morrison in 1891. The latter had a top speed of 14 mph (22.5 km h^{-1}) and was powered by a 24-cell lead–acid battery that required recharging every 50 miles (80 km); see Figure 1.14(d). In 1898, the Riker Motor Vehicle Company produced a two-seat three-wheel car, and the Electric Vehicle Co. started to make the famous Columbia cars. Henry and Clem Studebaker began their manufacturing career with an electric car which gave 40 miles (64 km) to a battery charge.

The main problem with electric cars was that one could go fast for a short distance or slowly for a greater distance, but one could not go both far and fast. And yet by the turn of the century, electric cars were everywhere. Large fleets of electric taxis were plying the streets of London and Paris. In 1900, battery vehicles occupied a third of the space at the exhibition of the Automobile Club of America and 40% of the nation's cars were powered by steam, 38% by electricity and only 22% by fuel. In the UK, certificates of competence to drive electric vehicles were issued by the Electromobile Company (an example is shown in Figure 1.15) but, in truth, the electrics were probably easier to drive than either the steamers or the internal-combustion-engined vehicles (ICEVs) of the time. In America, the electric car became primarily a town runabout, a woman's car that was feminine in upholstery and décor; see Figure 1.14(e). By contrast, in Edwardian England the vehicle became a gentleman's carriage that was chauffeured by a liveried servant; it was as decorous as it was silent. Thomas A. Edison had become a strong proponent of electric vehicles, especially as a means to create a market for his patented nickel–iron battery. Soon after the battery went on sale in 1904, however, reports of leaks and other problems prompted Edison to discontinue its production. He dedicated the next few years and spent more than US$1 million of his personal fortune on improving the battery, which he reintroduced in 1909 and subjected to highly-publicized endurance trials; see Figure 1.14(f). Although it was claimed to extend the driving range to 100 miles (160 km) between charges, the technology failed to attract most electric vehicle manufacturers, who preferred to continue making cars with lead–acid batteries.

In spite of isolated successes, the electric car was unable to compete in the long run with the ever-improving ICEV, the development of which accelerated as a result of World War I. In particular, the

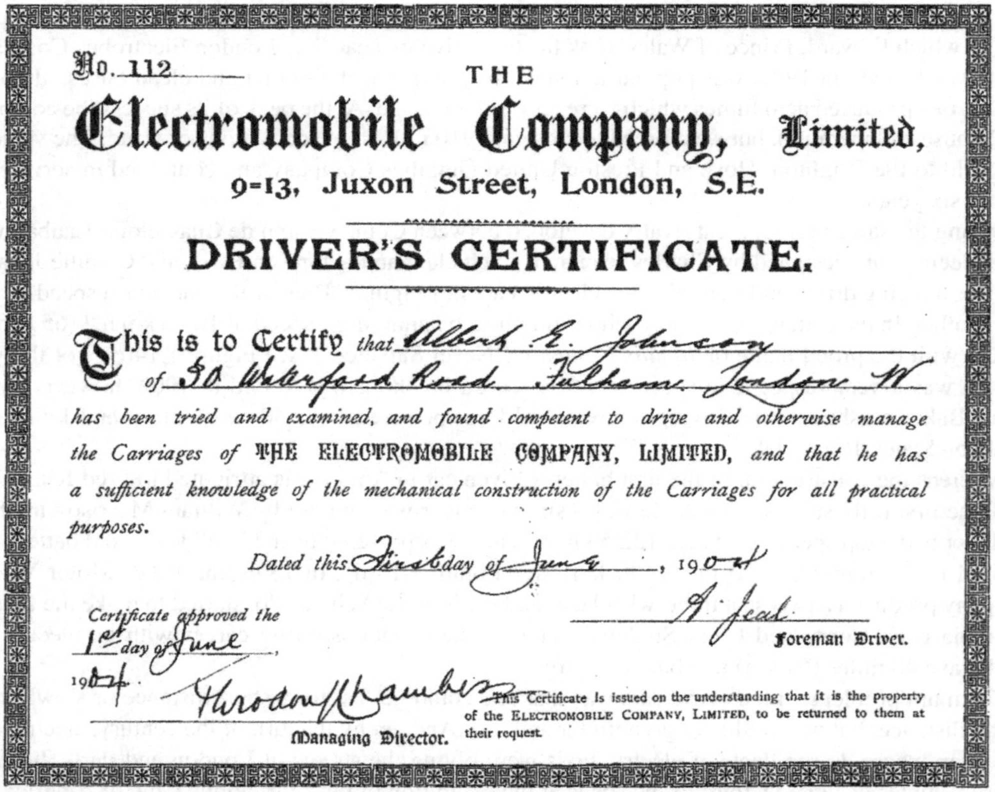

No. 112

THE

Electromobile Company, Limited,

9-13, Juxon Street, London, S.E.

DRIVER'S CERTIFICATE.

This is to Certify *that* Albert E. Johnson

of 30 Waterford Road Fulham London W

has been tried and examined, and found competent to drive and otherwise manage the Carriages of THE ELECTROMOBILE COMPANY, LIMITED, *and that he has a sufficient knowledge of the mechanical construction of the Carriages for all practical purposes.*

Dated this First *day of* June, 1904

A. Jeal

Foreman Driver.

Certificate approved the 1st *day of* June, 1904.

Throdon Chambers

Managing Director.

This Certificate is issued on the understanding that it is the property of the ELECTROMOBILE COMPANY, LIMITED, to be returned to them at their request.

FIGURE 1.15

1904 certificate of competence to drive an electric vehicle

relative primitiveness of the battery remained an insurmountable problem. By contrast, the greater driving range of petrol vehicles, faster refuelling and the burgeoning oil supply infrastructure caused worldwide attention in engine improvements to move from electrical propulsion to internal-combustion technology. Like steam cars, the electric form of road transport was virtually extinct by the mid-1920s, save for delivery vans for which speed was of little importance compared with easy starting/stopping and silence.

1.3.2.2 Buses and trucks

In America, where to venture outside the city limits could be a hazardous operation given the poor state of the roads, electric vehicles enjoyed a much longer security of tenure, and were employed in all sizes from light two-seat runabout cars to buses and urban delivery trucks; see Figure 1.16(a)–(c). Notable electric car companies included Waverley Electric, Pope-Waverley, The Woods Motor Vehicle Company, and Century Electric. Nevertheless, by 1925, most American electrics were no longer on the market, although the Anderson Electric Car Co. continued to offer their celebrated *Detroit Electric* against special order until 1939.

Special mention should be made of the hybrid power systems that were introduced by Tilling-Stevens in 1906 – particularly given the present-day interest in all types of hybrid electric vehicle; see Figure 1.16(d). The popularity of the company as an early builder of buses was largely due to the development and use of a 'petrol-electric' transmission rather than the conventional 'crash' gearbox of the period. The adoption of the simplified gear changing was significant because, at that time, very few bus drivers were accustomed to driving motor vehicles. The engine was connected to an electricity generator and the current produced was passed to a motor that drove the rear wheels. Unfortunately, the petrol-electric chassis was not considered suitable by the army for use during World War I because many men had been trained to drive on vehicles with conventional gearboxes. This led to a decline in popularity of the Tilling-Stevens system and by the 1930s chassis were being produced with conventional petrol/diesel engines, gearboxes and transmission.

Towards the end of the 20th century, battery electric buses were once again being produced given the mounting concern over the degradation of air quality in certain cities. Battery electric buses of various seating capacities for different services have been deployed in several countries, notably Australia, China, France, Germany, Japan, South Korea, the UK and the USA; examples are shown in Figure 1.17. Their limited range may be acceptable for vehicles that are used only in the morning and afternoon rush hours, as recharging can then be conducted during the day. For more general use, however, fleet managers have shown a preference for hybrid electric buses, which have somewhat smaller (therefore less costly) batteries and no range limitation; see Section 5.2.6, Chapter 5.

The development of battery electric trucks has also seen only limited progress. The door-to-door delivery of milk was an early and successful application. This service typically involves short journeys (20–40 km, or 12–25 miles) with frequent stops – a driving schedule that is ideal for electric vehicles, but would make heavy demands on an internal-combustion engine. Moreover, milk is delivered early in the morning when customers appreciate the silence of the electric milk cart. At one stage in the mid-20th century, about 40 000 'milk floats' were operating in the UK (see Figure 1.18), but now there are far fewer as a result of milk sales through supermarkets.

There have been attempts to introduce the use of battery electric trucks for postal deliveries where the duty cycle is similar to that for milk delivery. Extensive trials have been carried out in Austria, Germany and the USA (see Figure 1.18(b)), but have failed to lead to the implementation of such vehicles on a national scale for largely logistical reasons, e.g. delivery of mail to difficult locations.

FIGURE 1.16

(a) Electric sightseeing buses, New York, 1904; (b) Edison battery bus in the Lancaster Corporation fleet, UK, taking charge, 1916; (c) electric trucks, General Motors Corporation, 1912–1917; (d) Tilling-Stevens petrol–electric buses operating in Melbourne, Australia, *c.* 1915 (left) and Kent, UK, 1908 (right)

(Part b from A. Thomas, 2007, Buses: The Evolution of Passenger Road Transport, *London: Courtesy of Bounty Books)*

1.3.2.3 Bicycles and scooters

Shortly after the invention of the bicycle in the late 19th century, battery-driven bicycles and tricycles appeared in Britain and France, e.g. see Figure 1.14(a) above. These were among the first forms of battery-powered personal road transport and, in recent years, have proved to be by far the most

FIGURE 1.17

Battery electric buses: (a) Seoul, South Korea; (b) St Helens, UK; (c) Stockton, California, USA; (d) Beijing, China

(All images sourced from Wikipedia and available under the Creative Commons License)

FIGURE 1.18

Milk delivery vehicle ('milk float') in the UK, 1980s

(Image sourced from Wikipedia and available under the Creative Commons License)

successful, particularly in China, where many millions have been sold (see Section 5.3.3.3, Chapter 5). Because the battery must not be too large or too heavy, these are mostly electrically-assisted bicycles and the role of the electric motor is simply to make pedalling easier for the rider. Typically, the battery provides 150–250 W of power. The machines are known as 'pedelecs', or 'e-bikes'. There are also more robust versions with batteries of 250–350 W where little or no pedalling is required and a range of 30 to 40 km (19–25 miles) in electric mode is possible. Collectively, these two types of bicycle are referred to as 'electric two-wheelers' (E2Ws).

Electric scooters have also been developed and are promoted as being more stable than bicycles for riding in cities. Nevertheless, sales have so far not matched those of electric bikes. Examples of an electric bicycle and an electric scooter are shown in Figure 1.19(a) and (b), respectively.

1.3.2.4 The eclipse of the battery road vehicle

Gradually, during the early part of the 20th century, ICEVs won out over steam and electric cars even though the latter two technologies had far simpler power sources. In the early days of motoring, the disadvantages of the ICEV were the complexity and unreliability of the engines, as well as the need to start them manually. Hand-cranking required physical strength and could prove hazardous when the engine backfired. Ironically, it was the invention of the battery-operated self-starter that solved this problem and thereby is said to have sounded the final death knell of the electric car as a prime form of personal transport; see Section 4.1, Chapter 4. Other major factors militating against the BEV were the mass and size of the traction batteries, the requirement for their frequent maintenance by 'topping-up' with de-ionized water, the long recharging ('refuelling') time and, above all, the limited vehicle range between charges.

From time to time, there have been occasions when a resurgence of enthusiasm for the BEV might seem possible, most noticeably at moments when supplies of crude oil have appeared to be threatened or legislation has been proposed to restrict the sale of ICEVs on environmental (air quality) grounds. To date, however, even though most of the major car manufacturers are able to offer 'sustainable' models, the prospects for large fleets of BEVs do not look strong, except perhaps for small vehicles

(a)   **(b)**

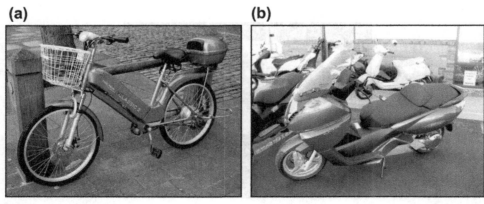

FIGURE 1.19

(a) Electric bicycle; (b) electric scooter

(Images sourced from Wikipedia and available under the Creative Commons License)

dedicated to city use. Better and cheaper batteries are still required before the number of BEVs on the road will grow substantially.

1.3.3 Off-road battery vehicles

Notwithstanding their disadvantages, BEVs have found acceptance in industry for specific applications where their attributes (silent operation, lack of exhaust emissions, ease of operation, good acceleration) have outweighed their limitations. Examples of off-road duties include haulage operations in mines and indoors transport in warehouses, factories, hospitals, etc. Most of these vehicles are forklift trucks, platform trucks or tractors that are employed for handling and conveying goods. Elsewhere, some airport authorities have chosen electric tractors to convey disabled passengers within terminal buildings, to transfer luggage to and from aircraft, and sometimes to 'push out' planes from their stands.

Battery electric vehicles also find use in recreational activities. Notably, electric golf carts and buggies are found on courses throughout the world. Because of their fume-free operation, BEVs are becoming increasingly popular at other sporting events; for instance, to allow journalists, cameramen and officials to follow horse or bicycle races without disturbing the competitors.

Off-road deployments are all niche applications, however, and do not hold the potential to upset the balance between the employment of ICEVs vis-à-vis BEVs in the road transport sector.

1.3.4 Mains electric traction

From quite early in the 20th century, it was recognized that electric vehicles could avoid all the shortcomings of batteries by using mains electricty. Consequently, two types of mains-powered vehicle have seen extensive service in public transport—namely, the tramcar and the trolley bus.

1.3.4.1 Tramcars

The earliest mains-electric road vehicles were tramcars (streetcars) that ran on steel rails embedded in the road. Tramcars were a logical development of the horse-drawn tram of the 19th century. The electricity was supplied through overhead wires which were supported by stanchions mounted on the pavement/sidewalk. The drivers of tramcars had no steering to do, but occasionally had to dismount to change the points at forks in the road. This was not, however, the only limitation. Overtaking was not possible without first lowering the power take-off arm of the lead tram from the wire. The vehicles were noisy, with steel wheels running on steel track, and with loud clanging bells to advise other road users of their approach. Tramlines were also dangerous, particularly for cyclists, who could be thrown off if one of their wheels became caught in the track. Further, as the twin tramways were generally located in the centre of the road, there was limited space for cars to overtake and customers were at risk when they crossed between the pavement and the tram, or vice versa. The upper deck was often open to the elements and so travel was not the most pleasant or comfortable of experiences. Modern tramcars tend to be single decked or to have enclosed upper decks. A few cities have learned to live with the disadvantages of trams given the environmental benefits they confer, but far more have reverted to buses or trolleybuses. Examples of old and modern tramcars are shown in Figure 1.20.

A variant of the tramcar is the cable car that is used in hilly cities. As described earlier in Section 1.2.5, traction power is supplied via a cable embedded in a third track between the wheels that serves

FIGURE 1.20

Development of trams in the UK: (a) 1892 South Staffordshire, one of the earliest photographs of this new form of transport; (b) 1903 Camborne, curtains on windows still remain; (c) 1906 Portsmouth, decorated with illuminations to celebrate a local event; (d) Leicester 1920s tram from the Tramway Museum, Crich, UK, courtesy of Dr Neil Clifton (e) Dresden, Germany; (f) Melbourne, Australia

(Images e and f sourced from Wikipedia and available under the Creative Commons License)

to haul the vehicle along its route, including up steep hills. Whereas in the first systems the cable was powered by steam, modern cable cars make use of mains electricity. The San Francisco cable car remains an iconic emblem of the city even today.

Regardless of their shortcomings, new tram systems are still being installed in some cities in the early years of the 21st century.

1.3.4.2 Trolleybuses

Most of the disadvantages associated with tramcars are overcome in trolleybuses; see Figure 1.21. These vehicles still use an overhead electrical power supply but they run on conventional wheels with rubber tyres so that the noise and points problems associated with tracked vehicles are avoided. In addition, a trolleybus has superior braking power to that of a tram and it is better at climbing hills since rubber tyres have more grip than steel wheels on steel rails. The steering is similar to that of conventional buses and thereby allows flexibility of movement in traffic and enables the vehicle to pull into the side of the road to pick up passengers. A notable technical difference between trams and trolleybuses is that the latter require two overhead wires to complete the electrical circuit, whereas trams require only one wire with the current returning through the track.

Trolleybuses progressively replaced tramcars in many cities during the 1940s and 1950s so that, eventually, there were some 900 systems in service around the world. A large share of these were withdrawn, however, during the 1960s and 1970s as transport options moved to private cars and diesel buses. Currently, over 300 cities in 45 countries still operate trolleybus systems.

1.4 The age of the motor vehicle – from dream to necessity
1.4.1 The internal-combustion engine

As outlined above, the early decades of the 19th century saw the introduction of a number of steam-driven road vehicles. Steam traction, however, suffers the particular disadvantage of being an inefficient user of fuel because an external-combustion engine is involved – that is, the fuel is burned outside the engine itself. Engineers realized well enough the advantages to be gained by burning the fuel inside the engine, despite the fact that it would impose a constraint on the type of fuel that could be used (liquid or gas). It was not until 1859, however, that Jean Joseph Étienne Lenoir, a Belgian engineer who, like so many other inventors, was completely self-taught, built the first successful internal-combustion engine; see Figure 1.22(a). In keeping with the early steam engines, it was designed as a static engine. The fuel was 'illuminating' gas (the coal gas used for street lighting, which had developed as a 'spin-off' from the need for coke for steel-making) mixed with air. The machine employed one double-acting cylinder; that is, there was a combustion chamber each side of the piston. As the piston moved along the cylinder, the combustion mixture was sucked into one of the chambers. On reaching the end of its run, a spark ignited the mixture and the piston was forced back in the other direction, which at the same time sucked fresh mixture into the chamber on the other side, and so on. Obviously, it was not very efficient – the 18-litre machine delivered less than 2 hp at 100 rpm and this represented only about 4% of the potential energy in the gas. Nevertheless, the engine did work. Lenoir patented his invention in 1860 and several hundred units were sold.

That same year, 1860, a newspaper report of Lenoir's machine was read by Nikolaus August Otto, a 28-year-old travelling salesman from the Rhineland; see Figure 1.22(b). He immediately started to

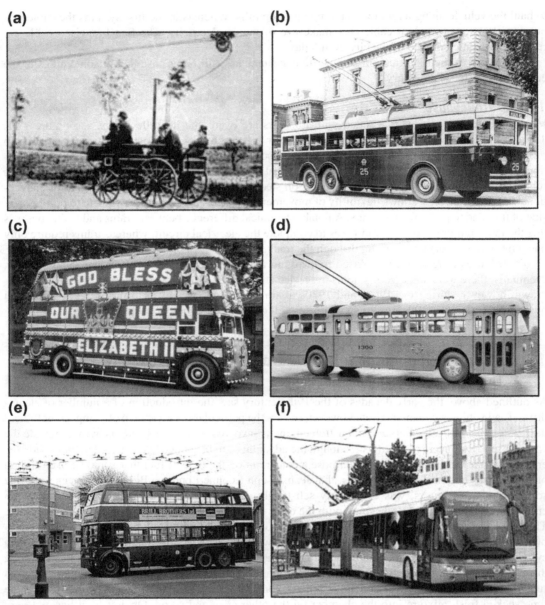

FIGURE 1.21

Evolution of the trolleybus: (a) 1882 *Elektromote*, Berlin, Germany; world's first electric vehicle; (b) 1935 Hobart, Australia; (c) 1953 Portsmouth, UK; decorated for coronation of Queen Elizabeth II. With kind permission from Middleton Press; (d) 1947 Cincinnati, USA; (e) 1966 Reading, UK; negotiating a turning circle; (f) 2009 Lyon, France

(Images a, b, d, e and f sourced from Wikipedia and available under the Creative Commons License)

(a)

Jean Joseph Étienne Lenoir
1822 – 1900

(b)

(c)

Nikolaus August Otto **Karl Friedrich Benz**
1832–1891   1844–1929

(d)

(e)

Gottlieb Wilhelm Daimler
1834 –1900

Rudolf Christian Karl Diesel
1858–1913

FIGURE 1.22

(a) Lenoir and his motor preserved in the Musée des Arts et Métiers, Paris; (b) Nikolaus Otto; (c) Karl Benz and his 1885 three-wheeler; (d) Gottlieb Daimler and being driven in the first four-wheeled car; (e) Rudolf Diesel

(Images sourced from Wikipedia and available under the Creative Commons License)

experiment on his own account and in 1876, after many mechanical detours, produced and patented the first engine to operate with a four-stroke cycle. The subsequent success of this system depends on the fact that the fuel–air mixture is compressed before firing (see Section 4.3.1.2, Chapter 4). On the first down-stroke of the piston, the mixture is drawn into the cylinder through a valve. On the following up-stroke, the mixture is compressed. At the top of that stroke, spark ignition is applied so that the next down-stroke is the power stroke, in which the piston is forced down by the expanding gases. The final up-stroke drives the used gases out through another valve. The system is called, after its inventor, the 'Otto cycle'.

In fact, the principle of the four-stroke cycle had already been expressed in 1862 by Alphonse Eugène Beau de Rochas, a French civil engineer, through both a patent and a pamphlet, which he distributed to the press. Otto was able to take out his own patent, however, since de Rochas's description of the cycle lacked sufficient detail and also the accompanying patent had lapsed through failure to pay the annuity. Yet, later in the 1880s, when Otto brought an action against French engine producers on the grounds that the activity infringed his patent, they successfully pleaded that Beau de Rochas had anticipated him. The overthrow of the Otto master patent obviously had a profound effect on the development of the motorcar, as all manufacturers were now free to use the four-stroke principle.

The question of who made the first car becomes crucial in the 1880s. Nicholas Cugnot (*v.s.*) rules undisputed as the man who, in 1769, made the first vehicle move under its own power. Following Otto's invention, much effort was devoted to finding a practical way of making the internal-combustion engine drive a vehicle. It had to be a light power source that was capable of a great number of revolutions a minute and that could run on a liquid fuel, since gas would require huge containers if reasonable journeys were to be achieved. The man generally credited with success was Karl Friedrich Benz, who in the spring of 1885 drove his three-wheeler round a cinder track at his factory in Mannheim, Germany; see Figure 1.22(c). His machine ran on petrol, which fed a single-cylinder four-stroke engine that produced 1.5 hp. The vehicle achieved 10 mph (16 km h^{-1}) and since history records that he was accompanied on the circuit by his workers, with his wife clapping enthusiastically as they ran, one can only assume that either they were all very fit, or that he slowed down out of kindness. Benz applied for a patent in early 1886 (Benz Patent DRP 37435, 29 January 1886) and thereby, according to the official history issued by the modern Mercedes company, 'paved the way for the motor vehicle as a complete unit'.

At Bad Cannstatt, some 60 miles (96 km) away, two of Otto's former assistants had set up their own business. They were Gottlieb Wilhelm Daimler and Wilhelm Maybach, and later in the same year (1885) they successfully drove the first two-wheeler to feature an internal-combustion engine; see Figure 1.27(a). The vehicle was an extremely crude affair and was only ever intended as a mobile test bed for their 264-cc four-stroke single-cylinder power unit. The little air-cooled engine had two fly-wheels, one each side of the crankshaft, all enclosed in a cast-aluminium crankcase. Starting was by crank handle. With a weight of 90 kg (198 lb), the machine produced 0.5 bhp at 750 rpm, and the two gear ratios gave speeds of 3.5 and 7 mph (5.6 and 11.2 km h^{-1}), respectively. Other notable features included an evaporating carburettor, heated tube ignition and almost conventional handlebars. Following a series of encouraging test sessions, first in the garden of his house and then in the streets of Cannstatt (where the longest journey was some 3 km), Daimler patented his design on 29 August 1885. Although this was the first true high-speed internal-combustion engine (the Benz tricycle achieved only 250 rpm), Daimler soon realized not only that the engine was not sufficiently powerful, but that his *Petroleum Reitwagen* ('Riding Wagon') or *Einspur* ('single track'), as he alternatively named it, was not easy to use. It was both difficult to balance and a real boneshaker, thanks in no small part to the dire state of the roads at that time. So instead, he concentrated his efforts upon the development of motorized carriages – a decision obviously strengthened by the resounding success of Karl Benz. Nevertheless, through his construction of the *Reitwagen*, some historians have ascribed to Daimler the title of 'father of the motorcycle'. As discussed in Section 1.4.3, however, this accolade may well be misplaced. Moreover, it is unfortunate that Maybach's contribution has been ignored.

On 8 March 1886, Daimler converted a four-seater phaeton to hold his engine and thus he designed the world's first four-wheel motorcar; see Figure 1.22(d). Shrewdly, Daimler ensured that the firm of coachbuilders was ignorant of fact that its carriage was to be engine driven; he said that it was a birthday present for his wife. When his old firm, Otto and Langen – in which he still had shares – turned down the offer of manufacturing rights he set up on his own to make cars.

Benz, meanwhile, was anxious to promote his motor vehicle commercially. An upgraded version was put on the market in 1888. It caused much interest but found few buyers. Business was soon to improve for Benz, however, following an action by his wife, Bertha, that has come to be regarded as a key event in the technical development of the motorcar. On 5 August 1888, without telling her husband and without permission from the authorities, Bertha and her two sons Eugen and Richard, 15 and 13 years old, set out from their home in Mannheim in one of the cars to visit other members of the family in Pforzheim, a distance of about 65 miles (105 km). The story goes that on the way they bought motor spirit from a chemist, had blacksmiths take up the slack in the chains, and persuaded cobblers to reline the crude brake shoes with leather. Frau Benz herself cleared a clogged fuel-line with a hairpin and used one of her garters as insulating tape to cure an electrical short. This expedition – the first long car-run in history – received a great deal of publicity (which was undoubtedly Bertha's intention) and resulted in steady sales of Benz vehicles.

It is quite remarkable that, although they worked only 60 miles (96 km) apart, Daimler and Benz never met. On 28 June 1926, long after Daimler's death, their firms formally merged to become Daimler-Benz AG and produce cars under the brand name of 'Mercedes-Benz'. That the names of the two men became linked in one of the most respected car companies is one of the most notable ironies of history. Three years after the merger had begun when the 85-year-old Benz lay bed-ridden in his home, a motorcade of vehicles from all over Germany paraded past his house in an Easter salute with the slogan 'Do Honour to Your Master', and a flight of Baden Flying Club planes dropped a victor's wreath near his Villa Benz. A few days later, on 29 April 1929, the 'father of the motorcar' died.

The compression-ignition engine, which needs no spark, was built at Augsburg in 1897 by yet another German engineer, the 39-year-old Rudolf Christian Karl Diesel; see Figure 1.22(e). This was based on the principle that if air is compressed sufficiently (far more than in petrol engines) it will get so hot, from the mere fact of compression, that when the fuel is then sprayed into the cylinder the heat of the air will ignite it. The diesel engine has the disadvantage, however, that it will do far fewer revolutions a minute than the petrol engine. On the other hand, it has two great advantages: it will burn very low-grade fuel (inherently cheaper because it avoids the refining processes of petrol), and it is far more fuel-efficient than the best petrol engines.

A limitation on the efficiency of all combustion engines is imposed by the 'Carnot cycle', which expresses the thermodynamic limitation on the fraction of heat supplied to an engine that can be used to perform useful work. Electric motors do not suffer from any such loss in efficiency, but this attribute has still not been sufficient for electric vehicles to be generally preferred over ICEVs.

1.4.2 Cars

Many different types of ICEV were built in small numbers in the late 19th and early 20th centuries, particularly in Europe and the USA. By way of example, a small French car of 1903, the de

(a)   **(b)**

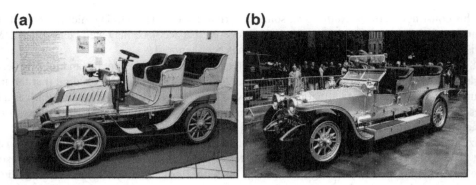

FIGURE 1.23

Examples of early cars: (a) 1903 de Dion & Bouton *8hp*; (b) 1906 Rolls Royce *Silver Ghost*

(Images sourced from Wikipedia and available under the Creative Commons License)

Dion & Bouton *8hp* (6 kW), and the iconic Rolls Royce *Silver Ghost* introduced in 1906 are shown in Figure 1.23(a) and (b), respectively. The latter was a huge, expensive car with a six-cylinder, 7-litre engine, but still only producing 40–50 hp (30–37 kW).

In the early days, the motorcar was the rich man's plaything – but not for long. In 1903, a farmer's son from Michigan set up a car factory. His name was Henry Ford and he saw that the car would become not a luxury, but a necessity. The early motorcar companies, for the most part, produced each vehicle as an individual, custom-built machine. Ford was anxious to introduce other ideas, which involved the cutting of costs by standardizing the production of each component and the complete assembly of the vehicle. He believed that as the price was brought down the market would expand and profits would increase so that, in turn, this would enable the price of the motorcar to be reduced again. This cycle would produce cheap motorcars and generate a huge profit. Ford openly declared:

> *I will build a car for the great multitude. It will be large enough for the family, but small enough for the individual to run and care for. It will be constructed of the best materials, by the best men to be hired, after the simplest designs that modern engineering can devise. But it will be so low in price that no man making a good salary will be unable to own one – and enjoy with his family the blessing of hours of pleasure in God's great open spaces.*

Ford's contemporaries regarded such notions as folly. Undeterred by such views, however, he set up the Ford Motor Company in 1903, where he produced his first car, the *Model A*; see Figure 1.24(a). Five years later, Ford launched his seminal 'car for everyman', the inexpensive *Model T*, which became affectionately known as the *Tin Lizzie* (the origin of the nickname is a subject of some conjecture).

There was one main assembly-line for the *Model T* chassis and engine unit; see Figure 1.24(b)–(d). The body unit was put together separately and once the engine was tested it was brought down a chute to be lowered on to the chassis and bolted into place. Once this was done, the car could be driven away. The innovations that Henry Ford introduced into his Piquette Plant in Detroit paid off handsomely and thereby had two far-reaching social consequences. They revolutionized the whole basis of vehicle production, and brought the motorcar within the financial reach of the average person.

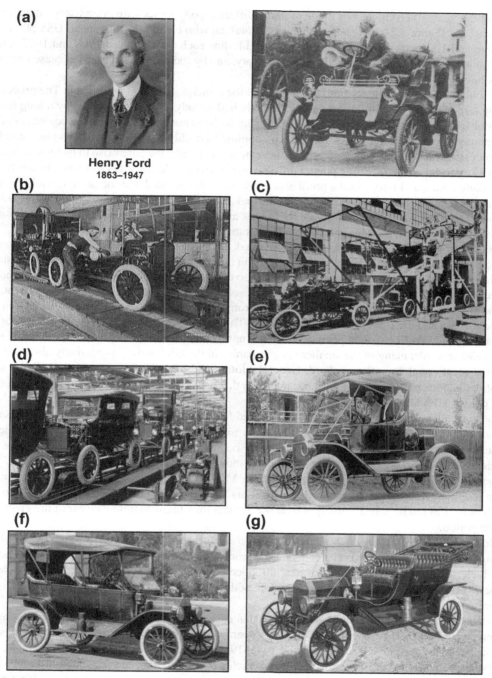

FIGURE 1.24

(a) Henry Ford and in his *Model A*, 1903. (b)–(d) scenes from the *Model T* assembly-line. *Model T* body styles: (e) roadster; (f) coupé; (g) tourer

(Images sourced from Wikipedia and available under the Creative Commons License)

Over the years, the vehicle was made in many different body styles with completely interchangeable parts; see Figure 1.24(e)–(g). By 1913, any American who could afford to spend US$ 500 could buy a car, and 1000 cars were coming off the assembly-line each day. Between 1908 and 1927, a total of 15 million *Model T* Fords emerged from the factory, and by the time this classic car ceased production its price was down to $290.

It should be noted that Henry Ford did not invent the principles of mass production. The process of creating large numbers of similar products efficiently had already been implemented by a long line of industrialists that started with Eli Whitney and his standardized gun components at the beginning of the 19th century. George Eastman was another who applied assembly-line methods in his photographic-film processing factories. Ford's particular achievement was to take the enterprise to its logical conclusion in a large-scale manufacturing plant. Ford also increased the salaries of his workers while reducing the working hours, and introduced a profit-sharing scheme. He insisted that the profits should be used to enable the company to expand. This was an unusual idea at the time and contrary to the wishes of the other stockholders, who wanted to divide the profits among themselves. Ford disliked the opposition and eventually bought out all his stockholders to assume sole control of the company.

The *Model T* only ceased production because the demands of the market were changing. Customers wanted extra comforts and the latest model designed in the current fashion to advertise their social standing. Never again would there be motorcars so cheap, so durable, and so available across a wide economic spectrum. The market dictated the ethos of built-in obsolescence, and even Ford was forced to adapt.

The success of Ford's production techniques prompted many other manufacturers in continental Europe, Britain and the USA to follow his example. They competed on quality, style and technological innovation, but to do likewise on price was difficult until a comparable mass production was achieved. From 1930 onwards, many of the smaller car companies in the USA were progressively absorbed into the three great conglomerations – Ford, General Motors and Chrysler. To a lesser degree, the same consolidation took place in Europe as it became clear that only companies with high production outputs could deliver affordable cars.

Those smaller companies that were not consumed focused on niche markets – prestige cars, sports cars, etc. One such British company was MG Cars, formed in 1924, which manufactured a series of two-seater sports cars in the 1930s that were highly popular. Like the *Model T* in America, these MG models have since become classic cars and are still much sought-after by enthusiasts. The MG *J-type Midget* (1932–1934) with an 847-cc engine and the MG *TD* (1950–1953) with a 1250-cc engine are shown in Figure 1.25(a) and (b), respectively. The cars had a top speed of around 65 mph (104 km h^{-1}) and 80 mph (128 km h^{-1}), respectively, which was typical for a high-performance car of the given production period.

Most early cars, including the MGs, were built on a substantial steel chassis to which the non-load-bearing body was then attached. The concept of a monocoque structure, in which the external body of the vehicle supports the structural load, was introduced much later; see Section 4.2, Chapter 4. Many of these early chassis-based cars had wooden bodies that were clad in light-gauge steel or in aluminium.

In the USA, the focus was on much larger sedan cars which were capable of accommodating six passengers and all their belongings. Styling was deemed all-important for sales. Consequently, some extravagant styles appeared during the period 1945–1965 and incorporated much chromium plate, two-tone colour schemes and whitewall tyres; four vehicles from the 1950s are presented in Figure 1.26. These cars were fuelled exclusively with petrol. Diesel engines were confined to trucks and buses.

FIGURE 1.25

MG sports cars: (a) *J-type Midget* (1932–1934); (b) MG TD (1950–1953)

(Both images sourced from Wikipedia and available under the Creative Commons License)

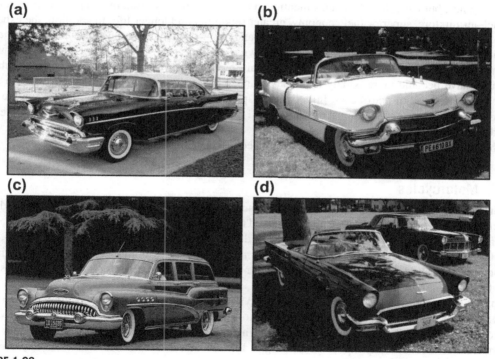

FIGURE 1.26

American cars of the 1950s: (a) 1957 Chevrolet *Bel Air* coupé; (b) 1956 General Motors *Cadillac Eldorado Biarritz.* (c) 1953 Buick *Roadmaster Estate Wagon;* (d) 1957 Ford *Thunderbird*

(All images sourced from Wikipedia and available under the Creative Commons License)

After the oil crisis of 1973 and the consequent rise in fuel prices, some Americans turned away from the over sized, eight-cylinder, 'gas guzzling' monsters they had been accustomed to driving and were prepared to consider a car of more modest size with four or six cylinders. At the same time, improvements were being made in engine design and efficiency so that just as much (or more) power could be obtained from the smaller engines. This was seized upon as an export opportunity by European and Japanese car manufacturers, who had a long history of building smaller cars. Japanese companies (Toyota, Honda, Nissan), in particular, expanded their outputs and addressed vigorously the opportunities afforded by the US market and, in due course, by the European and then world markets. European manufacturers had some success in penetrating the same market, especially with their more exclusive and expensive models (Mercedes, BMW, Ferrari, etc.), but were less successful than the Japanese companies when it came to mass sales. Later, in the 1990s, manufacturers from other Asian countries, e.g. Korea and Malaysia, entered the sector with models that competed on price. Finally, the concept arose of designing a car that would be acceptable in all countries. Following the Japanese lead, this idea was taken up in the USA by Ford, who claimed that their *Mondeo* model was the first 'world car', and named it accordingly.

From the 1980s onwards, some European manufacturers, especially the French, started to offer their cars with a choice of petrol or diesel engines. Diesel engines are rather more expensive to produce than petrol engines, but have the advantages mentioned above, notably: ignition by simple compression of the fuel–air mixture, superior fuel economy, greater reliability and longer life. In addition, the exhaust gases are less corrosive towards the exhaust system and, in an accident situation, diesel fuel is less liable to ignite than petrol. Diesel engines have been known to last for up to 300 000 miles (480 000 km) with minimal attention, which is one reason why they are preferred for use in taxis. In recent years, these attributes have become more widely appreciated and sales of diesel cars in Europe have grown at a faster rate than those of petrol cars. Initially, the Japanese manufacturers, with their focus on the US market (where diesel cars are almost unknown), did not offer diesel engines in their cars, but some are now doing so.

The engineering development of ICEVs is treated in greater detail in Chapter 4.

1.4.3 Motorcycles

The invention of the motorcycle could not take place until the two key elements – the motor and the bicycle – reached a stage of refinement that allowed them to be combined into a practical machine. The parallel development of the bicycle, with its strong diamond pattern frame (see Section 1.1 above), and the internal-combustion engine reached fruition at an ideal moment to allow the creation of a viable motorized cycle. Enthusiasm for two-wheeled transport swept through Europe and America as the invention captured the public imagination.

In fact, it is impossible to bestow on any individual the credit for having 'invented' the motorcycle. There was much preliminary development to be accomplished before any kind of motor vehicle could prove practical. Following the pioneering work on engines by Otto, Benz, Daimler and Maybach, the idea of adapting bicycle designs for the new internal-combustion engine clearly occurred to many inventors and engineers across Europe at around the same time, namely, the late 1880s. It is therefore somewhat ambitious to attribute the birth of the motorcycle solely to Daimler. Moreover, the *Reitwagen* actually had four wheels – history seems to have conveniently overlooked the two stabilizers; see Figure 1.27(a).

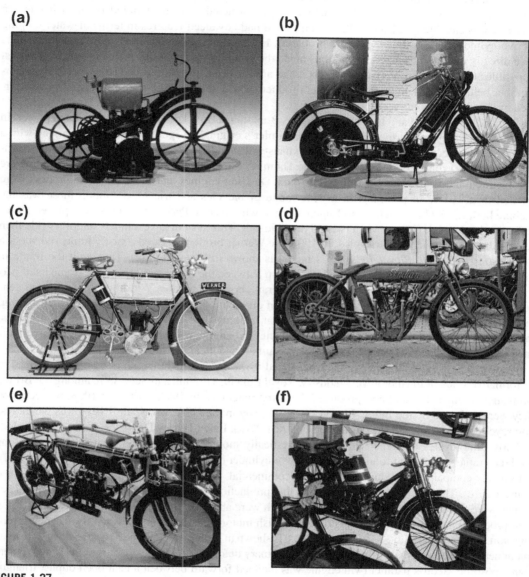

FIGURE 1.27

Early evolution of the motorcycle: (a) 1885 Daimler and Maybach; (b) 1894 Heinrich Hildebrand and Alois Wolfmüller. By courtesy of Deutsches Zweirad und NSU Museum, with many thanks to Ms. Dumas & Ms. Grams; (c) 1904 Michel and Eugen Werner; (d) 1914 Indian; (e) 1910 FN four-cylinder; (f) 1913 Scott two-cylinder

(Parts a and c-f sourced from Wikipedia and available under the Creative Commons License)

In 1894, Heinrich Hildebrand and Alois Wolfmüller designed the first commercially-built motorized two-wheeler; see Figure 1.27(b). The machine was constructed in Munich and also, under licence, in France. It was capable of 24 mph (38.6 km h^{-1}) and had advanced features in terms of water-cooling and two-cylinders, but also some severe problems. For instance, the connecting rods from the pistons drove directly on to the back wheel and thus made the drive far from smooth at low speeds, and almost impossible at less than 5 mph (8 km h^{-1}). As it had no flywheels, the pistons had to be forced by means of rubber belts to complete the whole cycle. Production was short-lived because the design was over-taken by developments in France.

Count Albert de Dion and his partner Georges Bouton had been keen enthusiasts of steam power. In 1884, however, they built a motor based upon Daimler's original design. It was a 120-cc engine with a vacuum inlet valve, a mechanically operated exhaust valve and electric ignition. It was capable of turn-ing over at 1800 rpm, i.e. double that of the Daimler. Subsequently, different sizes of engine were manufactured and sold. Over the following years, De Dion engines and those of their imitators were installed in a variety of positions in a wide range of bicycles and tricycles. Among these was the machine built in 1897 by Michel and Eugen Werner, who lived in Paris. The engine was placed above the front wheel which it drove via a twisted rawhide belt. This system gave a much smoother ride than the gear drive of the De Dion tricycles. In 1901, the Werner brothers created a vastly improved version of their so-called *Moto Bicyclette*; a 1904 model is shown in Figure 1.27(c). They split the frame in front of the pedals and bolted the engine into the gap between the two down-tubes. The frame itself was strengthened by adding a horizontal member that ran above the engine. With the engine located cen-trally and low, the new layout improved the weight distribution and resulted in much better handling. The redesigned bikes also boasted electric ignition, a spray-type carburettor, and wheel-rim brakes. Thus the new Werner had moved beyond the bicycle-and-engine concept. These machines ably domi-nated the international races of 1902. The first genuinely practical motorcycle had arrived.

The Werners also had a factory in Coventry and it was British manufacturers who took the lead in the formative years of the motorcycle industry. Royal Enfield, Triumph and Scott were among the British companies to enter the field and produced their first machines in 1901, 1902 and 1908, respectively. They were followed by the Indian Motorcycle Company in the USA, which until 1914 was the largest motorcycle manufacturer in the world; see Figure 1.27(d). Later, Harley Davidson assumed this role.

Early motorcycles operated with a single vertically mounted cylinder and a chain drive. Designs developed rapidly and led to four-cylinder and twin-cylinder models. Fabrique Nationale de Herstal (FN), a Belgian company, introduced the first commercial four-cylinder motorcycle in 1907; see Figure 1.27(e). The machine was so advanced that production continued for over 20 years, though later models were fitted with a clutch and gearbox, which were absent in the initial design. A year later, Alex Scott produced his twin-cylinder two-stroke, a British motorcycle that is still revered for its superb run-ning, soft purr and flexibility; a 1913 Scott 550 cc is shown in Figure 1.27(f). The new layout was a revela-tion in neatness and originality. Instead of the customary diamond or triangular frame and 'New Werner' engine position, Scott's parallel twin engine was inclined forward between a neat open duplex triangu-lated frame, which strongly recalled the Hildebrand and Wolfmüller configuration of 14 years earlier.

Before World War I, motorcycles had achieved speed and stamina as a result of intensive develop-ment for racing. The frame became heavier, the seat was lowered, the front forks incorporated springing to protect the rider from road shocks, and the engine was mounted where the pedal cranks had been previously. Though in some cases the engine still turned the wheels by means of V-belts running in pulley wheels, up-to-date machines already had a gearbox that provided two or three gear ratios between the engine and the rear wheel, a friction clutch, and power transmission by means of roller chains. The final refinement was a kick-starter, a single pedal crank that enabled the rider to start the engine without having to push the machine. By 1914, the constituents of modern motorcycles had been used with varying degrees of success.

Between World Wars I and II, the basic arrangement and componentry were both improved in detail and performance, and the British continued to dominate the motorcycle industry with well-constructed machines of conservative design; the Germans and Italians were their main rivals. The most popular type of machine had a one-cylinder engine; attempts to market more advanced versions with four-cylinder engines and shaft power transmission met with less success. This period witnessed the advent of Norton as one of the main manufacturers of high-powered motorcycles. The attractiveness of British motorcycles on the world market was enhanced by a string of major racing successes throughout the 1920s and 1930s.

By the start of World War II in 1939 there was a huge selection of manufacturers in Britain and fierce competition between them. Two companies in particular, BSA and Royal Enfield, produced motorcycles in large numbers for military use. Most of these were fairly elementary, single-cylinder models with no great refinements.

After 1945, there was consolidation among British motorcycle manufacturers and many well-known brands disappeared. Nevertheless, the country remained the major producer of machines, except for the USA, where there was still a strong attachment to larger and higher-powered motorcycles of the type produced by companies such as Harley Davidson. There was also a period when a proportion of motorcycles were fitted with sidecars. This was seen as a comparatively cheap form of transport for families who could not afford to buy a car.

As they started to rebuild their war-shattered economies, German and Italian manufacturers concentrated instead on the considerably lower-powered scooter with smaller wheels and improved weather protection; see Section 1.4.4. Gradually, however, these companies returned to the motorcycle arena with machines that first rivalled, and then exceeded, the more traditional British motorcycles in both technical sophistication and performance.

Later, in the 1970s and beyond, Japanese manufacturers came to command the international motorcycle market, just as they would later the markets for cars. Their emerging industry had concentrated first on simple models that could be produced very cheaply, and then on machines of increasing sophistication. Assisted by the benefits of mass production, the strategy resulted in high-quality, reliable products at competitive prices for world markets. Among the well-known motorcycle brands were Honda, Suzuki, Yamaha and Kawasaki. These companies have introduced many technical and styling innovations to their machines which have proved popular. Recently, some European manufacturers have started to fight back, notably the re-launched Triumph brand, BMW and the Italian Moto Guzzi Company. Similarly, in the USA, Harley Davidson has retained a large share of the market.

There has been very little competition for motorcycles from machines that use alternative motive power systems.

1.4.4 **Motor scooters**

The motor scooter has its roots in the public need for inexpensive personal mobility. The machines have been manufactured for around 100 years and clearly have their origins in the children's toy that consists of a platform mounted on two wheels and supporting a pair of handlebars. The earliest models were little more than motorized versions of such playthings.

Whilst World War I had terrible consequences, a positive outcome was the rapid advances made in technology. Engineering underwent major expansion during the conflict and several businesses started to produce scooters as part of the war effort. These enterprises were joined at the end of hostilities by others who were without military work and with spare manufacturing capacity. In fact, the resulting 'scooter craze' was largely precipitated by an American wartime austerity machine called the *Autoped*, which had been introduced in 1916; see Figure 1.28(a). This scooter was seized upon avidly as a cheaper form of mobility than even the lightweight motorcycle. Its role was that of a handy 'runabout', with the most basic of mechanics and control, for local travel rather than long-distance work; a driving range of 10 miles (16 km) was claimed.

Other notable models were the UK-built *ABC Skootamota*, which first appeared in 1919 and probably sold more examples than all its competitors put together (see Figure 1.28(b)), and the *Unibus* (UK, 1920), which was the first totally enclosed scooter and must be considered as the true forerunner of all modern machines. Unfortunately, however, most of the other products of this first generation gave very poor performance. Good, bad or indifferent, the breed soon foundered on its own limitations, but reappeared over a quarter-century later in more efficient form.

During World War II, simple scooters to mobilize parachute regiments and ground forces were manufactured by British, German, Italian and American companies. The resulting improvements in design, coupled with strong public demand for a reliable and affordable means of road transport, led to a proliferation of second-generation scooters. Notably, the Italian company Piaggio was not allowed to continue making its aeroplanes, and so instead created a small monocoque scooter in 1945. From this evolved the *Vespa* ('Wasp'). The design was so advanced that its shape and engineering principles formed the basis of all the *Vespa* scooters sold up to the mid-1990s.

Meanwhile, Innocenti in Milan launched its 'transport for the masses' in 1947 with the *Lambretta M* (later renamed the *Model A*). Unlike the *Vespa*, the machine was open-framed with tubular construction and it did not offer much weather protection. *Model B* was launched in 1949 and was equipped with proper sprung suspension that had been lacking in its predecessors. From this point on, the *Vespa* and the *Lambretta* vied for market share with steady improvements in performance and increases in engine size. For instance, the *Vespa* progressed from its original 99-cc capacity through 125 cc, 150 cc and later 200 cc. For the most part, the *Vespa* never deviated from its concept of a monocoque bodyshell with the engine/gearbox unit mounted inside/adjacent to the rear wheel, whilst the *Lambretta* stayed true to its tubular frame origins with either shaft drive or totally enclosed chain drive.

The boom in sales throughout the 1950s was not to last. The entry into the market of small cars like the *Fiat 500* and the *Mini* soon enticed an ever more prosperous consumer away from the rain, cold and relative lack of protection afforded by two-wheel travel. By the mid-1960s, scooter sales were in serious decline, supported only by the sub-cultures and fashions that had grown up among the scooter clubs that they had spawned. Nowadays, however, motor scooters are proving popular in Europe (particularly Italy and the Mediterranean), Japan and Taiwan, but not in the USA. They are even more popular in most parts of the developing world. Parking, storage and traffic issues in crowded cities, along with the easy driving position, make them a convenient form of urban road transport. In many nations, sales of scooters and low-powered motorcycles exceed those of automobiles, and a motor scooter is often the family transport. Motorized bicycles, so-called 'mopeds', are also being used widely; a modern example is shown in Figure 1.28(e).

FIGURE 1.28

Evolution of the motor scooter: (a) *Autoped*, USA, 1916; (b) *ABC Skootamota*, UK, 1919; (c) *Biella MP5 Piaggio Paperino, Italy, 1944*; (d) *Model 150d Lambretta*, Italy; (e) *VéloSoleX*, France, 2008

(All images sourced form Wikipedia and available under the Creative Commons License)

1.4.5 Trucks

For centuries, horse-drawn wagons were used to convey goods. The 'village carter' was a well-known character in 19th-century rural England. In the 1920s, however, motor-trucks were progressively introduced as a complement to horse-drawn wagons for local delivery services. Even as late as World War II, when strict petrol rationing was in force, horses were used to pull brewery drays, to convey goods to and from railway stations, and to deliver milk, bread and coal to domestic addresses. Railways were used for long-distance transport of freight.

The first motorized truck is said to have been built by Gottlieb Daimler in Germany in 1896; see Figure 1.29(a). This had a 4 hp (3 kW) two-cylinder engine, with a displacement of 1.6 litres, and a belt drive with two forward speeds and one reverse. Strictly speaking, this was a converted horse-drawn cart whose chassis featured transversely mounted fully elliptic leaf springs at the front and coil springs at the rear. This complex suspension was important not only because of the poor road conditions at the time but also because of the engine's distinctive sensitivity to vibrations. In 1898, Daimler made his commercial vehicle – a delivery van with a payload of 600 kg (12 cwt) which he sold to an American firm for delivering goods around New York. The towing industry had its origin in 1916 in Tennessee, USA, when a manufacturer began making tow trucks for sale to garages to rescue broken-down cars.

Early English trucks were small and used mostly for local deliveries; see Figure 1.29(b). They were referred to as 'lorries' – a term still used in the UK today, but rarely heard in the USA. As time went by, trucks were designed for different duties and various sectors of the market for such vehicles developed.

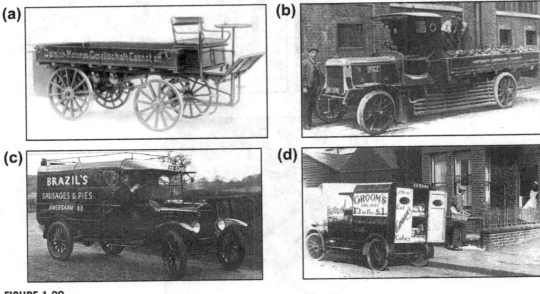

FIGURE 1.29

(a) Daimler motorized truck, 1896; (b) British lorry, 1914 (image from E. Watts, 1987, *Fares Please: The History of Passenger Transport in Portsmouth*, Portsmouth, UK: Milestones); (c) *Model T* Ford delivery van, 1917; (d) *Trojan* delivery van, 1920

(Images (c) and (d) sourced from N. Faith, 1995, Classic Trucks: Power on the Move, *London: Boxtree)*

Light trucks and vans (payload ~ 500 kg, or 10 cwt) are based on car designs with the rear seats, passenger windows and the boot/trunk removed and converted to an enclosed loading platform. Conversions of the *Model T* Ford became widely used as medium-weight vans throughout the USA and overseas; see Figure 1.29(c). In Britain, the *Trojan* van built in the 1920s by Leyland at its London factory was greatly appreciated by small shopkeepers because it was cheap and reliable; a baker's van making deliveries is shown in Figure 1.29(d). In official statistics, cars and light vans are often united under the banner 'light duty vehicles'. Larger vans, with payloads in the range of 1.0 to 1.5 t, have strengthened platforms to support the load.

Freight in developed countries is now generally transported by road (rather than by rail) in heavy goods vehicles (HGVs). The most common type of vehicle is the tractor-trailer (or articulated lorry), in which the freight-carrying facility (the trailer) is a separate unit that can be readily attached/detached from the tractor. Some trailers are designed to accommodate one or more standardized shipping containers that are used for the import/export of goods by sea. For long-distance road travel, the driver's cab may be fitted with sleeping quarters and limited cooking facilities. There are strict regulations concerning the maximum weight of HGVs, the maximum loading per axle, the number of wheels and the size of tyres fitted. For example, the maximum laden weight for a five-axle truck is 40–44 t in most European countries. Drivers of such vehicles must hold an HGV licence. There are also firm rules concerning how many hours a driver may be at the wheel without rest and sleep.

There is an important distinction between the designs of heavy trucks as used in North America and in Europe. North American trucks typically have the driver seated well behind the engine and front axle, as in a car; see Figure 1.30(a). This long-nose, conventional design has advantages of ease of access to the engine and improved safety for the driver. The disadvantage is that the vehicle is longer and therefore less readily turned in tight circles. In Europe, where many of the roads are twisting with sharp bends and there are more restrictive regulations over vehicle length, trucks are usually of the 'cab-over-engine' or 'flat-nose' design, as shown in Figure 1.30(b), in which the driver sits above the engine and front axle. Access to the engine is then achieved by tipping the cab forward.

In addition to these general freight-carrying vehicles, there are trucks for specialized uses, e.g. for conveying frozen goods, bulk liquids, ready-mixed concrete, tree trunks. Dumper trucks are employed for bulk quantities of rocks, soil and sand. Refuse collection is undertaken in custom-designed trucks with an onboard crushing facility, while trucks for carrying pallets (e.g. of bricks) are fitted with cranes for loading and unloading. Recovery trucks can now have a powerful hydraulic boom which can be extended under the vehicle to support the front axle. This underlift design is far superior to the high-mounted jib; loads of up to 12 t can be lifted. There is also a wide variety of military trucks – from the humble Jeep to the massive tank transporter – each with an important role to play. Although not designed to carry a load, mobile cranes should also be included in the list of heavy-duty road vehicles.

Some of the largest trucks are found in Australia, where the distances are long and the population sparse. A typical 'road train' consists of three trailers, is 53 m long and capable of hauling 115 t; see Figure 1.30(c). The vehicles cross the continent; for example, through the outback between Darwin in the north of Australia and Adelaide some 3000 km (1864 miles) to the south. All manner of materials are transported; in particular, livestock, fuel, mineral ores and general freight. Road trains can only be operated on lightly-used roads as overtaking such a long vehicle can be hazardous in the face of oncoming traffic.

Almost all of these forms of road transport currently consume hydrocarbon-based fuels, some of them in large amounts, and produce undesirable emissions. Much remains to be done to improve the efficiency and reduce the exhaust from trucks in the quest for sustainable road transport.

FIGURE 1.30

Heavy goods vehicles: (a) American design; (b) European design; (c) Australian road train

(All images sourced from Wikipedia and available under the Creative Commons License)

1.4.6 Buses

The term 'omnibus', a 'bus for all', was first coined during the age of horse-drawn carriages to describe vehicles that could carry a group of fare-paying passengers. Over the years, the name was abbreviated to 'bus' and, gradually, horse-drawn buses were phased out and replaced by motorbuses early in the 20th century. At first, there were a few battery electric buses and, rather more, steam buses. By the 1920s, however, most buses were ICEVs-either single deck (see Figure 1.31(a) and (b)) or double deck. Initially, double-deck buses followed early tram design with an open top and an external staircase, but first the top was covered and then later the stairs were enclosed, as shown in Figure 1.31(c)–(e). Access to the vehicle was now by means of an open platform at the rear from which the stairs led to the upper deck. Such buses had conductors who supervised the passengers, took the fares and issued the tickets. Single-crew buses were introduced in the 1960s, in which the driver had also to administer the fares and tickets. This necessitated a redesign of the bus so that both the driver and the ticket machine were incorporated in the passenger compartment and located adjacent to the entry door; see Figure 1.31(f). This scheme was introduced as an efficiency measure at a time when car ownership was becoming more widespread and travel by bus was declining.

FIGURE 1.31

Single-deck buses: (a) Bristol, UK, 1926; (b) London, UK, 1929. Double-deck buses: (c) London, UK, 1920; (d) London, UK, 1922; (e) Reading, UK, 1935. (f) Portsmouth, UK, 1966

(Images sourced from Wikipedia and available under the Creative Commons license)

The vast majority of buses worldwide are ICEVs, although there are an increasing number of hybrid electric buses in service on city routes; see Section 5.2.6, Chapter 5.

1.5 Growth of the petroleum industry

The fuel for the evolution of 'unsustainable' road transport has largely been based on oil. Indeed, modern civilization, as we know it, is almost entirely dependent on oil. During the 20th century, the petroleum industry grew rapidly as its products were increasingly required for transport, for the lubrication of engines, for the generation of electricity, for the central heating of buildings, and for the manufacture of chemicals, plastics and fertilizers. In recent times, natural gas has been progressively replacing petroleum for many of these applications, particularly electricity generation, central heating and the manufacture of fertilizers. Nonetheless, petroleum continues to reign supreme as the fossil fuel for powering road, sea and air transport, as well as a significant proportion of railway locomotives.

Native petroleum, tar and asphalt have been known since antiquity in parts of the Middle East. As early as 3500 BC, the Sumerians (who lived in present-day Iraq) used asphalt as an adhesive for bricks and to caulk the seams of boats. From 2200 BC onwards, the Babylonians followed the same practice to seal bridges, walls and tunnels. Bitumen, another fraction of native petroleum, was similarly employed. Later, the Romans burned crude oil for their lamps, and also used it as a fumigant to kill infestations of insects.

The Chinese first drilled for oil in 200 BC and penetrated the earth to a depth of 3500 ft (1067 m) – no mean feat for the time! Oil is widespread geographically and for centuries people who had access to deposits exploited the resource for domestic heating and lighting. In AD 1272, for example, Marco Polo noted that oil lamps were in general use in the town of Baku on the shore of the Caspian Sea and consequently there was a flourishing local trade in oil.

The modern petroleum industry originated in Pennsylvania following an observation that oil seeped to the surface in a valley known as Oil Creek. On 28 August 1859, Edwin Drake struck oil at a depth of 70 ft (21 m) in Titusville and found that it gushed freely to the surface. In the same month, oil was discovered near Petrolia in southwestern Ontario, Canada; a significant oil-refining industry started up in the town a few years later.

The petroleum industry began as a plethora of low-volume producers in the 19th century and, during the 20th century, was consolidated into a relatively small number of major companies that now operate on a global scale. Most of the enterprises began life in the USA or in Europe. A foremost exponent of the move to consolidation was John D. Rockefeller, who founded the Standard Oil Company of Ohio (SOHIO) in 1870. His strategy was to merge with competitors or, failing that, to eliminate them and his efforts led to the establishment of the Standard Oil Company of America (ESSO), later renamed EXXON. Technical factors favoured the consolidation of oil companies. Petroleum refining became increasingly complex – from simple distillation it was developed to progress through various processes such as catalytic cracking, hydro-cracking, reforming and hydro-desulfurization. These more sophisticated operations led to larger refineries with many specialized units that were more expensive to build. On the other hand, with greater outputs from larger companies came the associated economies of scale that led to cheaper products. Today, the petroleum industry essentially consists of an upstream sector (exploration, extraction from the ground and conveyance to refineries) and a downstream sector (refining and converting into products).

Although oil is widespread geographically, by far the largest inventories are found in the Middle East. Over 60% of the known global reserves are concentrated in just five countries, namely: Saudi Arabia 25%, Iraq 11%, Kuwait 9%, UAE 9% and Iran 8%. New oilfields are being discovered, for instance, in the countries of the former USSR and off the coast of West Africa, but these are unlikely to compare in size with those of the Middle East and will not significantly change the overall picture. Since the Middle East oilfields are far removed from the major markets in North America, Europe and, increasingly, Asia, bulk tankers are needed to ship the oil to refineries that are mostly located near the markets. The size of the operation is impressive; in 2011 the total world oil production was 84.8 million barrels per day, 30% of which were shipped from the Middle East. The largest of the tankers can transport around 2 million barrels of crude oil. Smaller vessels are employed to carry refined products to ports near the markets, from where the cargo is transferred to road tankers for onward distribution. These vehicles can convey 21 000 to 34 000 litres (4600–7500 gallons) of refined fuel or chemicals.

1.6 Development of roads

The inexorable growth in the use of ICEVs has been supported by major investment in the road systems on which most of them run. Before 1900, road travel was limited; few roads were metalled and even these were narrow and simply led from one town or village to the next. It was only in the early 20th century that 'A'-class roads fit for long-distance travel by motor vehicles were constructed, but these went straight through the centres of towns and thereby resulted in congestion and slow journeys; the concept of bypassing towns was a later development. Contrast that picture with the scene today, where most of the population in developed countries has its own road transport and the national road system is highly developed. People regularly expect to be able to travel 500 km (310 miles) or more in a day by car or by bus on a sophisticated motorway/autoroute/expressway network.

1.6.1 Road construction

Moving by land was slow and difficult until the Roman Empire constructed its vast network of military roads; indeed, the Empire was held together by its roads. At its peak under the rule of Emperor Trajan (AD 98–117), the international Roman road system covered an estimated 85 000 km (52 800 miles), from Hadrian's Wall in the north to the Euphrates in the east and the Sahara in the south. The main purpose of the network was to serve the needs of administration and trade, though this naturally involved the ability to move garrison troops quickly if the need arose.

Ordinary Roman roads were 5 to 6 m wide (16–20 ft), but the main highways could be as much as 10 m across (33 ft). The typical construction consisted of a metalled (e.g. gravel/pebbles) or stone-paved surface above a cambered solid foundation ('agger') that was formed by laying down successive layers of rubble, crushed stone, volcanic concrete and sand on a compacted sand/earth footing; see Figure 1.32(a). The structure was flanked by shallow drainage ditches that also served as boundary markers. The most remarkable feature of the design was its great thickness, often well over a metre – by

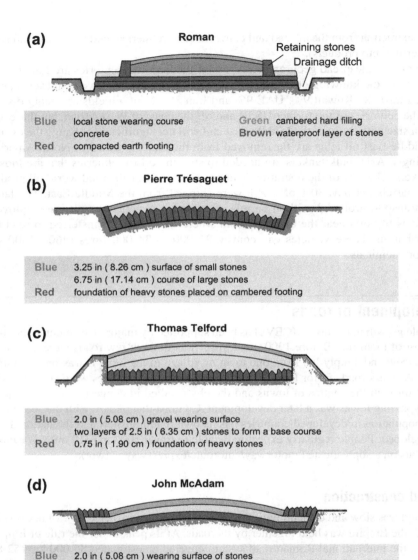

(a) **Roman**

Retaining stones
Drainage ditch

Blue	local stone wearing course	Green	cambered hard filling
	concrete	Brown	waterproof layer of stones
Red	compacted earth footing		

(b) **Pierre Trésaguet**

Blue   3.25 in (8.26 cm) surface of small stones
        6.75 in (17.14 cm) course of large stones
Red   foundation of heavy stones placed on cambered footing

(c) **Thomas Telford**

Blue   2.0 in (5.08 cm) gravel wearing surface
        two layers of 2.5 in (6.35 cm) stones to form a base course
Red   0.75 in (1.90 cm) foundation of heavy stones

(d) **John McAdam**

Blue   2.0 in (5.08 cm) wearing surface of stones
        2.0 in (5.08 cm) base course of stones
Red   cambered footing of compacted earth

FIGURE 1.32

Notable road builders, Roman to modern day

contrast, modern roads designed for very heavy motor traffic are rarely more than half this thickness, even though they are subject to greater stresses. The high standard of the Roman roads meant that vehicles could be exploited to the full; nearly 30 types of wheeled conveyance were regularly in use, from heavy agricultural carts with large wheels to exquisitely-decorated ceremonial carriages. There was so much traffic that the Romans had to establish a form of Highway Code that imposed certain limitations. At one time, for instance, loaded carts were not permitted to enter Rome for the first 10 hours after dawn.

No such road system existed outside the Roman Empire, though this does not mean that there were no well-developed and widely used lines of overland communication and commerce. The Chinese were constructing relatively broad and firm roads in 4000 BC, the Egyptians and Babylonians paved their main roads, and the Persians built the Royal Way from the capital of Susa to Sardis in Asia Minor, some 1700 miles (2700 km) away. The Silk Road, which helped to lay the foundations for the modern world, was forged as early as 130 BC. At the height of Chinese and Indian trading in AD 200, it ran for 3980 miles (6400 km) from Shanghai and Xian in China to the Levant, and via the sea to Cadiz (then Gades) in southern Spain. In most places, however, land transport was confined to approximate routes rather than roads and these were generally beaten tracks which changed course when over-worn and attracted or required little formal maintenance.

After the fall of the Roman Empire, road-building declined into a long period of neglect. Good roads were actually considered dangerous, in that they permitted enemies to advance more rapidly. The Silk Road became increasingly unsafe and its importance decreased with the opening of a sea route between Europe and India. (Nowadays, somewhat ironically, oil is piped along some 3000 km of the old road from Kazakhstan to fuel vehicles in China.) Trade still continued throughout the known world but it was mostly conducted by way of narrow and poorly negotiable tracks that had neither firm foundations nor ditches to drain them. Wattle bound together with flexible branches was laid in marshy places, but was renewed only sporadically. Although Charlemagne (AD 742) ordered the building of new roads and the Vikings established the Varangian Road, which was a long trade route that linked the Baltic with the Middle East via Russia, the extremely poor quality of roads severely hindered major improvements in the design of vehicles.

By the early 18th century, the growth of commerce and industry, together with the increase in freight traffic between towns and cities, stimulated a demand for better roads. The first leading pioneer of modern road-making was the French engineer Pierre-Marie-Jérôme Trésaguet (1716–1796), who became inspector general of roads and bridges in France in 1775. He devised a method of constructing a well-drained road with a strong 25-cm foundation of flat stones hammered in edgeways to take the weight of the traffic, a slightly raised crown and a surface of broken stone; see Figure 1.32(b). Through his efforts, by 1800 France had one of the finest road systems in Europe. Trésaguet's direct contemporary in England was John Metcalf (1717–1810), who, though blinded by smallpox in childhood, engineered roads in the north of England. He used a system of ditches to provide drainage and laid brushwood foundations in swampy areas (as had been the practice in mediaeval times, noted above).

The outstanding representatives of the next generation of road-builders are the great self-taught engineer and bridge-builder Thomas Telford (1757–1834) and John Loudon McAdam (1756–1836). In his road-building operations, Telford adapted Trésaguet's method in most of its essentials but emphasized the use of high-quality stone; see Figure 1.32(c). McAdam had his own theories on how the method could be improved. He saw the surface, rather than the foundation, as taking the brunt of the wear and tear of traffic. In 1783, he commenced experiments in road-making at his own expense. Later, he was able to continue these in Cornwall under government contract, but it was not until after being appointed the surveyor general to the city of Bristol in 1804 that he was able to put his ideas to the test on a large scale. The roads surrounding Bristol, then a rapidly growing port, were in a bad state of repair.

McAdam drew up a specification for a road with its bed raised slightly above the level of the surrounding land to facilitate drainage; see Figure 1.32(d). Ditches edged each side, and the roadbed consisted of a compact layer of large pieces of crushed rock. The road was 18 feet (5.49 m) wide and the crown in the middle rose 3 in (7.6 cm). The all-important surface layer consisted of broken angular pieces of granite or greenstone – no piece heavier than 6 oz (0.17 kg). Successive layers of stone were then laid as each preceding surface was compressed by the traffic. In this way, under pressure, the surface became virtually self-sealing. The iron rims on the wheels of carts and carriages, which had previously reduced natural surfaces to a chaos of ruts and potholes, were now in effect put to work in compacting the surface ever more tightly. The main disadvantage of the new type of road was that in dry weather an excessive amount of dust was thrown up to cover travellers from head to foot. On the other hand, the practical advantages were obvious and a parliamentary inquiry into road-building in 1823 led to the design being officially adopted. It was also quickly taken up by other countries, notably the USA, and a new verb, 'macadamize', entered the English language.

McAdam's specification remained the standard method of road construction until the arrival of motorcars with pneumatic tyres in the early 20th century. The rapidly rotating and flexible rubber did not tamp down the road surface as was the case with heavy solid wheels; on the contrary, inflatable tyres tended to tear the small stones in the road surface from their bedding. This made it necessary to seal the road surface with a protective layer of broken stone mixed with tar – so-called 'tar macadam' – that was spread with a uniform thickness and well rolled. The stones were added to prevent skids, whereas the top dressing provided additional waterproofing and avoided dust in dry weather, though the surface could melt in hot sun.

It should be noted, however, that in the 19th century tar was already known to stabilize road construction, but it was little used. The first modern road to be layered with bituminous material was the Vauxhall Bridge Road, London, in 1835. Asphalt, a mixture of bitumen and stone, was first used in Rue Saint-Honoré, Paris, in 1858. From then on, with the expansion of public transport systems and the enormous increase in traffic following the advent of the motorcar, the development of roads and highways was never-ending. Notably, although it had been experimented with in the 1890s, reinforced concrete only began to be used extensively in the 1930s through the introduction of fast-setting mixes. One advantage of concrete is that it is very rigid, so that the weight of heavy vehicles is evenly spread; tar, by contrast, is slightly plastic.

Current plans for future designs and networks of roads foresee little further change to the physical structure of the road itself, but rather anticipate the incorporation of increasingly sophisticated means of communication with individual vehicles while they are in transit.

1.6.2 Super highways

Most of the developed nations in the world have now constructed multi-lane super highways. Indeed, it might almost be said that the existence of such highways is one of the characteristics that define a

FIGURE 1.33

(a) Multi-lane freeway, Los Angeles, USA; (b) Gravelly Hill interchange, Birmingham, the original 'spaghetti' junction; (c) Interchange in Los Angeles, USA

(Images sourced from Wikipedia and available under the Creative Commons License)

'developed nation'. These highways go under various names in different countries, e.g. motorways in the UK; autoroutes in France; autobahns in Germany; freeways, expressways, turnpikes and through-ways in the USA; an example is shown in Figure 1.33(a). With one exception, the UK motorways are toll-free and their cost is met by general taxation, but in many countries a toll is charged. The super highways can have from two to six or more lanes in each direction, with three lanes being the most common.

The construction of super highways constitues a major feat of 20th-century engineering and they occupy large areas of land, especially at complex interchanges, such as the so-called 'spaghetti' junction that was opened near Birmingham, UK, in 1972; see Figure 1.33(b). The Judge Harry Pregerson

Interchange in Los Angeles, USA, is shown in Figure 1.33(c); completed in 1993 to join two of California's busiest freeways, it was described as one of the 'biggest, tallest, most-costly traffic structures yet built'. The erection of multi-lane suspension bridges over large tracts of water also represents a major undertaking. Because of the high cost of construction, such bridges usually attract a toll.

Generally, the traffic flows smoothly on super highways, although sometimes congestion can occur for no apparent reason. Agent-based models, which simulate the actions and interactions of autonomous agents with a view to assessing their effects on the system as a whole, have been developed to describe the formation of 'phantom jams'. These have demonstrated how small disturbances (a driver hitting the brake too hard, or getting too close to another car) in heavy traffic can become amplified into a major self-sustaining traffic jam. Speed limits on super highways vary from country to country but are generally around 70 mph (112 km h^{-1}), although top speed is unrestricted on certain autobahns in Germany. The roads tend to have an excellent safety record, thanks to the uniform traffic flow and the exclusion of pedestrians, cyclists, and low-powered mopeds and scooters.

The investment in and the convenience of 'super highways' are so great that any future changes that are proposed to road transport in pursuit of sustainability will almost certainly need to continue the use of these roads.

1.6.3 Traffic regulations and road accidents

When motor vehicles were first introduced, great concern was expressed over road safety. In 1875, fears were expressed at the US Congress that 'horseless carriages propelled by gasoline engines might attain speeds of 14 or even 20 miles per hour'. In Britain, this concern was allayed in 1865 when 'The Locomotive Act' ('Red Flag Act') was introduced (*v.s.*). As discussed above (Section 1.2.3), the Locomotives on Highways Act of 1896 exempted vehicles under 3 tons weight from the 'red flag' requirement and raised the speed limit to 14 mph (22.5 km h^{-1}). How far the road transport scene has advanced in little over a century!

The first motoring fatalities occurred in Britain in the 1890s, and in 1896 a law was passed requiring motorists to have an audible warning and to show lights at night. The Motor Act of 1903 required all vehicles to be registered with the local authority and to carry number plates. At the same time driving licences were introduced, but did not become mandatory until 1934 when driving tests became compulsory. Speed limits changed more than once, and in 1935 a limit of 30 mph (48 km h^{-1}) in urban areas was first imposed. As road networks became more complex and as the amount of traffic increased, more road signs were erected and became standardized when the Highway Code was first issued in 1931. Subsequent updates of the Highway Code amended and extended road signage.

In the early days of motoring, the number of accidents was quite disproportionate to the number of vehicles on the roads. For instance, in 1934 there were only 1.5 million vehicles in the UK but 7000 people were killed in road accidents. The great improvement since then in relation to the amount of traffic is a result of many factors, for example: greater pedestrian awareness, superior driver training, better road layout, comprehensive road signs, speed cameras, and improved safety features on vehicles. Over the past few years, the annual number of fatalities on UK roads has been fairly stable at 2000–3000 per year when there are almost 33 million registered vehicles, and the average mileage travelled by each vehicle is far greater than in 1934. In 2012, the UK recorded the lowest number of road deaths in the EU at 28 per million inhabitants, - nonetheless, there is no room for complacency so long as 3000 people are dying each year.

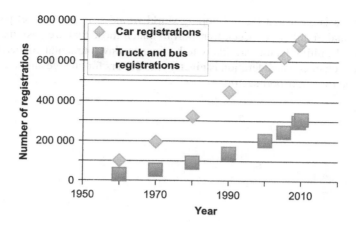

FIGURE 1.34

Historical trends of worldwide vehicle registrations 1960–2010 (thousands). Car registrations do not include US light trucks (sports utility vehicles, minivans and pickups) that are used for personal travel. These vehicles are accounted among trucks

(Image sourced from S.C. Davis, S.W. Diegel and R.G. Boundy, June 2011, Transportation Energy Data Book, Edition 30, *Office of Energy Efficiency and Renewable Energy, US Department of Energy)*

1.7 Growth of the automotive sector

The growth in manufacture and sales of road vehicles during the 20th century has been quite remarkable and is an urgent reminder that the greater the number of ICEVs in the world, the greater is the challenge to the achievement of sustainability. Starting from a base of almost zero in 1900, global production of passenger cars rose to 39.8 and 54.9 million in 1999 and 2007, respectively. By 2009, there was a small decline to 52.0 million as a result of recession and saturation of the Western market for private motorcars, but by 2011 the production figure had increased to 80 million. The consequent growth in world-wide vehicle registration from 1960 to 2010 is equally startling; see Figure 1.34. In 2010, there were 700 million cars and 300 million trucks and buses on the road.

In particular, the market for cars in China is growing vigorously as the country becomes more prosperous (see Section 2.2.2, Chapter 2). Bearing in mind that China and India together account for a population of about 2.6 billion (2.6×10^9) out of a global population approaching 7 billion, it is clear that, given rising prosperity, the scope for expanded car production is enormous. Today, there are over 1 billion vehicles in the world and on a 'business as usual' scenario (improbable), this number could double within a few decades. In practice, the rate of expansion may well be restricted, both by limitations of economic growth and the consequent inability of people to purchase cars, and by the lack of capital required to develop the road infrastructure to accommodate the vehicles.

The market for trucks is, of course, much smaller than that for cars. Nevertheless, as international trade expands there is an increasing need for more freight vehicles. In 2010, the 10 leading manufacturing groups delivered more than 2 million trucks of greater than 16-t gross vehicle weight (GVW). This is around 4% of the car production. To this must be added the much larger number of light commercial vehicles.

From these data, it is clear that the developing world would struggle to meet people's legitimate aspirations for personal mobility and truck transport to match those that are customary in most of the developed countries. Limiting factors are likely to be finance, environmental concerns, and petroleum resources. In the next chapter, consideration is given to the drivers for change and the likely impact that they may have on the road transport sector.

Drivers for Change

2.1 Challenges for new-generation road vehicles

Certain broad trends may be identified in the way that the world has developed in the late 20th and early 21st centuries. In part, these are political (the spread of democracy, advocacy of human rights), in part social (more informal dress and forms of speech, less deference to authority), and in part economic (the rise in living standards for many, especially in the developed world). Rural society is giving way to urban growth as people migrate into the cities. Concomitant electrification and rising transport aspirations are leading to ever-increasing use of energy. Consequently, attention is now being directed towards greater efficiency of energy production and consumption, together with increased diversification in the sources of energy employed that includes the realization of clean, low-carbon energy. The transport sector is facing a requirement to introduce novel, advanced technology to improve its efficiency of fuel use and to limit the emissions of both pollutants and carbon dioxide.

Petroleum use and greenhouse gas emissions are increasing steadily throughout the world due to the seemingly inexorable growth in demand for passenger and freight transport by land, sea and air. The challenge facing society is first to mitigate this growth, and then to reduce fuel consumption and greenhouse gas emissions. In the medium term, there will be an incentive to move towards unconventional fuels (see Chapter 3) and, ultimately, to renewable sources of energy. In parallel, there is much that can still be done to improve the efficiency of conventional internal-combustion-engined vehicles that run on petrol or diesel (see Chapter 4), and to introduce vehicles that operate with varying degrees of electrification (see Chapter 5). Later, fuel cell vehicles may find a role (see Chapter 8).

There are currently several *drivers for change* in the world transport scene, as follows:

- growth in the world population, coupled with the desire of people in developing and under-developed nations to attain the freedom of movement enjoyed by those in developed countries through the ownership of their own personal means of transport;
- expansion in global trade, with goods transported over long distances from factories to distribution centres and then onwards to retail outlets;
- concerns over the political vulnerability of oil supplies, which are the basis of road transport today, given that a large fraction of the world's conventional petroleum is located in just a few Middle Eastern countries and with many of the oil tankers passing through the narrow Strait of Hormuz;
- worries that known reserves of petroleum are being depleted at a pace that exceeds the rate of discovery of new oil fields;
- the pressing requirement to reduce atmospheric pollution, especially in urban environments;
- the impact of climate change and the conviction that this phenomenon is caused in major part by increasing levels of greenhouse gases in the atmosphere, especially carbon dioxide stemming from the combustion of fossil fuels.

Together, these considerations provide a powerful incentive to review and question the sustainability of the current transport scene – as regards road vehicles – in the decades to come.

2.2 Demographics and vehicle ownership

Key objectives of most politicians are to defend the realm from external attack, to maintain law and order, and to be re-elected at the next election. To achieve the last-mentioned goal, it is deemed necessary to promise and deliver economic growth and increased prosperity for the electorate. Consequently, the politicians are encouraged by professional economists who, basing their predictions on anticipated future scientific discoveries, new technology and smarter methods of working, do not appear to accept any limit to growth and thereby anticipate an enhanced material standard of living for all. Until now, these economists have been proved right. One has only to look at the greatly improved lifestyle for most people since, say, 1950 to acknowledge that shorter working hours, greater leisure and higher consumption of material goods are based largely on the spread of mass production, and the development and commercialization of new products and processes.

Not everybody, however, has subscribed to the above optimistic view. As long ago as 1971, the 'Club of Rome' used large computer models to forecast the economic future of the world. The subsequent best-selling report entitled *The Limits to Growth* warned that unbridled, indefinite growth was not feasible. Rather, it was concluded that the world would run out of petroleum by 1992. Clearly this did not happen, predominantly as a result of the discovery of new oil fields and advanced techniques for exploration and production in difficult environments. Even so, the unexpected outcome of greatly increased growth in oil production does not necessarily invalidate the general conclusion that there is an ultimate limit to economic growth. It is simply a matter of timescale. Is the world now approaching that limit based on ever-greater demand for petroleum to service a growing population? Many people think that it is.

2.2.1 World population statistics

The world population has increased enormously during the past few centuries. In 1800, it was around one billion (10^9); today it is around seven billion. The increase has not been linear. From 1800, it took almost 130 years to reach two billion (1927), then a mere 30 years to reach three billion (*c.* 1960), and a further 40 years to reach six billion (*c.* 2000). The final billion was added in just 11 to 12 years. The rapid increase in population since World War II may be attributed to many factors, which include: enhanced agricultural productivity; fewer young children dying from infectious diseases such as measles, diphtheria and scarlet fever, as well as malnutrition and diarrhoea; the eradication of smallpox; better control over tuberculosis, malaria and (recently) acquired immune deficiency syndrome (AIDS); fewer women dying in childbirth and men in battle. Improvements in water purity and in sanitation have also led to lower numbers of deaths. The great pharmaceutical companies of the world have developed many new drugs in the fight against disease and these, together with advances in surgery and nutrition, have been largely responsible for the increase in longevity.

Population growth has not been uniform across the world. The wealthy industrialized nations have gained the most from the above-mentioned advances in health and agriculture, but their population growth has been limited by a comparatively low level of fertility, which has resulted in an ageing society. By contrast, the developing and under-developed nations have benefited less from advances in medicine and nutrition, for both economic and educational reasons, but have had higher fertility rates that led to a population skewed towards younger people. As welfare in the poorest nations progressively improves, it is to be anticipated that life expectancy will grow. Fortunately, this factor is offset by a marked decline in fertility rates as prosperity has risen in developing nations, especially in Asia. This is as much a cultural and social issue as an economic and educational one, although clearly all these factors play a role. Whereas it is not possible to be certain of the outcome, there are grounds for optimism following the precedent set by the developed nations. It seems that a growing aspiration for a higher standard of living coupled with increased life expectancy and better education result in people limiting the size of their families.

The prime responsibility for gathering and collating global population statistics and predicting future trends lies with the United Nations (UN), although the US Census Bureau is also active in the field. Inevitably, there is great uncertainty when predicting ahead as far as the end of the 21st century. The UN graph for the world population from 1800 to 2005, which gives three predictions up to 2100 for

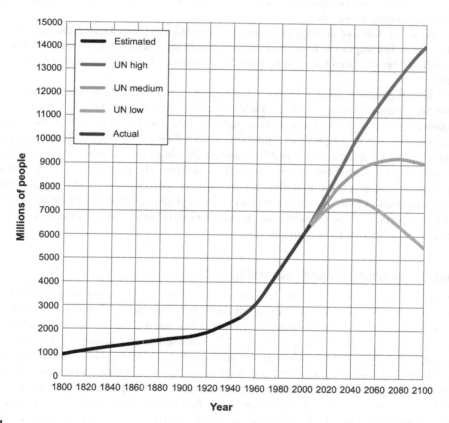

FIGURE 2.1

World population: 1800 to 2005, with extrapolations by the United Nations (UN) to 2100

(Permission to use is granted under the terms of the GNU Free Documentation License)

high, medium and low growth, is shown in Figure 2.1. The forecast for 2100 shows a high of 14 billion people to a low of 5.5 billion. Such a wide spread is almost valueless for planning purposes, although many authorities now settle on a figure of around 9 billion for 2050; thereafter, the curve is expected to flatten for the remainder of the century, reaching a plateau of 10–11 billion by 2100. Factors that may limit the rise in world population include:

- improved education;
- social trends;
- agriculture productivity and the number of people that can be fed properly;
- the impact of large numbers of people on the natural environment;
- the availability of strategic materials;
- climate change;
- the supply of fuels.

These factors, together with the advance of nations to developed status, seem likely to curtail unbridled population growth in the longer term.

What is the significance of demographic changes for road transport? During the past two decades, there has been growing migration from the countryside into urban communities on a worldwide scale; see Figure 2.2(a). This has been matched by growth in urban traffic at the expense of rural driving; see Figure 2.2(b).

Of the 7 billion people in the world today, 4.2 billion (60%) live in Asia and 1.0 billion (15%) in Africa – the two continents with the least vehicle ownership per 1000 inhabitants. China and India are the two countries with the greatest populations, namely, 1.35 and 1.20 billion, respectively. Both are developing countries where people have aspirations to own their own transport. Thus there is an enormous opportunity for the automotive industry, coupled with a huge challenge to the petroleum industry, to sustain such a vast potential expansion in vehicle ownership. The prospect also poses a major concern with respect to atmospheric pollution and greenhouse gas emissions. Already, the indigenous automotive companies in both countries are gearing up to produce vehicles appropriate to their home markets. By comparison, African nations are mostly less-developed and it will take longer for their personal transport objectives to be realized.

2.2.2 Vehicle production and ownership

Passenger cars are defined as motor vehicles with at least four wheels that are used for the transport of passengers, with no more than eight seats in addition to that of the driver. Other classes of road vehicle are light commercial vehicles, heavy trucks, buses, coaches and minibuses. In most countries and for most years, passenger cars have constituted at least 70% of vehicle production or sales; the percentage can, however, rise to over 90% in specific countries and years. Worldwide annual sales increased steadily until 2007, declined during 2008 and 2009 in the face of a universal recession, but bounced back in 2010 due to an improvement in economic circumstances and, presumably, to the retirement of older vehicles that were not replaced in 2008 and 2009. In 2011, the total number of new vehicles of all types exceeded 80 million for the first time, i.e. a 3% rise on 2010. The production statistics for 2000–2012 are recorded in Table 2.1. These are approximately correct, although there are slight variations in the data from different sources. The decline during 2008 and 2009 may be attributed not only to the economic recession, but also to improvements in the reliability and lifespan of motor vehicles.

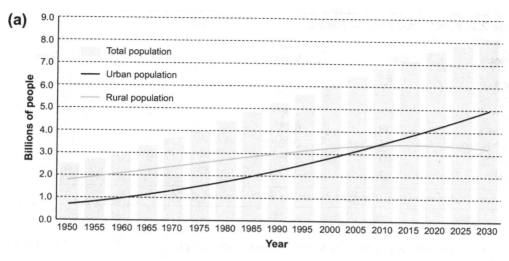

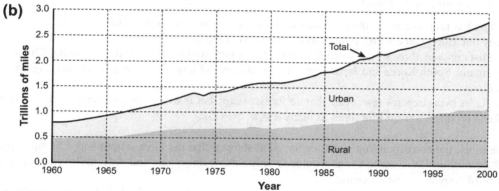

FIGURE 2.2

(a) Worldwide trends towards urbanization; (b) steady increase in urban travel

(Reproduced, with permission, from 'The future of transport: sustainable, affordable and effective', Dame Professor J. King)

Table 2.1 World Vehicle Production 2000–2012 (in Millions)

Year	Passenger Cars	Commercial Vehicles	Total
2012	63.07	21.03	84.10
2011	59.95	20.16	80.11
2010	58.34	19.36	77.7
2009	47.77	14.02	61.79
2008	52.73	17.79	70.52
2007	53.2	20.06	73.27
2006	49.92	19.3	69.22
2000	41.21	17.16	58.37

Data from Organisation Internationale des Constructeurs d'Automobiles (OICA).

Table 2.2 Vehicle Production in 2012 by Country (in Millions)

Country	Cars	Commercial Vehicles	Total	% Cars
China	15.52	3.75	19.27	81
USA	4.11	6.22	10.33	40
Japan	8.55	1.39	9.94	95
Germany	5.39	0.26	5.65	95
South Korea	4.17	0.39	4.56	91
India	3.29	0.86	4.15	79
Brazil	2.62	0.72	3.34	78

Data from Organisation Internationale des Constructeurs d'Automobiles (OICA).

The output of vehicles in 2012 by the top seven countries is presented in Table 2.2. The data reveal several interesting facts, namely:

- China has now overtaken the USA and Japan as the top vehicle manufacturer;
- only one European country (Germany) is included in the list;
- the list contains three of the so-called BRIC developing nations (Brazil, Russia, India, China);
- Germany, South Korea and Japan concentrate on cars and produce comparatively few commercial vehicles;
- the USA manufactures fewer cars than in former years, but is strong on commercial vehicles (CVs);
- China produces almost four times as many cars as the USA.

Remarkably, the corresponding statistics for 2000 showed that the USA output was 5.54 million cars and 7.26 million CVs, while China produced only 0.60 million cars and 1.46 million CVs – a total reversal of activity in just 10 years!

Most of the major vehicle manufacturers are now international in outlook and have assembly plants dispersed around the world. Consequently, there is no correlation between the home-base of a company and the vehicles made in that country. For example, Nissan, Honda and Toyota (all of Japan) and Tata (India) have plants in the UK; Ford and General Motors (USA) are present in Germany, the UK and elsewhere. Many companies have facilities in the USA. Also, it must be recognized that these operations are essentially assembly plants and that the various component parts of a vehicle are sourced from around the world. Vehicle production is truly a global industry.

Vehicle ownership varies widely across the world in line with the prosperity of different nations. The number of light-duty vehicles (LDVs; i.e. passenger cars and light vans) per 1000 of population ranges from over 700 in Australia, Canada, New Zealand and the USA, to 500–600 in the most highly developed countries of Europe, but just a few (< 10) in the least-developed sub-Saharan states. In China, the ratio for 2012 was approaching 100, with a comparable number again of commercial vehicles and motorcycles. Over the past few years, the annual growth rate of vehicles in China has averaged around 18% and it is predicted that the number will continue to increase substantially to give a total fleet of 500–600 million by 2040–2050. If this projection is at all realistic, the implications for petroleum consumption, urban pollution and greenhouse gas emissions are dire. Some of the smallest of the developed countries (e.g. Iceland, Luxembourg, Monaco, Lichtenstein) are among those with a high ratio of cars to population. Examination of the statistics for all the nations reveals that there is a non-linear relationship between car ownership and gross domestic product (GDP) per capita. It appears that the purchase of passenger cars takes off at the point where the middle classes attain a certain income and standard of living. This is precisely the present situation in China. The data reported in Table 2.2 clearly indicate that China has built up and continues to expand its automotive industry to meet the latent demand from its huge population as the privileged are educated to university level and become professionals, and as rural peasants drift to the city to become urban factory workers. India, too, is rapidly developing a large middle class as it industrializes. It is pertinent then to ask whether there will be sufficient fuel to support the burgeoning world fleet of private cars.

2.3 **Petroleum production and consumption**

Throughout most of the 20th century, petroleum played a key role in sustaining the world economy. Not only were most forms of transport dependent on this fuel but, additionally, petroleum was also widely employed in agriculture; in the construction industry; as a raw material for the manufacture of chemicals and fertilizers; to generate electricity; and for the central heating of offices, shops and homes. In the last quarter of the century, the price of oil rose dramatically from a very low point in the 1960s. At the same time, more and more natural gas was discovered and adopted as a substitute for petroleum in applications where this was practical, e.g. electricity generation, central heating and the manufacture of chemical products. Now, in the 21st century, the energy scene has reached a point where petroleum is largely conserved for use in internal-combustion engines (ICEs), particularly for transport applications where it is almost irreplaceable (except in the case of electric trains).

Twelve of the most important oil-producing countries are listed in Table 2.3, together with their annual output from 2007 to 2011 (in units of million barrels per day, 10^6 bbl d^{-1}),[1] and their percentage contribution to the world total. Together, these twelve nations provided 68% of the world supply in 2011; the remainder came from a multitude of smaller players. It is noticeable that all of the three foremost producers increased their respective contributions over the five years recorded, although only by modest amounts. Overall, world totals remained fairly stable. Clearly, world production and world consumption (see Table 2.4) must be broadly in equilibrium as the petroleum companies will only refine as much oil as the market demands. Higher outputs would depress the wholesale price and drive the marginal suppliers out of business.

A comparison of the data in Table 2.3 and Table 2.4 reveals that seven of the top twelve oil-producers are also among the top twelve consumers. In terms of petroleum use per capita, Saudi Arabia is the most prolific at 33.7 barrels per year (as well as being the top producer in most years), followed at quite a distance by Canada and the USA. France and Germany use about half as much petroleum per capita as the USA and Canada. In North America, the consumption of oil fell by 6.6% between 2005 and 2010 as a result of rising prices and the world economic situation, as well as growth in the supplies of natural gas. The corresponding fall for Europe was 4.2%.

---

[1] A barrel of oil is approximately 159 litres. Its mass depends on the density of the crude liquid. Different sources of production statistics vary somewhat according to what precisely is included or excluded, but the data in Table 2.3 are broadly correct and at least show the relative importance of the different countries. Nigeria and Iraq are other major producers with outputs in the region of 2.5×10^6 bbl d^{-1}.

Table 2.3 Major Oil-Producing Countries (Million Barrels per Day)

Country	2007	2008	2009	2010	2011	2011 Share of World Total (%)
Saudi Arabia (OPEC)	10.23	10.78	9.76	10.52	11.15	12.80
Russia[a]	9.88	9.79	9.93	10.15	10.23	11.75
USA[a]	8.48	8.51	9.14	9.69	10.14	11.64
Iran (OPEC)	4.04	4.17	4.18	4.25	4.23	4.86
China	3.90	3.97	4.00	4.31	4.30	4.94
Canada[b]	3.36	3.35	3.29	3.43	3.60	4.13
Mexico[a]	3.50	3.19	3.00	2.98	2.96	3.40
United Arab Emirates (OPEC)	2.95	3.05	2.80	2.81	3.10	3.56
Kuwait (OPEC)	2.61	2.74	2.50	2.45	2.68	3.08
Venezuela (OPEC)[a]	2.67	2.64	2.47	2.38	2.47	2.84
Norway[a]	2.57	2.47	2.35	2.13	2.01	2.31
Brazil	2.28	2.40	2.58	2.71	2.69	3.09
World totals	**84.26**	**85.36**	**84.32**	**86.75**	**87.10**	

Data from US Energy Information Administration; includes natural gas plant liquids. OPEC = Organization of the Petroleum Exporting Countries.
[a]Peak production of oil has already passed in this country.
[b]Conventional production, excluding oil sands.

Table 2.4 Major Oil-Consuming Countries (in 2011)

	10^6 bbl per Day	Population (M)	bbl per Year per Person
USA[a]	18.84	314	21.9
China	9.85	1345	2.7
Japan	4.46	127	12.8
India	3.43	1198	1.0
Russia[a]	2.72	140	7.1
Germany	2.40	82	10.7
Brazil	2.79	193	5.3
Saudi Arabia	2.99	25	43.7
Canada	2.26	33	25.0
South Korea	2.23	48	17.0
Mexico[a]	2.13	109	7.1
France	1.79	62	10.5

Data from the US Energy Information Administration.
[a]Peak oil production already passed in this country.

The overall transport sector is responsible for almost 60% of oil consumption by the members of the Organisation for Economic Co-operation and Development (OECD – a consortium of 34 developed and developing countries). Transport is fuelled predominantly by oil, with over 95% being either petrol (gasoline) or medium-distillate products such as diesel, kerosene or jet fuel. Other energy sources, such as liquefied petroleum gas (LPG), compressed natural gas (CNG), electricity and biofuels, still play only minor roles in the overall world transport scene, despite the prevalence of electric trains in certain regions and the recent enthusiasm for biofuels. In North America, around 70% of road transport is powered by petrol and 30% by diesel, whereas in Europe the ratio is almost reversed (35–40% petrol, 60–65% diesel). This reflects the far larger number of private cars in Europe that use diesel, in addition to trucks and buses.

A significant fraction of the road transport fuel that is consumed is actually wasted. The principal cause is traffic congestion in cities; vehicles wait in long tailbacks with their engines running. Outside cities, congestion is caused by accidents, highway repairs and excessive holiday traffic at peak seasons. Modern 'start-stop' technology in cars (see Section 5.2.1, Chapter 5) is beginning to ameliorate the problem. Other sources of fuel waste are hold-ups at traffic lights and the tendency of some drivers of commercial trucks and buses to keep their diesel engines running for long periods when stationary. This practice is more understandable when the ambient temperature is very low. In the USA, it has been estimated that in recent years congestion has accounted for around 4 billion gallons of extra fuel per year.

The refining of crude oil is a highly sophisticated industry that was developed throughout the 20th century. Oil refineries are large and complex installations. The first stage in the process is fractionation. This involves heating the oil to a high temperature in the absence of air. The volatile components, predominantly paraffins (saturated hydrocarbons), distil off and ascend a 'fractionating column'. This column effects a first separation, with the lighter components (one-to-six carbon atoms, $C_1–C_6$, in the molecule) rising to near the top of the column, and successively heavier fractions condensing and being taken off lower down. The process is shown schematically in Figure 2.3, which also indicates the applications for the various petroleum fractions. For road transport, the focus is on petrol ($C_5–C_8$) and diesel ($C_{14}–C_{20}$) and it is possible to increase the proportions of these by further processing of other fractions. Specifically, this involves 'cracking' of the heavier molecules by passing them over a suitable catalyst at high temperature to yield lighter molecules. Numerous other catalytic processes also take place in an oil refinery, e.g. dehydrogenation to convert paraffins to olefins (thereby improving the octane number; see Section 4.3.5, Chapter 4) and hydrodesulfurization (to remove sulfur from the fuel).

Petrol for road transport is a very sophisticated product and a great deal of research and development has gone into perfecting it for use in spark-ignition engines. By contrast, diesel is more straightforward to produce because the performance of diesel engines is less demanding of a precise fuel specification. In recent years, however, this situation has changed somewhat with the requirement for diesel with ultra-low sulfur content to meet environmental regulations.

2.4 Conventional petroleum reserves

A key question concerning the future of road transport is: how many years of petroleum reserves are still available to exploit? This is exceedingly difficult to answer, although many 'experts' have promulgated views. The difficulty arises from numerous uncertainties, as follows:

(i) How much oil is there still to be extracted from existing known reservoirs?
(ii) How many new reservoirs remain to be discovered, particularly in areas that are difficult to access such as the polar regions and deep offshore waters?

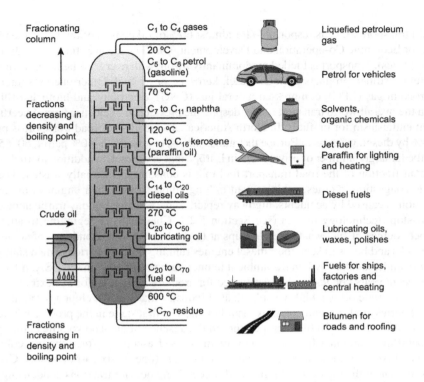

FIGURE 2.3

Distillation products of crude petroleum

(Reproduced with permission from The Royal Society of Chemistry from R.M Dell and D.A.J. Rand, 2004, Clean Energy, Cambridge, UK)

(iii) What further advances will be made in drilling and production technology?

(iv) How do production volumes and sales depend upon the international market price of crude oil?

(v) Will oil sands and oil shale make a significant contribution?

(vi) To what extent will the development of alternative fuels influence the situation?

(vii) Will political factors intervene to restrict supply?

(viii) What will be the impact of legislated targets for reducing pollution and greenhouse gases?

(ix) How many extra vehicles will be manufactured in the years ahead to satisfy market demand, especially in China and India?

(x) What will be the global requirement for petroleum in the light of advances in automotive engineering that may lead to improved efficiency?

The first six questions lie within the realm of the petroleum industry, questions (vii) and (viii) are political in nature, while the remaining two relate to the automotive industry. Answering each question individually is in the nature of crystal-ball gazing – collectively, they pose a formidable problem for forecasters. There is a view among many geologists and petroleum engineers that virtually all of the mega-large oil fields are already known and worked, and that no more comparable in size to those in the Middle East are likely to be found.

Table 2.5 Proved Reserves in 2000 and 2010 of the 10 Major Oil-Producing Countries (Expressed as 10^9 bbl Oil) and Ratio of Reserves to Current Production Rate (R:P)

Country	2000	2010	R:P
Saudi Arabia	262.8	264.5	72.4
Venezuela	76.8	211.2	> 100
Iran	99.5	137.0	88.4
Iraq	112.5	115.0	> 100
Kuwait	96.5	101.5	> 100
United Arab Emirates	97.8	97.8	94.1
Russian Federation	59.0	77.4	20.6
Libya	36.0	46.4	76.7
Nigeria	29.0	37.2	42.4
Canada[a]	18.3	32.1	26.3

Note: these data are little more than estimates and should be treated with caution. Nevertheless, they provide a semi-quantitative view of where the bulk of the reserves resides.
[a]*Excluding oil sands.*

The proved reserves in 2000 and in 2010 of the 10 countries currently believed to have the largest amounts of oil yet to be recovered are given in Table 2.5. The final column shows the ratio of proved reserves to the current production rate per annum (R:P). This ratio indicates the number of years that a given country might continue producing at the current rate. It is notable that some 60% of the known world reserves of oil are found in just five countries in the Middle East. This domination by one geographic region causes concern for the future security of oil deliveries, mitigated somewhat by the knowledge that these nations need the revenue as much as the rest of the world needs the oil.

The second remarkable statistic shown in Table 2.5 is that in every case (bar one) the proved reserves in 2010 were greater than in 2000, despite all the oil that had been produced and sold during the intervening decade. Furthermore, in many countries the reserves were substantially greater. Why should this be? There are at least two factors at work here. One is that continuing exploration and improved extraction technology lead to more oil fields being discovered and exploited. The other is in the nature of statistics. 'Proved reserves' relate to oil that is economic to recover at a given market price. As this price has increased over the past decade, i.e. from around US$30 per barrel in 2000 to around US$100 per barrel today, it has become economic to extract more oil from a given reservoir before closing it down and moving elsewhere. Also, more reserves that are located in marginal fields have been upgraded from the speculative category into the 'proved' category. There are many oil fields that were not economic to operate 10 years ago, but are profitable at present-day oil prices. These considerations emphasize that proved reserves and R:P ratios are somewhat vague, economically dependent measures. If the price of crude oil were to rise to US$200 per barrel, yet more oil would become economic to recover. This leads to the realization that the world will never 'run out' of oil; it will simply become more and more expensive and consumption will fall as its use is restricted to premium applications. Moreover, as prices rise, unconventional sources of petroleum (oil sands, oil shales) become competitive and, beyond that, there are alternative fuels that might be adopted (see Chapter 3). The wild fluctuation in the price of crude oil over the past 40 years is shown in Figure 2.4.

Nevertheless, individual oil fields do become depleted and once they are shut down it is often impossible or uneconomic to return at a later date and extract oil that has been left behind. The North Sea is one region that has passed its peak production and is running down. So also are many of the oil

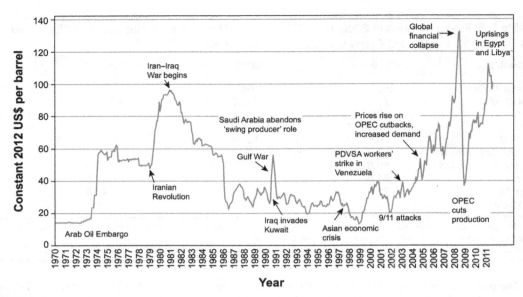

FIGURE 2.4

World crude oil price and associated events: 1970–2011

(Figure adapted from the US Energy Information Administration, 2011, What Drives Crude Oil Prices?)

fields in the USA. In 1971, M. King Hubbert predicted that world production of oil would peak in the period 1990–2000, that the total ultimate recoverable reserves would lie between 1.4 and 2.1 trillion barrels, and that 80% of these reserves would be processed in about 60 years from 1970 to 2030. While it is now clear that this forecast was unduly pessimistic, its error is only a matter of degree and time-scale. By re-plotting King Hubbert's data on a scale of four centuries, as shown in Figure 2.5, it is clear that fossil oil is but a transitory phenomenon that is not sustainable in the long term. It seems likely that petroleum may become relatively scarce within the lifespan of children alive today.

Since the 1970s, the idea of 'peak oil' has been taken up by numerous energy analysts and various prognostications have been made as to when any particular country or oilfield will reach its maximum production rate and go into decline. Road transport is unlikely to be the first casualty of a depleting oil supply as other sectors of the economy are more readily converted to natural gas or other fuels. Already, the use of oil for central heating and for electricity generation has gone into sharp decline. Ultimately, road transport will suffer, if not from limited supplies then from ever-rising prices. For true sustainability, it is necessary to look to the 22nd century and beyond.

2.5 Atmospheric pollution

Automotive exhaust consists, typically, of carbon dioxide and water, together with unburnt hydrocarbons, partially combusted products (carbon monoxide and aldehydes) and oxides of nitrogen (designated collectively as NO_x). Diesel engines also liberate fine particles of soot. Much progress has been made with suppressing these pollutants from vehicle exhausts. The work was stimulated by the

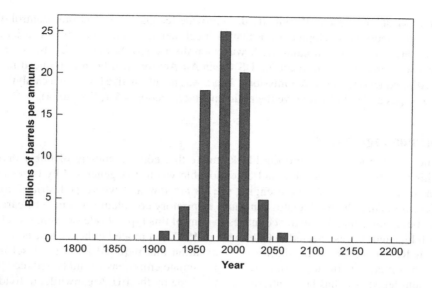

FIGURE 2.5

Estimated petroleum reserves: Hubbert's Peak

(Reproduced with permission from The Royal Society of Chemistry from R.M Dell and D.A.J. Rand, 2004, Clean Energy, Cambridge, UK)

increasing levels of atmospheric pollution that developed in many cities of the world. Degradation of air quality is particularly prevalent when there is a combination of (i) a city situated in a basin surrounded by hills and (ii) a natural temperature inversion, which leads to entrapment of pollutants in the basin. As the fleet of internal-combustion-engined vehicles increases in size, the pollution worsens and people develop symptoms of running eyes, coughing, chest pains, etc. The classic example has been Los Angeles, which as early as the 1940s began to experience significant atmospheric pollution. Other cities to be likewise affected have included Athens, Rome and Tokyo.

In 1998, the UK Committee on the Medical Effects of Air Pollutants estimated that up to 24 000 people died prematurely each year in the country as a direct result of air pollution. Similar findings have emerged from international research. According to the World Health Organization, up to 13 000 deaths per year among children (aged 0–4 years) across Europe are directly attributable to outdoor pollution. It was further argued that if pollution levels were returned to within the limits imposed by the European Union (EU), more than 5000 of these lives could be saved each year. Much of this atmospheric pollution is attributed to road vehicles, especially in urban areas.

Under conditions of strong sunlight, a photochemical reaction occurs between hydrocarbons and NO_x to give rise to ozone (O_3) and other undesirable products such as peroxyacetyl nitrate, a strong lachrymator irritant. The overall phenomenon is known as 'photochemical smog'. This differs from the urban smog (a term derived from 'smoke' and 'fog') that formed in earlier years, before the automotive age. The latter was not photochemical in origin, but contained sulfur compounds rather than NO_x that arose predominantly from the combustion of coal in open fireplaces. When vehicle exhaust gases first became a serious problem, engine modifications were made to provide some reduction of emission levels, in particular the development of 'lean-burn' engines in which the air-to-fuel ratio was increased. The development of oxygen sensors in the 1970s facilitated control of the air-to-fuel ratio. These were placed in the exhaust gas, close to the manifold, to provide feedback for electronic control of the carburettor. Another important development was that of direct fuel injection, where it was easier to control the air-to-fuel ratio than with carburettors. It was soon discovered, however, that these engineering improvements were insufficient to meet the US Clean Air Act that had been introduced in 1970 and mandated a 90% reduction in exhaust emissions. There was no alternative but to fit a catalytic device to treat the exhaust gas and thereby remove the pollutants (see Section 4.5.1, Chapter 4).

2.5.1 Emission regulations

Diesel engines are notorious for emitting black smoke that consists mostly of finely-divided soot particles. This is especially noticeable and objectionable when it is generated by buses and heavy trucks in an urban environment where engine speeds are low and where pedestrians congregate. Because of concerns that the particulates can cause respiratory complaints and may be carcinogenic, legislation is being introduced in many countries to control this type of release from diesel vehicles. The first such legislation was enacted in California in 1987 and related to heavy trucks. In 2008, the California Air Resources Board passed further legislation that required all road diesel trucks and buses to be retro-fitted with devices to ensure that particulate emissions would be reduced by at least 85%. Equivalent legislation has been introduced in Europe by the EU. Meanwhile, individual cities around the world have set their own standards for diesel emissions. Retro-fitting of buses and trucks is required in Hong Kong, Tokyo and Mexico City, for example, while London and Milan impose stiff fines on vehicles entering the central 'Low Emission Zone' if they do not comply with regulatory requirements.

In Europe, EU directives have been instrumental in reducing emissions of carbon monoxide, NO_x, hydrocarbons and particulate matter of less than 10 μm in size (so-called 'PM10'). First introduced in 1992, the legislation forms a set of rolling regulations that are designed to become more stringent year on year. Currently, limits for new cars and light-duty vans must conform to 'Euro V' standards; see Table 2.6.

The effect of tighter EU standards on vehicle emissions has been to accelerate the introduction of 'greener' vehicle technologies. For petrol cars, this has been achieved in part through the use of the three-way catalytic converter and the move to fuel injection systems. For diesels, NO_x and particulate emissions have been reduced through the development of direct injection engines and diesel particulate filters (DPFs). These technological advances, together with the cleaner fuels that made these developments possible, have led to a dramatic reduction in regulated pollutants; so much so, that a car manufactured today produces 20 times less emissions than a 1970 model. Car manufacturers are well aware that future vehicles will have to conform to increasingly stringent regulations – indeed, although Euro V only came into force in 2010, it is to be replaced in 2015 by Euro VI, which embodies an even tighter set of standards; see Table 2.6. The regulations in the USA are more complex, but broadly similar in intent and scope.

It should be noted that since diesel vehicles are more economical than their petrol counterparts (typically, 20–30% more km per litre of fuel, for engines of similar power), their emissions are less. Nevertheless, the reduction in the amount of carbon dioxide emitted is not correspondingly greater since diesel has a higher carbon-to-hydrogen ratio than petrol. In the UK, vehicles are now taxed on the basis of carbon dioxide emissions, which are measured in g km^{-1}. This legislation favours small cars and, to some extent, diesel cars.

Table 2.6 EU Emission Standards for Passenger Cars (in g km^{-1})

	Euro Standard	Implementation Date[a]	CO	HC[b]	NO$_x$	HC + NO$_x$	PM[b]
Diesel	Euro I	Jul-93	2.72			0.97	0.14
	Euro II	Jan-97	1.00			0.70	0.08
	Euro III	Jan-01	0.64		0.50	0.56	0.05
	Euro IV	Jan-06	0.50		0.25	0.30	0.025
	Euro V	Sep-10	0.50		0.18	0.23	0.005
	Euro VI	Sep-15	0.50		0.08	0.17	0.005
Petrol	Euro I	Jul-93	2.72			0.97	
	Euro II	Jan-97	2.20			0.50	
	Euro III	Jan-01	2.30	0.20	0.15		
	Euro IV	Jan-06	1.00	0.10	0.08		
	Euro V	Sep-10	1.00	0.10	0.06		0.005
	Euro VI	Sep-15	1.00	0.10	0.06		0.005

[a]Market placement or first registration date, after which all new engines placed on the market must meet the standard.
[b]HC = hydrocarbons; PM = particulate matter.
Source: European Union directives.

Small quantities of sulfur-containing compounds are another undesirable component of both petrol and diesel fuel. These impurities cause the production of minor amounts of sulfur dioxide, which, like NO$_x$, is an acid pollutant and therefore is detrimental to the life of the catalytic converter, as well as damaging to the environment. Consequently, the major petroleum companies have developed and brought to the market low and ultra-low sulfur brands of petrol and diesel. The action to reduce dramatically sulfur levels in motor fuel has also been made in response to approved regulations in the USA and Europe. In the USA, the Environmental Protection Agency has set a requirement for diesel fuel to have less than 15 ppm sulfur (ultra low sulfur diesel, ULSD). The EU introduced a mandatory requirement in 2005 for all road vehicle fuels to contain less than 50 ppm sulfur and for fuels of < 10 ppm sulfur to be introduced progressively. Good progress towards these goals is reported from a majority of the member states. Legislation is also in place in most developed countries to control the emission of carbon dioxide, see Section 2.7 below.

2.6 Fuel and vehicle efficiencies

In 1993, the US government launched its Partnership for a New Generation of Vehicles (PNGV) with a target to deliver, by 2004, production-ready vehicles with a fuel economy of 34 km per litre (= 2.94 litres per 100 km = 96 miles per UK gallon = 80 miles per US gallon). It was soon realized that this was only likely to be achieved with direct-injection diesel engines, which were not commonly used in passenger cars in the USA. Nonetheless, the automotive companies are well on the way to reaching the technical target, but not in the timescale envisaged, through the use of smaller, turbo-charged engines. This progress has not been matched by any government-backed initiatives to wean US motorists from petrol-engined cars and sports utility vehicles (SUVs) that consume large quantities of cheap fuel. In Europe, five-seat passenger cars equipped with direct-injection diesel engines are currently capable of returning around 25 km per litre.

Traditional engines and transmissions are notoriously inefficient. Typically, the mechanical energy available at the wheels to propel the vehicle is around 20–25% of the chemical energy liberated on combustion of the hydrocarbon fuel. Sales brochures for vehicles usually state efficiencies in terms of fuel consumption (litres per 100 km; or miles per gallon, mpg) rather than as a percentage of the chemical energy of combustion. The former is of more use to the motorist, while the latter is of more fundamental interest to the scientist or engineer. Fuel consumption is vehicle-specific. A heavy vehicle, such as a bus or dumper truck, that returns high litres per 100 km (low mpg) may still be efficient – it is simply that more energy is required to move it along. Fuel usage also depends on the driving environment and in many countries it is conventional to quote values for urban driving, extra-urban driving and a combination of the two modes, with each situation measured according to a standard driving cycle. Increasingly, sales literature also provides an indication of the mass of carbon dioxide emitted per km.

The overall efficiency of converting crude oil to traction at the wheels of a vehicle is composed of the following two factors:

(i) the efficiency of extracting petroleum from the ground, transporting it to the oil refinery, processing it to products, conveying the refined fuel to the petrol station and dispensing it. This is referred to as 'well-to-tank' efficiency and may have a value of 85–90%, i.e. the operations involve an energy loss equivalent to 10–15% of the calorific value of the fuel;

(ii) the efficiency of converting the chemical energy of the fuel into mechanical energy at the wheels. This parameter is the 'tank-to-wheels' efficiency (20–25%).

Multiplying the two efficiencies together yields an overall 'well-to-wheels' efficiency in the range 17–22.5%.

The low 'tank-to-wheels' efficiency is, in part, a fundamental thermodynamic consequence of operating an energy-conversion process in a non-reversible manner and, in part, reflects friction losses in the engine and transmission. The energy that does reach the wheels is used to propel the vehicle, but some is lost in aerodynamic drag and in tyre friction (so-called 'rolling resistance'), as shown schematically in Figure 2.6.

The question facing engineers today concerns the extent to which the comparatively low 'tank-to-wheels' efficiency can be improved by the redesign of vehicles, engines, transmissions and tyres so as to maximize power output and minimize drag and frictional losses. These areas of development are being addressed vigorously and are discussed in Chapter 4.

2.7 Emissions and climate change

In all the debate over climate change (global warming) and its origin, three facts seem incontrovertible:

(i) the Earth's climate has changed and generally warmed over the past two centuries;

(ii) the concentration of carbon dioxide in the atmosphere has risen steadily since the Industrial Revolution, when fossil fuels started to be burnt in large quantities;

(iii) carbon dioxide is a greenhouse gas that absorbs infra-red radiation reflected from the ground and prevents its escape into space; see Box 2.1.

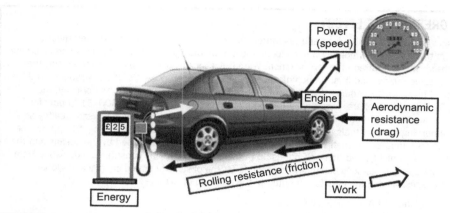

FIGURE 2.6

Schematic illustration of the relationship between energy, work and power in a passenger car

(Reproduced with permission from The Royal Society of Chemistry from R.M Dell and D.A.J. Rand, 2004, Clean Energy, Cambridge, UK)

Most authorities link these three facts and conclude that carbon dioxide derived from fossil fuels is largely responsible for the observed change in climate. Is this, however, the complete story?

The factors that control the climate are very complex and only partly understood. Carbon dioxide is released in vast quantities from natural processes, notably by volcanoes; forest fires; and loss from oceans, lakes and rivers. Other natural sources include the respiration of animals, as well as the decay of plants and animals on land and plankton in the oceans. These releases are approximately balanced by a similar quantity taken up by photosynthesis (the process whereby living organisms use light to produce complex substances from carbon dioxide and water) and by re-absorption in the oceans, so that historically (that is, before mankind started to burn fossil fuels) the climate was in equilibrium.

Anthropogenic releases of carbon dioxide from combustion processes in, for example, power stations and motor vehicles are small by comparison. For instance, the burning of fossil fuels has been shown to contribute only ~ 6 gigatonnes (Gt) to the 150–200 Gt of carbon taken up by the troposphere each year, (i.e. 3–4%). With extensive de-forestation taking place, however, one of the principal sinks for carbon dioxide is being removed and this is compounding the effect of such combustion. Thus, in terms of the annual balance, fossil fuels, and to a lesser extent deforestation, are held to be entirely responsible for the growing concentration of carbon dioxide in the atmosphere.

Other commentators (so-called 'climate sceptics') reject this view and claim that climate change is entirely attributable to numerous natural processes that involve, for example, the effects of water vapour, clouds, solar activity, volcanoes and ocean currents. Many books and research papers have been published on the subject of climate change and arguments have been advanced both for and against the contribution made by the combustion of fossil fuels. This book is not the place to rehearse these complex issues. Rather, the authors accept for now the scientific consensus that anthropogenic greenhouse gases are a major cause of climate change and that they must be curtailed by limiting the extraction and combustion of fossil fuels. The challenge that has been set is to cut global emissions of carbon dioxide by at least 50% by 2050. The transport sector will have to take its share of the task.

BOX 2.1 THE GREENHOUSE EFFECT

When sunlight strikes the earth it is absorbed and re-emitted as long-wavelength infra-red radiation (heat). Certain molecules in the atmosphere, notably water vapour and carbon dioxide (CO_2), are impervious to incoming visible light, but absorb the re-emitted infra-red radiation. Were it not for this absorption by water vapour, the world would be a far colder place than it is. Carbon dioxide is an essential component of the atmosphere; indeed, without it there would be no photosynthesis and no plant life, and therefore no animal life. Historically, the concentration of CO_2 in the atmosphere, in parts per million by volume, has been around 300–310 ppmv for millennia. Since the advent of the Industrial Revolution around 1800, this concentration has risen steadily as a consequence of the burning of fossil fuels in ever-increasing amounts and it is now approaching 400 ppmv; see Figure 2.7. It is estimated that roughly half of the CO_2 emitted by mankind is re-absorbed by the oceans and the terrestrial eco-system fairly quickly, and that half builds up in the atmosphere. This constitutes a deviation from the natural equilibrium and is thought to be largely responsible for the observed climate change and global warming of recent decades. Among the effects attributed to the build-up of CO_2 are:

* melting of the Greenland and Antarctica ice sheets;
* acidification of the oceans that leads to the destruction of a major part of the world's coral reefs.

The rate of change of CO_2 concentration and its effects over the past century are unprecedented in the geological record.

Water vapour and carbon dioxide are by no means the only greenhouse gases (GHGs), although they are the most important. Another significant GHG is methane (CH_4), which, molecule for molecule, has more than 20 times the warming potential of carbon dioxide. The quantity of methane released through human activity is comparatively limited, but major quantities arise naturally from decay of animal and vegetable matter, from release by the seabed and lakes, and from digestive processes in animals.

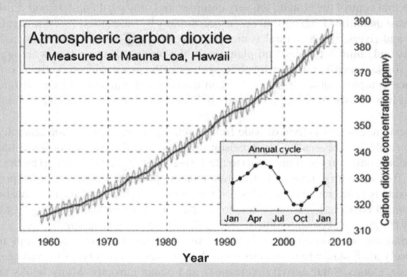

FIGURE 2.7

Steady increase in the concentration of carbon dioxide in the atmosphere over the past five decades

(Image courtesy of Narayanese, Semhur and the NOAA)

About 25% of worldwide anthropogenic emissions of carbon dioxide are attributable to transport and this is the fastest-growing source of the gas. Three-quarters or more of this amount stems from road transport, with the remainder from aviation and shipping. Given current trends ('business as usual' scenario), the International Energy Agency predicts that emissions from transport are likely to rise by 50 and 80% by 2030 and 2050, respectively. Clearly, this is unsustainable if the aim is to hold the average global temperature rise to less than 2 °C, as recommended by the Intergovernmental Panel on Climate Change (IPCC), and so avoid the worst consequences of climate change.

In 2009, the European Parliament passed new legislation on carbon dioxide emissions from light duty vehicles that imposes a cap of 130 g km^{-1} averaged over all new cars produced by each manufacturer by 2015. The limit will be the equivalent of 58 miles per UK gallon for diesel engines and 52 miles per UK gallon for petrol engines. The steps to reach this goal are being phased in over three years; 65% of the newly registered cars of each manufacturer had to comply by 2012, 75% by 2013, 80% by 2014 and 100% by 2015. An extended target is set to be an average of 95 g km^{-1} by 2020. Manufacturers that exceed targets from 2012 onwards will be required to pay a penalty for each car registered that amounts to €5, €15 and €25 for the first, second and third g km^{-1} over the averaged limit, and €95 for each subsequent gram. From 2019, stricter penalties will be introduced – every exceeding g km^{-1} will carry a €95 surcharge. These measures constitute a serious financial disincentive for the producer of large cars with powerful engines (e.g. SUVs), which is a most profitable market sector. If companies wish to continue to supply these vehicles, then they will either have to pay up or focus the bulk of their output on small cars with efficient engines. The inclusion of hybrids or pure-battery electric vehicles in their range of models would certainly help.

In the USA, the automotive manufacturers are required by the Corporate Average Fuel Economy (CAFE) regulations to meet certain standards for fuel consumption (measured as miles per US gallon), averaged over all their product line. (Note: fuel consumption is related to carbon dioxide emissions according to the carbon content of the fuel.). By 2016, the CAFE standard calls for an average of 42 and 26 miles per US gallon for cars and trucks, respectively. Again, a penalty will be imposed for failing to comply with these standards. The Environmental Protection Agency (EPA) and the National Highway Traffic Safety Administration (NHTSA) are working together to set new standards for light duty vehicles in model years 2017–2025 as regards both fuel consumption and greenhouse gas emissions.

China is the world's foremost liberator of carbon dioxide (the level overtook that of the USA in 2010, to represent one-fifth of global emissions) and is therefore also taking climate change seriously whilst building up its vehicle fleet. The legislation being formulated closely parallels that of the EU in terms of both fuel consumption per vehicle and emissions. A comparison is made of the evolution of regulations for carbon dioxide emissions in various countries in Figure 2.8. Starting in 2003, the mandates become increasingly stringent to 2020 and beyond for Europe, Japan and the USA. All three regions plan to work towards a situation where the corporate average for light vehicles will be well below 100 g CO_2 km^{-1}. It seems likely that China, Australia and other developed countries will follow suit. Indeed, as cars will be designed and built for the major world markets to meet these standards, other nations will inevitably fall in line when they purchase the vehicles.

In summary, it is clear that governments, together with automotive manufacturers, are now greatly concerned with improving fuel economy in order to mitigate greenhouse gas emissions. Governments impose targets that manufacturers strive to achieve. The technology being adopted in new model vehicles is described in Chapter 4 and Chapter 5. Despite the efforts by engineers, however, other steps will

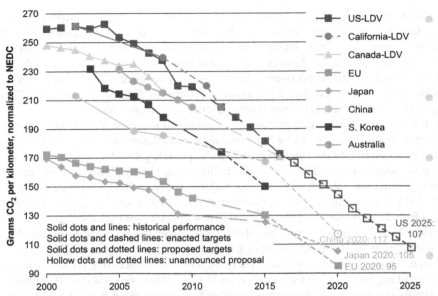

FIGURE 2.8

Progressive tightening of automotive emissions regulations: enacted and planned

(Data from the International Council for Clean Transportation and available under the Creative Commons License)

need to be taken if the world is to reach its global targets for reducing greenhouse gases in the atmosphere. For the transport sector, the measures must include both a reduction in travel, facilitated by the use of modern means of communication, and a shift towards greater use of public transport modes. To bring about the latter it will be necessary to employ both sticks and carrots. Sticks might include higher taxes on cars, congestion charges, and restricted and expensive parking places. Carrots would comprise public transport that is convenient, frequent, reliable, clean, comfortable, cheap and secure for passengers. This is the challenge faced by urban transit companies and also by long-distance bus and train operators.

2.8 Electricity and hydrogen as energy carriers

One of the long-term objectives of a new paradigm for road transport must be to reduce, and ultimately eliminate, its dependence on petroleum-based fuels. There are three reasons for this:

- to ensure security of supply and protect it from political or other disturbance, by diversifying the primary fuel base;
- to eliminate, as far as possible, the release of carbon dioxide to the atmosphere;
- to anticipate the advent of 'peak oil' and the decline in production, just as global demand for road transport is growing.

The first of these objectives can be met by opening up new oil fields in different parts of the world, for instance in Alaska or offshore Brazil, or by utilizing novel petroleum feedstocks such as oil sands from Alberta, Canada. Such initiatives address the security-of-supply issue, but make little or no contribution to resolving the other two concerns. The use of fuels of lower carbon content (compressed natural gas, liquid petroleum gas) for transport would play a part in reducing carbon dioxide emissions, as would the burning of liquid alcohols (methanol, ethanol). These are only interim, partial solutions and bring with them technical challenges of implementation on a large scale, as well as economic implications.

Ultimately, it will be necessary to derive road transport fuels either from renewable sources of energy or from carbonaceous fuels processed at large central plants furnished with carbon capture and storage facilities. In either case, the endeavour will require a vector to convey the energy from the primary production plant to the road vehicles. At present, this energy vector consists of road tankers that distribute processed petroleum from the refinery to the service stations. In the long-term future, a new energy vector will be needed and the two most canvassed candidates are electricity and hydrogen.

Converting the road transport fleet to electric drive imposes both technical and logistical problems, as discussed in Chapter 5. Also, the future sources of mains electricity are by no means certain as there are question marks over both coal and nuclear power as means of generating electricity. In the near term, natural gas is a more secure source of electricity, but this too requires the implementation of carbon capture and storage. Renewable energy, the final goal, is growing in importance but still makes only a minor contribution to our electricity supply, even with the help of government subsidies. Electricity supply issues are discussed in Chapter 6.

Finally, there is the option of hydrogen as a transport fuel, either as a cryogenic liquid or as compressed gas, burnt in an internal-combustion engine or converted to electricity on board the vehicle in a fuel cell. Hydrogen as a road transport fuel, discussed in Chapter 8, has difficulties of a technical and engineering nature that are now being addressed, as well as doubts about overall efficiency and cost.

Despite these reservations, it is necessary to evaluate thoroughly all possible options even though it is accepted that road transport is only one component, albeit an important one, of a constantly changing and evolving energy scene.

Unconventional Fuels

3.1 The need for 'unconventional fuels'

There is widespread awareness of the rising cost of crude oil as the more readily exploited sources are depleted and it becomes necessary to seek further reserves in progressively more difficult environments such as in deeper regions of the continental shelf. Over the past 40 years, the price of a barrel of crude oil has more than quadrupled. The consequent increase in the price of motor fuels 'at the pump' impacts everyone in one way or another.

There is considerable uncertainty over how much petroleum remains to be extracted and how long world supplies will last. Alleviation of these two global concerns is a complex technico-economic issue that depends on:

- advances in exploration and extraction technology in difficult and/or hostile environments and the need to avoid disrupting or polluting sensitive eco-systems;
- improvements in road vehicles and engine design that result in greater efficiency;
- the response of the market to ever-rising fuel prices;
- the growth of road transport – in Asia and Africa especially;
- how motorists adapt their lifestyles and how freight businesses modify their operations to reduce fuel consumption.

Increases in the efficiency of internal-combustion engines (reviewed in Chapter 4) and the electrification of power-trains (Chapter 5) will help to reduce petroleum consumption. It is also expected that the opening up and exploitation of unconventional sources of motor fuel – the subject of this chapter – will play a significant role in extending the ultimate lifespan of the internal-combustion engine as a power source for road vehicles. On the other hand, the anticipated escalating demand for products derived from petroleum as standards-of-living rise in the less-developed nations of the world is a negative factor. The only point on which there is a consensus, as noted in Section 2.4, Chapter 2, is that petroleum will never 'run out', but will become increasingly expensive as time goes by. The era of 'cheap oil' is now well past and society will have to exercise greater ingenuity and inventiveness in meeting its future requirements for transport fuels.

3.2 Raw materials

There are many alternatives to conventional petroleum that are capable of being employed in the manufacture of motor fuels. Some of these are already being exploited on a significant scale (oil sands, heavy crude oil, energy crops), while others remain as yet undeveloped and are awaiting favourable economic conditions for their commercialization.

3.2.1 Oil sands

Perhaps the most important of the alternative fossil fuels are the oil sands (formerly, but erroneously, known as 'tar sands'; tar is derived from the dry distillation of coal). Oil sands consist of loosely compacted sand grains that are coated by water and clay. The intervening pore spaces are filled with bitumen, which is an especially heavy semi-solid, form of crude oil that consists of a complex mixture of condensed polycyclic aromatic hydrocarbons. Often referred to as asphalt, bitumen is widely used hot, as a viscous liquid, for surfacing roads and driveways, as well as in roofing applications. The material may also be refined to produce hydrocarbons of lower molecular weight, viscosity and density.

Oil sands are found in major deposits around the world, the greatest of these being the Athabasca resource in Northern Alberta, Canada. This and other deposits in Alberta are estimated to contain 1.7–2.5 trillion barrels of bitumen – about one-third of the known oil reserves on Earth. The proven *recoverable* hydrocarbon reserves, a fraction of this total, are estimated as equivalent to 170–180 billion ($170–180 \times 10^9$) barrels of oil (~25 gigatonnes of oil equivalent, Gtoe), although clearly this depends on both the technology employed and the price of oil. The fraction that is deemed to be recoverable may well increase as new extractive technologies are developed and as the price of crude oil rises.

Some of the Albertan bitumen (~20%) may be recovered by open-cast mining operations. Surface mining is simple and relatively inexpensive, but has a major impact on the local environment. After the topsoil and overburden (up to 75 m thick) are removed, the sand is simply shovelled up by giant excavators and taken by truck to a central cleaning plant where it is treated with warm water and caustic soda to separate the bitumen; see Figure 3.1(a). Subsequently, the bitumen is upgraded to petrol, diesel and other useful products in an oil refinery; see Figure 3.1(b). An average of four tonnes of material has to be removed to separate the two tonnes of oil sands needed to produce a barrel of synthetic crude oil. After the oil has been extracted, there remain the problems of disposing of the contaminated water, returning the sand to the pit from which it was extracted, replacing the overburden and topsoil, and finally making good the landscape. These operations are energy-intensive, and thereby reduce the net gain in energy.

(a)  **(b)**

FIGURE 3.1

(a) Bitumen mining operations; (b) refinery for upgrading bitumen

(Images reproduced courtesy of: Suncor Energy, Canada)

Particulate and gaseous emissions, including greenhouse gases such as carbon dioxide and acidifying compounds such as sulfur dioxide, constitute deleterious by-products of the processing of oil sands. Furthermore, the extraction and recovery of useful products from oil sands are extremely water-consumptive and may reduce the level of local water tables, lakes and rivers. Hot waste water released during the cracking process may cause thermal disruption of ecosystems.

Deep-lying oil sands have to be treated *in situ* by injecting vast quantities of steam down vertical boreholes. This procedure heats the bitumen and reduces its viscosity so that it can flow. As the steam passes horizontally through the sand bed, it frees the bitumen such that a mixture of water, steam and heavy oil emerges from an exit borehole. The technique has less environmental impact on the surrounding countryside than does the open-cast scheme, but requires enormous quantities of steam, and this is expensive. Also the recovery rate of the bitumen is much lower (50–60%) than that achieved with surface mining (> 90%) and the energy consumption and greenhouse gas emissions are each approximately 25% higher. Again, huge volumes of contaminated water are produced and these have to be cleansed before discharge. Most operations to date have therefore been confined to the simpler method of surface mining. Newer production techniques, such as vapour-assisted petroleum extraction (VAPEX) or *in situ* combustion, can reduce some of the penalties associated with treating bitumen beds that are too deep for mining, but the relative economics of such schemes remain to be clarified.

Each of the two principal processes requires a substantial input of energy – for extraction *in situ* this is in the form of steam; for surface mining it is in the form of fuel for the excavators and transport trucks. When the bitumen has been recovered, a large amount of hydrogen (from natural gas) is required to upgrade the raw material so that it is suitable for treatment in a conventional oil refinery. The energy input in the form of natural gas, electricity and diesel to produce 1 GJ of motor fuel (approx 30 litres petrol) is illustrated in Figure 3.2. The diagram also gives the quantity of carbon dioxide emitted in each of the unit operations.

It is said that the energy needed to mine, process, refine and upgrade the bitumen, as well as to restore the landscape, is as great as the energy content of the resulting petroleum products. Therefore, on strict energy grounds, it is dubious whether any case can be made for the undertaking, but this is not the foremost consideration in the minds of the operators. The government of Alberta (and that of Canada more generally) is focused on the value of the petroleum produced and sees the enterprise as a source of wealth to be exploited, as well as a means of self-sufficiency from conventional petroleum in the fuelling of North American transport. The petroleum companies, especially Shell Canada and Suncor Energy, estimate that the production of petroleum from bitumen is a profitable operation now that crude oil prices on the world market are at a high level. Nevertheless, from an environmental viewpoint the true value of the oil sands remains questionable.

The production of conventional oil in Canada peaked around 1997 and declined over the following nine years. Forward projections are for a continued decline, which is more than compensated by growth in production from oil sands. Indeed, by 2030, the total output is expected to have more than doubled (see Figure 3.3).

To the extent that natural gas (which is abundant in Alberta) is used to generate the hot water, steam and hydrogen, the overall process may be viewed as an indirect means of using this low-value fossil fuel to produce valuable petroleum products. Seen in this light, the extraction of petroleum from bitumen may be regarded as an alternative to established 'gas-to-liquids' technology, whereby natural gas is converted to liquid hydrocarbon fuels; see Section 3.3.2.

The scale of the operations involved in the exploitation of oil sands is such that it has even been suggested that a dedicated nuclear power plant, operating in combined heat and power (cogeneration) mode, might provide the heat to generate the steam and hot water.

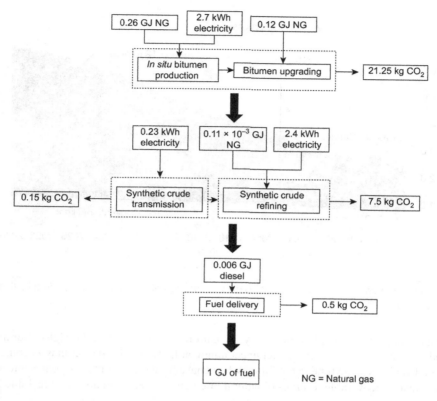

FIGURE 3.2

Energy input in the form of natural gas, electricity and diesel to produce 1 GJ of motor fuel from bitumen. The mass of carbon dioxide released in each operation is also shown

3.2.2 Heavy oil deposits

Heavy oils lie intermediate between light oils, which flow readily and constitute the 'petroleum' used as feedstock in oil refineries, and bitumen, which does not flow at all at ambient temperature. Although heavy and extra-heavy crude oils are widespread in 30 or more countries, there has been little exploitation of this resource because extraction, transportation and processing are more difficult than for light oils. Furthermore, heavy oils often contain higher levels of sulfur and other undesirable elements such as selenium and transition metals. For pipeline transport, a diluent is often added at regular intervals to facilitate flow.

The largest reserves of heavy crude oil in the world are located north of the Orinoco River in Venezuela, where it is estimated that the deposit represents a massive 1.2 trillion barrels, of which 250–300 billion barrels are thought to be economically recoverable. The reserves of heavy crude oil in Venezuela are of the same order of magnitude as the conventional oil reserves in Saudi Arabia; see Table 2.5, Chapter 2. The Venezuelan heavy oil differs from the Alberta oil sands product in that the heavy oil, unlike the semisolid bitumen, is not intimately bound up in the stratum. Much of the Venezuelan oil is deep lying and cannot be extracted by surface mining. Several decades ago, the oil company BP

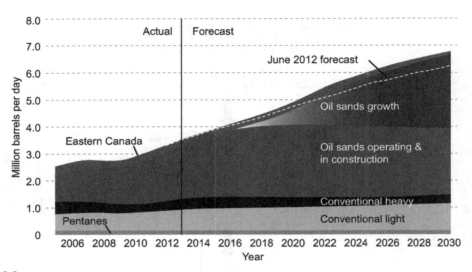

FIGURE 3.3

Canadian oil sands and conventional production. Based on an original from the Canadian Association of Petroleum Producers

formulated a major project to exploit this heavy oil. Unfortunately, however, the high sulfur and metals content made the processing to a clean fuel uneconomic at that time. Heavy oil may become a source of fuel for road vehicles some time in the future when conventional oil eventually comes into short supply. The technically recoverable reserves of natural bitumen and heavy oil are listed in Table 3.1.

3.2.3 Oil shales

Oil shale is again a misnomer as it contains neither oil nor bitumen. Rather, it is sedimentary rock that incorporates considerable quantities of kerogen, a solid mixture of organic chemicals. The deposits were formed millions of years ago by the co-sedimentation of silt or sand and organic debris, followed by compression and heating. In terms of its hydrogen-to-carbon ratio, oil shale lies partway between crude oil and coal. The material is processed by pyrolysis – that is, by heating in the absence of air to 450–500 °C, at which temperature the kerogen decomposes to form a liquid that can be vaporized and condensed to leave the solid matter behind. Major deposits of oil shale are found in western USA, Queensland (Australia), Russia, Brazil, China, Sweden and Estonia, with lesser quantities in many other countries. The world resources have been estimated at 3–5 trillion barrels of oil equivalent, of which perhaps one-fifth is technically and economically recoverable. It is anticipated that, on average, about 40 litres of oil may be recovered per tonne of shale processed, although this figure is quite variable from place to place. Clearly, oil shale is a major resource of world petroleum that is geographically widespread.

The history of the extraction of oil from shale reaches back to the 19th century. The first industrial-scale operation was in Scotland around 1860, but soon the process was taken up in other parts of the world. Several million tonnes of oil shale per year were mined right up until the early 1960s, after which the production of oil from this source could no longer compete economically

Table 3.1 Technically Recoverable Reserves of Natural Bitumen and Heavy Oil (Billion Barrels)

	Natural Bitumen		Heavy Oil	
	Recovery Factor	Technically Recoverable	Recovery Factor	Technically Recoverable
North America	0.32	530.9[a]	0.19	35.3
South America	0.09	0.1	0.13	265.7[b]
Africa	0.10	43.0	0.18	7.2
Europe	0.14	0.2	0.15	4.9
Middle East	0.10	0.0	0.12	78.2
Asia	0.16	42.8	0.14	29.6
Russia	0.13	33.7	0.13	13.4

[a]*Mostly Canada.*
[b]*Mostly Venezuela.*

with petroleum. Even as late as 2006, however, Estonia generated most of its electricity from shale oil.

Shale from the mine is crushed to a particle size that is appropriate for the particular method that is to be employed for the subsequent heat treatment. The decomposition of the kerogen and liberation of the oil can be undertaken by numerous procedures, of which the most common is performed in a horizontal retort. In addition to the oil, some combustible gas is released and the residual shale ash contains carbonaceous coke or char. These by-products may be progressively burnt to assist with heating the retort. An alternative process is to crush to a small size and heat the shale in a fluidized bed. As is the case with oil sands, there is also the possibility of *in situ* pyrolysis underground without mining the shale. The operation requires considerably more heat than the above two procedures, however, and is therefore less energy-efficient. Accordingly, it is not widely practised.

The materials-handling involved in the mining of shale and conveying it to a central processing plant, in the return of spoil to the pit, and in the retorting stage are all very energy-intensive operations. Unfortunately, the exploitation of oil shale also imposes severe environmental constraints. The shale often contains significant amounts of sulfur and nitrogen. The former element may be liberated as hydrogen sulfide and sulfur dioxide and the latter as ammonia; these gases must be sequestered to avoid polluting the atmosphere. Often, toxic heavy metal impurities and carcinogenic polycyclic aromatic hydrocarbons are present in the waste and may leach into surface and ground water if not contained; thus safe disposal is essential. Similarly, water that is required for the processing of the shale, at a rate of 1–5 barrels per barrel of shale oil produced, may also become contaminated and therefore require purification before release back to the environment. Nevertheless, despite all these difficulties, shale oil extraction and subsequent refining is considered to be an economic proposition given the current high price of oil.

The most complete estimates of oil shale availability have been published by the US Geological Survey, which confirms that deposits are to be found in many countries. The bulk of the world's known resources, however, is located in the USA, where there are estimated to be more than two trillion barrels (about 75% of the world total) of oil shale of medium quality. Today, few deposits can be economically mined and processed in competition with petroleum. It is also noteworthy that large-scale processing of oil shale would be expected to generate about five times the emissions of carbon dioxide per litre of fuel produced compared with conventional oil refining.

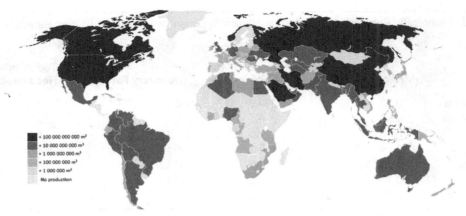

Natural gas production by country /10⁹ m³

Top 10 producers of natural gas (10⁹ m³ per year)	
Russia	669
USA	651
EU	168
Canada	160
Iran	146
Qatar	117
Norway	103
China	103
Saudi Arabia	99
Algeria	85

FIGURE 3.4

World map of natural gas production and annual output of top 10 countries

(Map sourced from: Wikipedia and available under the Creative Commons License)

3.2.4 Natural gas

Natural gas, which is a mixture consisting of methane with other hydrocarbons, carbon dioxide, nitrogen and hydrogen sulfide as minor constituents, is abundant and relatively inexpensive in many parts of the world (see Figure 3.4) but it must be compressed for storage on a vehicle. With only little modification, internal-combustion engines can be made to run on compressed natural gas (CNG). At acceptable storage pressures, the calorific value of CNG (per unit volume) is lower than that of liquid petroleum fuels and therefore the storage tank has to be significantly larger in volume than the fuel tank in a vehicle that uses petrol or diesel, otherwise the vehicle range is reduced.

Sedimentary oil deposits generally have natural gas associated with them. Regarded as a nuisance in many remote oil fields, the gas was traditionally flared off from production wells since there was no practical means of delivering it to market. This practice represented a flagrant waste of a valuable resource and

FIGURE 3.5

Tanker ship for conveyance of liquid natural gas

(Image sourced from Wikipedia, available under the Creative Commons License)

contributed to the burden of carbon dioxide in the atmosphere. In recent years, facilities have been built around the world to collect the so-called 'stranded' gas, liquefy it at a very low temperature (boiling point of methane is −164 °C), and then transport it in specially designed refrigerated ships to reception terminals in developed countries; see Figure 3.5. After landing, the liquid natural gas (LNG) is re-vaporized and fed into the country's gas grid for distribution. Japan, in particular, has no indigenous fossil fuels and is now heavily dependent upon the import of LNG for its energy supplies, but reception terminals are also being built in Europe to receive LNG from the Middle East and Africa.

Not all natural gas is associated with oil. For instance, large sedimentary basins that contain gas but no liquid petroleum are found in the southern part of the North Sea, in Algeria and in many of the American states. Where large markets exist that can be reached by pipeline, it is straightforward to collect the gas from the sedimentary basin and convey it to customers. The distances involved may be quite long, however – for instance, from Alberta to eastern Canada or from Siberia and the Caucasus to western Europe – and this might be seen as a potential instrument for exerting political pressure.

There are large deposits of shale rock, often deep-lying, that hold substantial amounts of natural gas. This vast resource, which is comparable in size with shale oil, is referred to as 'shale gas' and is released from the rock by hydraulic fracturing (now known as 'fracking'). High-pressure water, often with added chemicals to increase the fluid flow, is injected into a borehole to cause the shale to crack and expel the gas. The process has been made feasible by the development of horizontal drilling techniques that enable a significant area of shale to be fractured from a single borehole. Shale gas is being tapped extensively in the USA and by 2010 represented over 20% of the total national production of gas. Large reserves are also found in China and Australia. There has been criticism of fracking technology on account of the accompanying consumption of large quantities of water – a particular problem in desert regions – and the possible contamination of the supplies of drinking water by the chemicals that are injected during gas extraction. It is also reported that the procedure can give rise to earth tremors.

Sources of natural gas are not restricted to sedimentary basins. Comparatively small amounts are recovered from landfill sites where gas is generated by the slow decay of waste matter. Much larger quantities are present in coal seams in which the gas is adsorbed on the extended porous inner surface of the coal. Because of its large surface-area, coal can hold many times more methane than a conventional rock reservoir of similar volume, e.g. up to ~ 25 m^3 per tonne of coal. Unfortunately, coal beds also contain fractures ('cleats') that are often filled with water. This has to be pumped out to allow the 'coal-bed methane' to be desorbed and recovered. Disposal of the saline water that is removed from the cleats may present a significant problem. The largest resources of coal-bed methane are to be found in regions that are rich in coal, namely North America, Russia, Germany, China, Australia and South Africa. To date, these coal fields have not been tapped for gas in a major way but, as with most other unconventional fuels, they await the arrival of an economic climate that promises good profitability.

Methane is also tied up in the 'permafrost' of Siberia and under the ocean in the form of solid 'methane hydrate', $CH_4 \cdot xH_2O$, which is effectively a variety of ice that contains methane locked within a so-called 'clathrate' structure. This compound is only produced at low temperatures and high pressures, as encountered on or under the ocean bed, and decomposes to discharge methane when the pressure is reduced and/or the temperature is raised. The quantities of methane hydrate are thought to be prodigious, but no serious attempts have yet been made to harvest them as present supplies of natural gas from other sources are plentiful. Indeed, although world consumption has risen five-fold over the past few decades, the proven reserves of natural gas have grown steadily as more fields of the conventional type have been discovered and brought into operation. Given that methane is a far more effective (and therefore pernicious) greenhouse gas than carbon dioxide, concern has been expressed that global warming could lead to uncontrolled release of this 'locked-in' natural gas to the atmosphere that would result in a damaging positive feedback loop.

Because of its convenience of transmission, distribution and use, natural gas is widely employed for space heating and cooling, both domestically and in commercial buildings. It is also becoming the preferred fuel for electricity generation in combined-cycle gas turbines; see Section 6.3.2, Chapter 6. These units have higher efficiency than traditional coal-fired plants and emit substantially less carbon dioxide per megawatt-hour of electricity generated. The potential importance of methane as a transport fuel is not so much for its direct use (as CNG), although this is a niche market, but rather for conversion to liquid fuels via 'gas-to-liquid' technology; see Section 3.3.2 below.

3.2.5 Coal

Coal measures were laid down in the carboniferous period, around 300 million years ago. Trees and other plant material died, fell to the ground, and gradually were compressed. Under immense pressure, the decomposing vegetation was converted first into peat and then into coal. The latter is not a unique product and in fact exists in various forms with properties that depend upon the age of the deposit and the stresses imposed by temperature and pressure. Geological processes that took place over millennia converted peat first into lignite, then into sub-bituminous coal, then bituminous coal and finally into anthracite, which is the hardest form of coal with the least content of volatiles.

Coal deposits are distributed very widely throughout the world. Historically, coal was mined for domestic heating purposes and for raising steam to power engines used in industry, in agriculture and in transport. Today, the foremost application of coal is in electricity generation. This is quite remarkable given that it is regarded as a 'dirty fuel' because of the pollution that is emitted on combustion – in particular sulfur dioxide and nitrogen oxides (which give rise to acid rain) and carbon dioxide. Much may be done to minimize the release of sulfur dioxide to the atmosphere by equipping the exhaust stacks with desulfurization equipment. Nitrogen oxide emissions can be kept to a low level by controlling the combustion temperature. Countries without indigenous coal supplies tend to adopt other means of electricity generation such as natural gas or nuclear power.

The relevance of coal to road transport operations is that it, too, may be converted to petroleum by 'liquefaction'. This process is known as 'coal-to-liquids' (CTL) technology. In the past, CTL technology has been practised in situations where coal was abundant and petroleum in short supply. During World War II, for example, Germany introduced and developed CTL operations to provide fuel for its military. South Africa, which also has ample coal but no oil, later perfected a similar process during the apartheid period when sanctions prevented the import of petroleum. The first South African 'coal-to-gas' plant (Sasol I) began operations in 1955. Two further large facilities (Sasol II and Sasol III) were commissioned in the 1980s; each was capable of converting 50 000 tons per day of low-grade bituminous coal into petrol, diesel, jet fuel, and other petroleum and chemical products. These plants are still in operation and each is capable of producing 150 000 to 160 000 barrels of synthetic crude oil ('syncrude') per day.

The basis of the Sasol process is first to gasify the coal by reacting it with steam and air in a Lurgi gasifier to form 'water gas', a mixture of carbon monoxide and hydrogen. This output is then passed over a catalyst of iron, cobalt or thorium at elevated pressure (350 psi) and temperature ($\sim 350\,^{\circ}$C) – the Fischer–Tropsch synthesis – to produce a variety of hydrocarbons that are separated into a higher-boiling fraction for use in diesel engines and a lower-boiling fraction, which is further reformed catalytically to yield petrol.

The significance of the Sasol process lies in the fact that the world has huge reserves of coal and that the technology is well-developed and could be utilized should it become necessary. Countries such as China and India that have coal but little oil may well be among the first to import the technology. Liquid fuels will therefore be available for the foreseeable future, albeit at a higher price than that of current fuel from conventional crude oil. From a US government perspective, there is great comfort in the knowledge that North America has huge reserves of oil sands, shale oil, natural gas and coal – all of which can be converted to liquid transport fuels and thereby enhance security of supply. Indeed, it has recently been projected by the International Energy Agency (IEA) that over the next decade or so the USA will once more become self-sufficient in oil and gas, thanks largely to its huge shale deposits.

3.2.6 Energy crops

Energy crops fall into three broad categories, namely:

- crops containing sugars or starch, e.g. sugar cane, sugar beet, sweet sorghum, corn/maize, wheat;
- oilseed crops, e.g. cottonseed, flaxseed, rapeseed/canola, soybean, sunflower seeds;
- ligno-cellulosic raw materials, e.g. trees, shrubs, straw, grasses.

The starchy crops can be used to manufacture ethanol, a petrol substitute, by fermentation, whereas varieties of oilseed may be chemically processed to produce bio-diesel. Ligno-cellulosic feedstocks are mostly burnt as biomass to generate heat, or are employed in agriculture. Another unconventional motor fuel, not widely used as yet but perfectly feasible, is methanol, which can be produced from a variety of agricultural and other waste products.

A major industry has grown up, particularly in Brazil and the USA, to produce bio-ethanol on a multi-tonnage scale. In Brazil, sugar cane is grown and fermented directly; the residue (bagasse) is currently a waste product which is burnt to provide process heat. In the USA, the feedstock is mostly corn (maize), which has ears rich in starch that can be fermented. Again, the bulk of the crop is a waste residue.

In a 2011 report, the IEA recorded a fast growth in global production of bio-fuels, namely, from 16 billion to more than 100 billion litres between 2000 and 2010; see Figure 3.6. It is apparent that the major increase has been in bio-ethanol. Brazil has long been active in obtaining ethanol from sugar cane and the IEA reported that 21% of all its transport needs are met by bio-fuel. The USA was a latecomer to the field, but with its much larger transport fleet it now processes a greater tonnage of ethanol than Brazil, mostly from corn grown in the Midwest. Bio-ethanol currently accounts for around 4% of the transport fuel consumed in the USA. Almost half of the American ethanol plants are owned and operated by farmers. This practice provides a secure market for their produce, diversifies their activities beyond farming, adds value to their output and avoids the cost (and energy consumed) in transporting the crop to a central processing plant. It is reported that the US government's 'ethanol from corn' programme may be attributed, at least in part, to lobbying by farmers in the Midwest.

In Europe, with its greater proportion of diesel-engined vehicles and with a climate unsuited to growing sugar cane, the emphasis has been on producing bio-diesel from oilseed-bearing crops such as rape and sunflowers – the output constitutes around 3% of Europe's transport fuel. Both of the crops bear yellow flowers. In early summer, the fields of England are a sea of bright yellow where rape is being grown, whereas further south in France and Spain fields of sunflowers are everywhere to be found a little later in the year. Soybean, cottonseed, coconuts and palm nuts are other oilseeds that can be processed to yield bio-diesel.

In principle, energy crops constitute a sustainable source from which to produce petroleum substitutes. In practice, this attribute is degraded to the extent that fossil fuels are used to provide process heat or to drive agricultural machinery used in harvesting and transport. Because energy crops are largely sustainable, unlike fossil fuels, the carbon dioxide liberated on combustion is not counted towards the

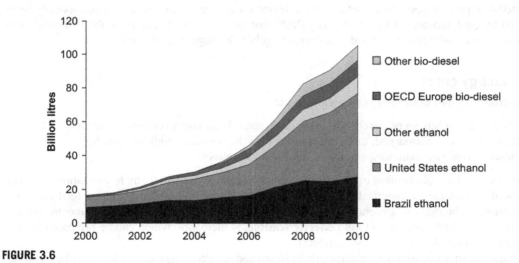

FIGURE 3.6

Global production of liquid bio-fuels 2000–2010. OECD = Organisation for Economic Co-operation and Development

(Figure based on one from International Energy Agency 2011 publication 'Technology Roadmap: Biofuels for Transport')

atmospheric burden of greenhouse gases; it is deemed to be recycled as new crops are grown. Other benefits of bio-fuels are that they reduce a country's dependence on imported oil, they introduce a new rural industry and an additional market for farmers, and they contain no sulfur or aromatic compounds.

A major switch to bio-fuels, however, will be accompanied by some disadvantages, of which the principal ones are as follows:

- biofuels have a considerably higher cost than conventional liquid fuels; this is especially true for bio-diesel, where the estimated production cost is as high as US$250 per barrel, in part due to a comparatively low agricultural yield per hectare;
- the amount of energy indirectly consumed in growing the crop (e.g. that used in the manufacture of fertilizer, pesticide), as well as the petroleum employed in harvesting and transporting it, may approach or exceed the energy content of the final product;
- the land utilized in growing energy crops might better be devoted to the production of food for humans or fodder for animals; this is a hot topic, both politically and morally, when so many of the world's population are under-nourished.

The energy consumed in manufacturing bio-fuels is in part petroleum-based (for operating farm machinery, transport by truck) and in part heat (for distillation of dilute ethanol, or for the chemical processing of oilseeds to bio-diesel). The extent to which the process heat derives from natural gas constitutes a saving in petroleum, but still contributes to greenhouse gas emissions. To be truly sustainable, the farm machinery should run on bio-diesel and the process heat should be obtained from waste agricultural products.

From a pure energy-balance viewpoint, the case for ethanol as a motor fuel is questionable. This is on account of the energy that is employed in growing and harvesting the crop and, especially, the considerable energy required to distil the dilute alcoholic solution. Obtaining ethanol from grain, rather than from sugar cane, is energy-intensive and typically yields a fuel that contains only 1.3 to 1.5 energy units for every unit of energy expended in its production. In addition, in the USA this external energy input is generally derived from coal or natural gas, thus making matters worse from a greenhouse gas perspective.

The debate over whether land is better employed in growing food or energy crops is extremely complex and is likely to vary from country to country as dictated by the size of the operation and the nature and fertility of the available land, together with the population to be fed. Ultimately, this question will be resolved by government policy, according to whether directives or incentives are given to farmers to grow food or, in other territories, subsidies are offered for bio-fuel crops. Taking account of these factors, the final determinant will be the relative price that the farmer can obtain for agricultural products vis-à-vis energy crops. Environmental factors must also be considered, e.g. the impact on wildlife, the long-term effect of growing the same crop without rotation and possible flow-off of nutrients to watercourses. Farmers have long experience of growing foodstuffs, but energy crops are still relatively new and it will therefore be necessary to proceed with caution.

Whether the crop be sugar cane or corn for fermentation to ethanol, or oilseed for conversion to bio-diesel, there is inevitably a great deal of waste product in the form of bagasse (from sugar cane), or stalks and husks from the other crops. The waste is composed largely of ligno-cellulose, a stable substance that cannot be readily processed to useful products and therefore is generally burnt.

Considerable research has been undertaken into the possibility of converting this material to ethanol. Four steps are required, as follows:

- physical and chemical pre-treatment to break up the structure and to separate the cellulose from the lignin;
- hydrolysis of the cellulose to form simple sugars such as glucose; the process is generally effected biochemically by means of enzymes;
- fermentation of the sugar to yield a dilute solution of ethanol;
- distillation of the alcoholic solution to produce concentrated ethanol.

A number of large-scale pilot plants have been constructed in the USA and elsewhere to develop an economic process for the conversion of ligno-cellulose to ethanol. Success would open up the possibility of utilizing fast-growing trees such as poplar, or herbaceous perennials such as miscanthus or switchgrass, for ethanol production. Forestry waste arising from coppicing might also be used. Commercialization of this endeavour would result in much greater productivity and could transform the economics of energy crops.

Although many uncertainties are still to be resolved around the broader environmental impact of growing energy crops, there is an increasing consensus that bio-fuel will eventually become a part of the transport fuel mix, although the size of its contribution in the long term is difficult to foresee. Much will depend upon whether ethanol from ligno-cellulose sources can be made profitable.

3.3 Motor fuels

A selection of potential substitute fuels, both liquid and gaseous, is available for combustion in internal-combustion engines. Each of these energy sources has its merits and demerits. One of the primary functions of the petroleum industry is to produce fuels that give rise to as little pollution as possible. This is accomplished by catalytic processes carried out in the refinery. The automotive industry contributes to reducing pollution through engine design and adjustment, as well as by fitting catalytic converters to the exhaust systems of vehicles. A new objective, which has arisen since the general acceptance of the role of greenhouse gases in climate change, is to minimize the quantity of carbon dioxide released to the atmosphere. This is achieved through the introduction of smaller and more efficient engines and the move to hybrid power-trains. Meanwhile, the petroleum industry is assisting by the development of fuels with higher hydrogen-to-carbon ratios, i.e. lighter fractions of lower molecular weight taken from the top of the distillation column. To a degree, it is possible for the petroleum engineer to adjust the yields of the various hydrocarbons formed, but account has to be taken of what will burn smoothly in an engine and what the motorist will purchase. The lightest practical motor fuel is liquid petroleum gas.

3.3.1 Liquid petroleum gas

Vehicles equipped with engines capable of burning liquid petroleum gas (LPG) are found in many countries. This fuel, which is the best-known and most widely used alternative to petrol and diesel, is a mixture of predominantly propane (C_3H_8) and butane (C_4H_{10}) in varying proportions. The boiling points of propane ($-42\,°C$) and butane ($-0.5\,°C$) are such that LPG is gaseous at most ambient temperatures and therefore to remain in the liquid state it has to be stored and delivered under pressure. This necessitates the use of pressurized storage tanks at the 'gas station' as well as on board the vehicle. The source of the component gases is either petroleum or natural gas. The catalytic cracking and reforming of petroleum yields some gaseous hydrocarbons, which constitute the lightest fraction of the distillate. The unsaturated olefins propene (C_3H_6) and butene (C_4H_8) are also likely to be present, although these compounds have greater value as chemical feedstocks for the manufacture of plastics, rubber and other petrochemical products. Natural gas is principally methane (CH_4), but also contains some ethane, propane and butane. The latter two hydrocarbons are condensed out as LPG before feeding the bulk methane to the national gas grid. The calorific value of LPG per unit volume (i.e. the 'energy density') is about 2.5 times that of LNG. This is a direct consequence of the higher molecular mass of its two components.

Conversion kits to enable petrol engines to operate on LPG are readily available for many different models of vehicle. The fuel is stored in cylindrical steel vessels and sold under the common name of 'Autogas'. Given that the vapour pressure of LPG rises rapidly with temperature, the containers have to be sufficiently strong to withstand the highest ambient temperatures that are likely to be encountered in use. An over-pressure relief valve is usually fitted. Liquid petroleum gas may be injected directly into the engine, i.e. similarly to petrol, or vaporized before being introduced into the inlet manifold. In the latter case, the required latent heat is provided by means of warm engine coolant passing through a suitably located coil.

A dispenser for pressurized Autogas is shown in Figure 3.7. Advantageously, LPG is a cleaner-burning fuel than petrol or diesel with much reduced emissions of NO_x and particulates. Further, on account of its higher hydrogen-to-carbon ratio than petrol, substantially less carbon dioxide is discharged. In this respect, the fuel may be seen as a first decarbonizing step on the road from petrol to pure hydrogen fuel. Autogas is available in many countries, albeit sometimes only at a minority of petrol stations. Although,

FIGURE 3.7

Dispenser for LPG

(Image sourced from Wikipedia and available under the Creative Commons License)

as stated above, it is the third most widely used motor fuel, as yet only a few percent of vehicles are designed or adapted to utilize LPG, and even then usually as dual-fuel vehicles. Because of its lower emission of pollutants, LPG is particularly favoured for buses and taxis that operate predominantly in cities.

3.3.2 Natural gas

Internal combustion engines that run on petrol can be adapted for natural gas. Dual-fuel vehicles permit the option of using either fuel with efficiency comparable with that of petrol. Engines dedicated solely to natural gas operation have compression ratios and efficiencies that are higher than those of diesel engines. World-wide, over 10 million vehicles are running on natural gas, with Pakistan, Argentina, Brazil, Iran and India among the principal users. Some major natural gas engine units (13-and 15-L capacity) have been developed in the USA for use in large trucks and earth-moving equipment.

The onboard storage of natural gas for use as a transport fuel necessitates the carriage of a heavy pressure vessel of cylindrical shape and this presents a challenge for vehicle designers. In the case of cars, this fuel tank is generally located in the boot (trunk) and thereby compromises the space available for luggage. Buses, on the other hand, can accommodate the gas storage more easily. A metro-bus operating on natural gas in Washington, DC, is shown in Figure 3.8; conventional cylinders are mounted in an enclosed pod on the roof. Pressurized methane (boiling point $-164\ ^\circ$C) remains in a gaseous state and therefore the energy-storage capacity of a given cylinder of natural gas is much less than that of a similarly sized cylinder of LPG, and consequently the driving range of the vehicle is reduced. On the other hand, methane is a cheaper fuel and is cleaner-burning. Thus, it may be regarded as a second step on the path from petrol to hydrogen in terms of decarbonization and reduced emissions.

In countries where natural gas is imported as a cryogenic liquid the possibility exists, in principle, of storing LNG directly aboard the vehicle and vaporizing it by heating before entry into the engine. In

FIGURE 3.8

Metrobus in Washington, DC, fuelled by natural gas

(Image sourced from Wikipedia and available under the Creative Commons License)

practice, such a procedure would impose major logistical problems with respect to distribution and local storage of a cryogenic fluid, as well as the need to equip vehicles with well-insulated fuel tanks. It is unlikely, therefore, that this approach will be adopted for vehicles generally, although tanker trucks employed to convey liquid air and similar cryogenic liquids might store LNG for their own use.

An alternative option for exploiting natural gas as a road transport fuel is to convert it chemically to petroleum. This may be accomplished by means of 'gas-to-liquid' (GTL) technology. The gas is reacted with steam and air (or oxygen) over a nickel catalyst at ~ 900 °C to yield a mixture of carbon monoxide and hydrogen – 'synthesis gas'. The mixture is then subjected to the Fischer–Tropsch synthesis, which was first developed in Germany in 1923 and is also used for the treatment of coal (*v.s.*). The synthesis gas is passed over an iron- or cobalt-based catalyst at a lower temperature (200–350 °C) and thereby converted to a variety of liquid hydrocarbons (syncrude) that can be processed in a conventional oil refinery. The lower-boiling fraction constitutes petrol and the higher-boiling fraction yields diesel. Major oil companies have invested heavily in this technology as conventional petroleum reserves have become harder to find and natural gas has proved to be more abundant.

3.3.3 Ethanol

Alcoholic drinks have been made from time immemorial by the enzymatic fermentation of the sugars contained in fruit juices (e.g. wine from grape juice). The enzymes are contained in yeast and this acts as the fermenting agent. The alcoholic content of the wine is limited by the concentration that the yeast, a living entity, can survive. Stronger, fortified wines (e.g. brandy) are produced by distilling the more dilute table wine. Pure ethanol (C_2H_5OH), as used in industry, is generally made from ethene (ethylene), C_2H_4, a product of petroleum refining, by catalytic hydration with a sulfuric acid catalyst.

As mentioned above, ethanol for application as a bio-fuel is produced by the fermentation of crops that contain sugars or starch; see Box 3.1. The scale of operation is large, i.e. comparable with that of a conventional oil refinery; see Figure 3.9. In 2011, Brazil and the USA produced 21 and 52 billion litres of ethanol, respectively. The European Union lagged behind, with a bio-diesel output of about 10 billion litres.

After fermentation is complete, the product solution is distilled in order to concentrate the alcohol. After reaching a maximum concentration of around 95%, the ethanol distils off along with the water. Various chemical methods may then be employed to remove the remaining water. A modern means of concentration, which avoids distillation (and the associated input of heat energy), involves the use of molecular sieves. These are filters with micropores through which water can penetrate, but not the larger ethanol

BOX 3.1 FERMENTATION OF SUGARS

There are many different sugars. One of the simplest is glucose, which is a monosaccharide of chemical formula $C_6H_{12}O_6$. Common sugar, as extracted from sugar cane or sugar beet, is a disaccharide (sucrose, $C_{12}H_{22}O_{11}$), in which glucose and fructose (another monosaccharide) are linked together with elimination of water. The first stage of fermentation of sucrose is hydrolysis using an enzyme contained in yeast (invertase) to break the glucose–fructose bond. The resulting monosaccharides are then fermented by zymase, another enzyme present in the yeast. The net result is that one molecule of sucrose yields four molecules of ethanol and four of carbon dioxide. Starch, an essential constituent of plant material that stores solar energy, is a more complex polysaccharide in which multiple monosaccharide units are joined together to form a giant molecule. This can be broken down in a similar way to sucrose, although more slowly, by enzymatic hydrolysis, and then subjected to fermentation as with the monosaccharides derived from other sugars.

FIGURE 3.9

Bio-ethanol plant in Iowa, the corn belt of the USA

(Image sourced from Wikipedia and available under the Creative Commons License)

molecule. Irrespective of the method chosen, the product is ethanol with around 1–2 vol.% water, at which level it is miscible with petrol in any proportion without separating into two liquid phases. The greater the concentration of ethanol in an ethanol–petrol mixture, the more is the water content that can be tolerated without phase separation. Water in the fuel is undesirable, however, not least because it reduces the kilometres per litre of fuel, but also because it enhances corrosion of metallic components.

Cellulose, which is also a long-chain polysaccharide, is yet more complex in structure. The basic skeletal structure of plants is composed of ligno-cellulose, a material formed from cellulose and lignin. This, too, can be decomposed and the resulting cellulose degraded enzymatically to monosaccharides and converted to ethanol, although with considerably greater difficulty than fermenting simple sugars or starch. The production of ethanol from woody and fibrous feedstocks is technologically difficult because of the need to optimize all the steps in the production process. Until now, the route to ethanol has not been developed to full production scale, although much experimental investigation has been carried out and pilot facilities have been constructed and operated. The process is perfectly feasible scientifically; the challenge is to make it economically viable. If a new cellulose ethanol industry does indeed develop, then it will greatly expand the quantity of ethanol that can be produced from a given area of farmland.

Other research is directed towards the possibility of producing all the required enzymes inside the reactor vessel, thereby enabling the microbial production both of the enzymes that break cellulose down into sugars and of those that ferment the sugars to ethanol. This consolidated bio-processing is seen by many as the logical end point in the evolution of biomass conversion technology.

Ethanol has a long history as a motor fuel that started in the late 19th century, when engines were designed and built to run on it. Early in the 20th century, a keen debate arose between the proponents of petrol and ethanol as to the respective merits of the two fuels, and car races were held to demonstrate which one gave the better performance. In the 1930s, the Cleveland chain of service stations in the UK sold a mixed petrol–ethanol fuel under the trade name 'Discol'.

Ethanol is a clean-burning, high-octane fuel. Its use as a blend with petrol tends to lower nitrogen oxide (NO_x) emissions but to raise the emissions of toxic compounds such as aldehydes formed by

Table 3.2 Calorific Value of Various Fuels

Fuel	Formula	HHV (MJ kg^{-1})[a]
Hydrogen	H_2	141.8
Methane	CH_4	55.5
Propane	C_3H_8	50.4
Butane	C_4H_{10}	49.50
Methanol	CH_3OH	22.7
Di-methyl ether	CH_3OCH_3	31.8
Ethanol	C_2H_5OH	29.7
Isobutanol	$C_4H_{10}O$	36.0
Petrol		46.4
Petrol with 10% ethanol		44.8
Diesel		45.8
Biodiesel		40.6

For liquid fuels, the values will be slightly higher if calculated from the vapour phase rather than the liquid phase.
[a]*The higher heating value (HHV) is the heat of combustion when the product water is in the liquid phase rather than in the vapour phase, i.e. as steam.*

oxidation of alcohol. Today, ethanol–petrol mixtures are given an 'E' number to indicate the percentage of ethanol in the mixture, e.g. E20 contains 20 vol.% of ethanol. From July 2011, all light-duty vehicles (LDVs) in Brazil were required to use E25 fuel. In the USA, over 90% of ethanol-containing fuel sold has 10–15 vol.% ethanol (E10–E15). Internationally, many LDVs are now manufactured with engines that are fuel-flexible and operate with pure petrol or with up to E85 in North America, and up to E100 in Brazil. Ethanol contains only about two-thirds the energy of petrol (see Table 3.2) and, therefore, with all other factors being equal, the distance travelled on a tank of pure ethanol would be only two-thirds of that achieved with petrol. In practice, however, pure ethanol is rarely used and the inherent reduction in range on blending it with petrol can be mitigated somewhat by adjusting the engine timing to correspond to the higher compression ratio of the alcohol mixture and then by turbo-charging as described in Section 4.3, Chapter 4.

Ethanol in high concentrations requires separate handling facilities as well as vehicle engine modifications. As such, a push to high-level blends, such as E85, implies the building of a parallel infrastructure, which will include the installation of tanks, pumps and nozzles that will not corrode when used with ethanol–water mixtures, and also modification of the existing vehicle fleet. Clearly this will take time, even if deemed desirable.

The economic case for bio-ethanol without subsidies is questionable. Nevertheless, despite reservations concerning the overall benefits, governments seem prepared to support bio-ethanol in order to lessen dependence on imported petroleum and also to reduce greenhouse gas emissions as a contribution towards meeting national targets. In the long run, however, a dependence on subsidies to ensure economic viability cannot be a sustainable situation.

An alternative alcohol fuel is isobutanol, $(CH_3)_2 \cdot CH \cdot CH_2OH$. This alcohol possesses several advantages over ethanol, namely: it is less volatile, mixes with petrol in all proportions, is less corrosive to pipework and can be used in an internal-combustion engine without any modification. The reduced corrosion rate eliminates the need for dedicated storage facilities and tanks. Isobutanol has

a higher calorific value (see Table 3.2) than ethanol and is closer to petrol in its properties. Like ethanol, it can be made by fermentation of starches and sugars, but by using a different enzyme. Fermentation plants have been constructed in the USA. In principle, isopropanol can be also be produced by
enzymatic breakdown of cellulose, but a naturally occurring enzyme to bring about the transformation has yet to be found. Consequently, research is in progress to genetically modify an enzyme for
such purpose.

3.3.4 Bio-diesel

The main production of bio-diesel takes place in Europe, because a far greater proportion of the LDVs
have diesel engines than elsewhere in the world. The preference for this technology is driven by the
good reliability and long life of diesel engines, and especially by the greater fuel economy in a region
where road transport fuels are heavily taxed. It is by no means uncommon for taxis with diesel engines
to cover 200 000 miles (320 000 km) with little maintenance. Diesel fuel is also far less inflammable
than petrol and thereby safer in a road accident situation.

The oil obtained from oilseeds is a mixture of esters, which are the glycerides of long-chain
aliphatic acids (the so-called 'fatty acids'). The oil extract is reacted with methanol or ethanol,
together with sodium or potassium hydroxide, to liberate glycerol from the molecule and replace it
by a simpler alcohol to form a new methyl or ethyl ester. The transformed oil constitutes 'bio-diesel'.
The glycerol is a useful by-product that finds application in the food industry as a sweetener and
preservative, as well as in the pharmaceutical and personal-care industries. Bio-diesel is not usually
employed in the neat form in vehicles, but is blended with conventional diesel in a 1:5 ratio that may
be fed to an unmodified diesel engine. The mixed fuel is said to help in keeping engines clean and
well-running, and to improve lubrication so that components last longer. The calorific value of bio-
diesel is somewhat less than that of the traditional diesel (see Table 3.2), but in terms of fuel consumption this shortcoming is partly offset by improved engine performance. Another positive
attribute is that bio-diesel is sulfur-free.

Europe has produced substantially more bio-diesel than any other region of the world, with output
rising to about 170 000 barrels per day by 2009; see Figure 3.10. While the bulk of the European output
is based on oil seeds (rape or sunflower), some is derived, on a smaller scale, from used frying fat and
surplus animal fat. The raw materials have to be purified first before undergoing hydrolysis and reaction
with methyl alcohol (esterification).

Another approach to making high-quality diesel-compatible fuel is the hydrocracking of vegetable
oils (e.g. palm oil, soybean oil) and animal fats. This technology is fully commercial. The most cost-
effective approach to the production of bio-diesel by this route may be to integrate the process into oil
refineries, where the necessary facilities and quality control mechanisms are already in place. Although
hydrocracking may help to make bio-fuels of higher quality from oil-seed crops, it does not remove the
basic drawbacks associated with these feedstocks. The objections to the bio-diesel industry are much
the same as those put forward for bio-ethanol as a substitute for gasoline, viz., the land used could be
better employed in growing food, the high cost of these fuels when unsubsidized, the lack of sustainability of agriculture without crop rotation, the energy required in the manufacture of fertilizers, possible ecological or environmental impacts and, in the case of rapeseed/canola, aversion to the 'unnatural'
appearance of fields of yellow flowers. Nevertheless, there is growing confidence that such concerns
may be overcome and, consequently, the use of bio-diesel is increasing.

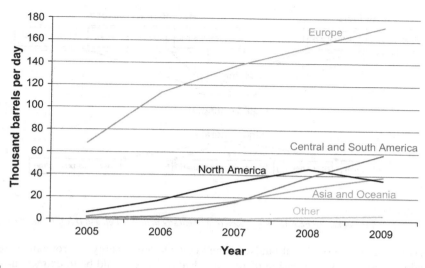

FIGURE 3.10

Growth in production of bio-diesel

(Graph from the Department of Energy, USA)

3.3.5 Methanol

Methanol (CH_3OH) is the simplest of the aliphatic alcohols, with just one carbon atom. In practical terms (although not chemically), methanol may be regarded as 'two molecules of hydrogen made liquid by one of carbon monoxide'. Seen in this light, methanol is one further step along the road towards the decarbonization of road transport fuels. The ultimate goal is hydrogen, but as a gas this presents all kinds of practical problems in use (see Chapter 8). Liquid methanol, for example, is much more convenient to handle. Moreover, in volumetric terms methanol contains over twice as much hydrogen as liquid hydrogen itself, in which there is a very low packing density of the molecules. Thus, 1 m^3 of liquid methanol contains hydrogen equivalent to 1660 Nm^3 of gas (measured at normal temperature and pressure) whereas 1 m^3 of liquid hydrogen contains 788 Nm^3.

Methanol is usually manufactured in industry via the GTL conversion of methane-derived synthesis gas (*v.s.*) with a copper-based catalyst at 200–300 °C. Commercial units can produce 5000 t per day, mostly for use in the chemical industry. The annual production of methanol world-wide is some 35–40 Mt. Methanol can also be derived from a wide range of low-grade fossil fuels, from agricultural waste products such as biomass, and from municipal waste that contains cardboard, paper and plastics. In fact, almost any carbonaceous material can be converted to methanol. As a fuel and energy store, methanol is very versatile – it can readily be transported and dispensed, much like petrol and diesel fuel. This utility has led to the concept of the 'methanol economy', in which the fuel can be derived from many different sources and put to many different uses; see Figure 3.11. While the present-day applications are mostly in the chemicals sphere, in the longer term, when oil becomes a scarce commodity, methanol has the potential to be the keystone of a sustainable energy future.

For true sustainability, however, it will be necessary to employ non-fossil (renewable) fuels to produce methanol. Biomass is ideal, but is only available in limited quantities. Also, the material is

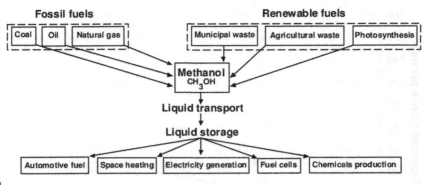

FIGURE 3.11

The methanol economy

(With permission from The Royal Society of Chemistry from R.M Dell and D.A.J. Rand, 2004, Clean Energy)

widely dispersed over the agricultural landscape and considerable energy (petroleum) is required for collection and transport to the processing plant. The ideal practice would be to extract carbon dioxide from the atmosphere and transform it to a fuel. This is the natural process of photosynthesis whereby plants, under the action of solar energy, convert atmospheric carbon dioxide to fresh biomass. Unfortunately, photosynthesis is complicated, inefficient in its use of solar energy, slow and has yet to be replicated in the laboratory. By contrast, the production of methanol is achievable via the reaction of carbon dioxide with hydrogen over a suitable catalyst:

$$CO_2 + 3H_2 \rightarrow CH_3OH + H_2O \qquad (3.1)$$

Pilot plants have been built in several countries to demonstrate this process at a sensible scale. There remain, however, the twin problems of where to source the carbon dioxide and the hydrogen. In the case of carbon dioxide, extraction from the atmosphere is difficult on account of its low concentration (0.04 vol.%), but collection from large industrial sources (e.g. the exhaust from power stations, cement kilns and natural gas wells) is both simpler and practical. The challenge with hydrogen is how to release it from water without employing primary fuels, as discussed in Chapter 8. Ultimately, this would involve large-scale electrolysis using renewable energy (e.g. hydroelectricity, solar cells or perhaps nuclear electricity), but this objective is far from being realized. For the present, electricity from renewable sources is better used directly rather than via conversion to hydrogen, except possibly in specialized locations such as in Iceland, where hydroelectricity and geothermal energy are abundant; see Section 6.4.1, Chapter 6. Provided that the electricity stems from renewable energy sources this is then a totally sustainable route to liquid fuels while, at the same time, not generating any new carbon dioxide. In terms of economics, though, this has to be seen as a long-term venture.

A nearer-term possibility is to derive the hydrogen from methane. When methane from natural gas is heated to 800–900 °C, it decomposes to carbon and hydrogen. This process has long been operated industrially to produce carbon black for use in tyres and as a pigment for printing inks and paints. The process can be re-engineered and scaled-up to focus on hydrogen as the desired product. The carbon would then be a by-product that could be stored, utilized (for example, in road building), or disposed to landfill. The beauty of this process is that most of the carbon in the methane is not liberated to the atmosphere as carbon dioxide. Of course, heat is required to decompose the methane and therefore a fraction of the natural gas would have to be combusted to raise the temperature of the reaction chamber. Consequently, this is not a fully sustainable process, but is a step in the right direction.

Methanol, like ethanol, can be employed directly as a motor fuel. It has a high octane number and is thereby suitable for use in engines with high compression ratios. Methanol is also a clean-burning fuel, but one problem is that its calorific value is only half that of petrol (see Table 3.2) and so a full tank goes only half as far. Other issues are its acute toxicity (unlike ethanol) and its miscibility with water, which can lead to corrosion of fuel tanks and feed pipes. For these reasons, its commercial application has been seriously limited. When used at all, methanol is generally blended into petrol at the 10–30 vol.% level.

There are other possible applications for methanol as a road transport fuel. Partial oxidation of methanol gives hydrogen and carbon dioxide. These gases can be separated to yield pure hydrogen for combustion in fuel cells (see Chapter 8). Methanol can readily be dehydrated to produce dimethyl ether, or DME ($CH_3 \cdot O \cdot CH_3$), which is a non-toxic diesel substitute with a cetane number of 55 (see Section 4.3.1.3, Chapter 4) and can be blended with regular diesel. Given that DME is a gas at ambient temperature, it can be stored in low-pressure cryostats as a refrigerated liquid at −25 °C, or in pressurized tanks at ambient temperature. Its calorific value is comparable with that of ethanol (see Table 3.2). In experimental tests, DME combustion produces very little nitrogen oxides or particulate matter. Current global production of the chemical is less than 0.5 Mt per year, however, and it is mainly used as an aerosol propellant.

There have been recent developments in the direct production of DME from synthesis gas, and from coal. China has been active in its manufacture in recent years and plans further expansion. The most significant barrier to the wider use of DME as a transport fuel is the absence of a distribution infrastructure. Given its low boiling point, it would have to be distributed as a pressurized liquid, like LPG. In addition, to compete with conventional diesel, it would need to be produced on a large scale, which requires heavy capital investment.

3.4 Summary

There are numerous opportunities to develop and utilize unconventional fuel sources and, indeed, new fuels for road transport if and when the supply of conventional petroleum does not meet the market requirements, and/or in response to concerns over greenhouse gas emissions. Some of these alternative fossil sources and new products are not economically viable at present without government subsidy, but will be when the retail price of petrol and diesel rises further in response to increased demand.

Early candidates for growth are the use of LPG in dual-fuel vehicles and bio-fuels blended into petrol and diesel. These are well-established procedures and implementation is straightforward, although there remain questions to be addressed over how much LPG the refineries can make available, and over how much bio-fuel can be produced in competition with growing food crops.

The technology for converting syngas, derived from natural gas or from coal, to petrol has also been demonstrated, although its further implementation will require the construction of large new chemical plants. Of the two feedstocks, natural gas is the simpler option, although it might be expected that countries with ample supplies of coal would follow the lead shown by South Africa and use their indigenous resource rather than importing gas to manufacture synthetic petrol. The widespread adoption of compressed natural gas as a road transport fuel presents infrastructure problems, and is less likely to be

Table 3.3 Typical Values of 'Well-to-Tank' Efficiencies

Fuel	Well-to-Tank Efficiency (%)
Diesel	85
Petrol	82
LNG	82
DME	70
Biodiesel	65
Methanol	63
Hydrogen (chemical)	52
Hydrogen (electrolytic)	35

attractive to the motorist given that the gas cylinders occupy more space in the vehicle than a petrol tank and that the driving range between refuelling stops is reduced.

Ethanol has become established as a petrol additive in Brazil and the USA and is likely to continue to be used now that the industry is in place. Moreover, the concentrations of ethanol added to petrol may well rise in an effort to reduce greenhouse gas emissions, although high levels of ethanol in petrol (e.g. E85) do introduce technical problems in distribution and utilization, and they also have a significant impact on vehicle range. Whether other countries with no established industry will adopt ethanol as a motor fuel is more problematic. Much will depend on climate, availability of land for growing suitable crops and the attitude of governments towards subsidies of bio-fuels.

In Europe, the bio-diesel industry is also established, although to a lesser degree, and is likely to continue for the present. Its growth may hinge on whether the recent swing towards diesel-engined cars accelerates as conventional fuels become more expensive and motorists seek to maximize fuel economy for their vehicles.

Methanol, although a highly versatile fuel in terms of the raw materials from which it can be manufactured, is not without shortcomings in road transport applications and therefore may not be an early candidate as an alternative fuel.

One important consideration when comparing different fuels is the relative efficiencies with which they can be extracted from the ground, processed to manufacture end products (motor fuels), and then conveyed to the retailer. This is generally termed the 'well-to-tank' efficiency. The actual numerical values can be somewhat controversial, but a set of representative figures, culled from various sources, is shown in Table 3.3. In order to obtain the overall efficiency of fuel usage, the 'well-to-tank' efficiency has to be multiplied by the 'tank-to-wheels' efficiency, which is a measure of the efficiency of the vehicle's power-train. This topic is discussed further in Section 8.8.2.1, Chapter 8, where hydrogen is evaluated as a road transport fuel. Such calculations are really rather esoteric for the typical motorist, who is principally motivated by availability, price and performance of fuels. Nevertheless, 'well-to-tank' efficiency is a key determinant of fuel price and is also important in terms of greenhouse gas emissions.

In summary, it can be predicted with some confidence that unconventional fuels will continue to be developed and those finally adopted will vary from country to country according to local circumstances. There is no shortage of feedstocks, of various kinds, and the technologies for conversion to acceptable automotive fuels already exist even if not yet exploited to their full potential.

Development of Road Vehicles with Internal-Combustion Engines

4.1 Early days of the motor industry

During the last decade of the 19th century, a variety of cars were built – by more or less gifted engineers and by enthusiastic amateurs. The first such vehicles to take to the roads were primitive in design and individually assembled by hand at considerable cost. Relatively few models of that time were sufficiently promising, technically or commercially, to encourage further development; fewer still can be regarded as laying the foundations of what was to become a major world industry.

By the turn of the century (1900), a profusion of small manufacturers of road vehicles had sprung up, in both the USA and Europe. Among the notable European companies were Panhard-Levassor and Peugeot in France, the brothers Opel in Germany, and Giovanni Agnelli's Fabbrica Italiana Automobili Torino (Fiat) in Italy. The two great German pioneers Gottlieb Daimler and Karl Benz had relatively brief careers in the early industrial phase of the motor car. The former died in March 1900 and the latter retired from the Benz Company in January 1903. Daimler, however, had grown his business steadily during the 1890s both as a car maker and as a builder or licenser of engines used by other firms. His brilliant designer Wilhelm Maybach was largely responsible for the introduction of the V-twin engine. This technology was a cornerstone of the company's production in the 1890s and was also used in Panhard, Peugeot and other cars. The great firm of Daimler Benz was created from a merger in 1926.

Only affluent people could afford to buy the first cars, particularly the so-called luxury marques. These early vehicles were built on a rigid chassis to which was bolted the body. The latter was commonly high and short (like many horse-drawn carriages of the time) and therefore was unstable, especially when cornering. The crude steering could easily be wrenched out of the driver's hand if the solid wheels hit a large stone. The many deficiencies of early cars included inefficient engines that gave few miles per gallon of petrol (or km L^{-1}), poor transmissions with no synchromesh gears, no self-starter (and hence the need to hand-crank the engine), magneto ignition,[1] low top speed (typically 30–40 mph, 48–64 km h^{-1}), feeble brakes (crude shoes pressing against hard steel-rimmed or solid-rubber-shod wheels) and acetylene lighting. Engine cooling depended on a tank from which the water boiled away at frequent intervals and had to be replenished. Radiators, if fitted at all, usually took the form of a serpentine coil of pipes underneath the car. Comfort was lacking, with largely open cabins and essentially no heating. Early motorists, exposed to the elements, wore overcoats, goggles, scarves and gloves. Safety considerations were not addressed beyond the use of a rigid chassis. There was no 'crumple zone' to absorb the energy of a crash, there were no seat belts to restrain the vehicle's occupants, and there were no airbags to protect them in the event of a collision. In general, then, most cars of this time had a specification that could appeal only to the most dedicated enthusiasts – they were too unreliable to be used confidently as an everyday means of transport.

---

[1] A magneto is a simple AC generator used as a high-tension source to start engines in the absence of a battery. The engine was hand-cranked to turn it over and also to rotate the magneto and so generate the spark.

Since those early days, the deficiencies have been progressively overcome and new technologies have been introduced to deliver the sophisticated models of today. This does not mean, however, that the car has reached the end of its development – far from it! Now that the problem of urban pollution from petrol engines has been largely overcome, thanks to the development of catalytic converters to treat the exhaust gases (see Section 2.5.1, Chapter 2, and Section 4.5.1 below), there remain the twin problems of increasing the efficiency of engines to conserve petroleum and of reducing the emissions of carbon dioxide, while retaining affordability. These are challenging issues.

The first breakthrough towards vehicle affordability came in the USA in October 1908 when Henry Ford (1863–1947) introduced the *Model T* Ford as the first 'people's car'. Many of the critical highly stressed parts (e.g. the crankshaft, forged front-axle, wheel spindles) were built of vanadium steel and therefore became sufficiently strong to withstand the buffeting that a vehicle had to endure on early pot-holed roads. Initially, the body was made of wood, but five years later sheet steel was introduced. The car was powered by a 2898-cc four-cylinder water-cooled engine that produced 20 bhp at 1600 rpm and a top speed of 40 mph (64 km h^{-1}). Ignition was by flywheel magneto that ran in an oil bath, and transmission consisted of an epicyclic gearbox with only two speeds, which were engaged by stamping on a pedal. Incidentally, the *Model T* was the first Ford with left-hand drive. Most importantly, the car was also the first to be made on a moving assembly line, which Henry Ford installed in his factory at Highland Park, Michigan, in 1913, along with power-driven conveyers for sub-assemblies. It was said that this innovation was inspired by the overhead trolley systems that meatpackers used for moving carcasses through a factory – instead of butchering an entire animal at a time, each worker performed a single operation on a carcass and thereby output per man more than doubled. By having a simple model with relatively few parts and by aiming for mass production, Ford brought the cost of the *Model T* within the reach of the average middle-class American. It was one of the most successful cars ever; between 1908 and 1928 over 15 million were built.

A major technical breakthrough was the invention of the electric starter motor and ignition system by Charles Kettering, an American, in 1911. This obviated the requirement for hand-cranking the engine, a task that was difficult, physically demanding, inconvenient and occasionally dangerous when the engine 'kicked back'. By 1920, most manufacturers included an electric 'self-starter' in their vehicles. Just imagine the confusion in today's congested urban traffic if every time an engine stalled it was necessary to leave the driver's seat, and go to the front of the vehicle with a large handle to hand-crank the engine. Society owes a debt of gratitude to Charles Kettering. Ironically, it was this battery-operated technology that finally led to the demise of the battery electric car, which could no longer compete once the self-starter had been introduced.

Soon competition emanated from other enterprises in the USA and a colossal growth in vehicle sales eventuated. By contrast, the rate of progress in European motor companies was painfully slow. In 1913, when 483 000 cars were made in the USA, the total made in Britain, for example, was only 23 238. The situation was little different in France, where Renault, the leading manufacturer, proudly announced in 1913 that it had, for the first time, delivered 10 000 cars in one year. The main Italian company, Fiat, completed about half that number. World War I almost halted the European output of cars, but it made little difference in the USA. During the 1920s, however, car production took off in much of the rest of the world and by the late 1930s there was a very wide choice, both of maker and model, in the USA and in Europe.

Many further improvements in vehicle design and construction were introduced during the course of the 20th and early 21st centuries. Some of these had to do with the capability of the car to perform different functions and hence there emerged a great variation in the size and capacity of different car bodies. Other factors concerned the desire of drivers and passengers to travel in comfort, with adequate storage space for luggage and other goods. The needs of the market also demanded that the car should not be too expensive for the majority of the population to afford. Thus it had to be built of the cheapest possible materials consistent with adequate strength and long life, and, at the same time, should be easy to repair and attractive in appearance. These requirements led to vehicles of much improved performance, lower cost, superior comfort, increased convenience and greater safety. The evolution of automotive design to meet these demands is outlined in this chapter.

4.2 Developments in vehicle body design

Leaving aside for the moment the propulsion system and all accessories, the prime goals of body design have been to provide the desired accommodation at the lowest possible cost, to enhance appearance and sales appeal, to reduce weight while maintaining strength, and to inhibit corrosion. Other aims were to heighten safety, and to reduce air friction (coefficient of drag) so as to improve fuel efficiency. This combination of objectives presented serious challenges to the designer.

From the 1940s onwards, a new method of car construction was introduced – the 'monocoque' construction (French 'single shell'). Instead of having a rigid chassis as the load-bearing structure with a cabin bolted to it, the load-bearing members are fully integrated into the bodyshell. The latter comprises the car's underbody (floor pan) suitably reinforced with lateral and longitudinal struts' the engine compartment, the body frames and the suspension units – all welded together. Most mass-produced passenger cars and light goods vehicles are now built in this way. The other basic components of the body – the load-bearing panels and the external panels forming the 'skin' of the car – are stamped out in their thousands by press tools. The monocoque design reduces the weight and cost of the vehicle, facilitates assembly, and considerably enhances safety by dispensing with the rigidity of the traditional chassis. It also permits the incorporation of 'crumple zones' at the front and rear of the vehicle to absorb much of the energy of a crash.

Many of the major manufacturers employ a single platform (i.e. customary design of floor pan, wheel base, steering, suspension, engine and power-train) for several models of car. Thus, basically the same car can appear as a four-door saloon, a two-door coupé, a three- or five-door hatchback, or a five-door estate car. Almost all light goods vehicles are based on a platform which is used in the corresponding saloon model but with different body panels.

Sometimes this sharing of major components, particularly platforms, extends across competitive automotive companies. The obvious advantage of such practice is to reduce the cost of manufacture, whereas the disadvantage is that there is limited scope for creativity of design and differentiation between rivals. Individuality is then only achieved in the design of body panels attached to the platform and items such as radiator grills, doors, lights, etc., as well as internal fittings. Sports cars and convertibles with soft tops lack the rigidity provided by the roof of the saloon car and this has to be compensated by extra structural members added to the platform in order to avoid body distortion or vibration.

Another important advance is in the area of body protection. In the past, the rusting of cars was a serious issue – particularly in communities that border the seashore, where salt spray aggravated the problem. Salt spread on roads during the winter to restrict the formation of ice also contributed substantially to corrosion, especially for those vehicle bodies made from recycled steel. The reclaimed metal contained impurities such as copper that set up galvanic cells and thereby accelerated the rate of attack. In the worst cases, serious rusting of bodies occurred in just a few years; areas such as the sills under doors and the wings/bumpers were prime targets. Initial attempts to overcome this corrosion involved coating the underside of the steel with a rubberized or bitumen layer. The procedure was only partially successful, however, because the coating soon became brittle and cracked. This degradation enabled water to enter and become trapped between the metal and the coating, so exacerbating the process. When the situation became unacceptable in terms of customer satisfaction, automotive companies were forced into more radical solutions. Today, most steel car bodies receive sophisticated surface treatment that includes galvanizing to ensure corrosion resistance, followed by a high-quality paint finish.

Heavy trucks and buses are in a different category to light vehicles. Because of their size, weight and the heavy loads that they carry, it is impractical to use a monocoque structure and such vehicles have a traditional chassis with the members bolted, riveted or welded together. The engine, transmission, suspension, axles, steering, braking system, electrics and the cab or bus body are then attached. In order to spread the load, heavy trucks have multiple axles usually with two large wheels at each end. Illustrations of typical European and American heavy trucks are shown in Section 1.4.5, Chapter 1.

4.2.1 Body size and shape

Car body designs are broadly targeted, as regards size and shape, at specific market sectors. In detail, the designer has many different goals to address, as follows.

Weight and strength. A bodyshell must be sufficiently strong and rigid to withstand the forces that tend to make it bend and twist when the car is in motion. These forces are due to the opposition between, on the one hand, the dead weight of the whole structure plus its occupants and, on the other hand, the vertical forces generated by uneven road surfaces and horizontal forces generated when cornering and braking. The strength and weight of the body are determined primarily by the gauge of steel sheet employed. If it is too thin, the cabin will not have adequate strength, will not be sufficiently rigid, will distort easily and will corrode through more quickly. Consequently, up until about 1940, cars had thick sheet-steel bodies that were very strong, but heavy and expensive. Most manufacturers today settle for a gauge of around 0.7 mm with reinforcing members up to 2.0-mm thickness. This is seen to provide adequate strength whilst being as lightweight and inexpensive as possible.

Much of the drive toward strengthening body structures has been in the cause of safety. In particular, laws passed in many countries have required designers to tackle the problem of protecting the occupants of cars that are involved in major incidents. For instance, cars must be able to withstand a crash at a prescribed speed without serious distortion of their structure, and without the steering wheel being pushed significantly back towards the driver's chest. Vehicles must also be able to roll over at speed without the roof being crushed, and to survive a serious impact from the side without the doors being pushed onto the passengers.

Visibility. Rear-view visibility is important. Rear pillars to support the roof are needed in standard saloon models, but if these are too wide they impede visibility and give rise to 'blind spots' such that overtaking vehicles may not easily be seen.

Internal space. Within the confines of the bodyshell, designers attempt to maximize the available useful space. Hatchbacks are particularly popular as they generally have a capacious boot/trunk and the rear seats may be lowered to permit the carriage of larger items. Designers exercise ingenuity in creating space within the passenger cabin – for instance, the utilization of door pockets to store safely the personal possessions of occupants. A recent development is that of flexible seating, whereby the configuration of the seats may be changed to accommodate differing passenger- or load-carrying requirements – for instance, the transport of several children or the conveyance of a wheelchair. Ease of access and egress is important, especially for elderly or disabled people. Even in quite large cars, it can sometimes be difficult for passengers with disabilities to enter and leave the rear seats.

Aerodynamic drag. Fuel efficiency (miles per gallon of fuel, or litres per 100 km) is determined by a number of factors, one of which is the aerodynamic drag coefficient (C_d). This is strongly influenced by the frontal shape of the vehicle. Most new designs of car are tested in a wind tunnel to determine the optimum streamlining to minimize the C_d. Any object that protrudes from the bodyshell (such as door handles, lights, mirrors, radio aerial, roof rack) significantly increases the drag coefficient and hence lowers the fuel efficiency. Some cars, particularly sports cars, have a 'spoiler' fitted to the boot. This is an aerodynamic device to deflect air so as to increase the downward force on the rear of the car and thereby improve grip on the road. It should be noted, however, that too much down force would aggravate drag.

Comfort. A key factor that determines the nature of the 'ride' is the stiffness of the suspension. Traditionally, large American cars of the 1950s and 1960s had soft suspensions that smoothed out the discomfort that arose as a result of imperfections (such as pot holes) in the road surface. Smaller European cars tended to have firmer suspensions. Now that the car market has become globalized, it has become customary to adopt a compromise in stiffness that is acceptable to all. Some consumers may have a preference for a firm ride and therefore manufacturers may fit more rigid suspensions in their so-called 'sports models'.

Noise and vibration are two further factors that detract from the comfort afforded to the occupants of a car. Noise stems mostly from the engine and the tyres, and is minimized by acoustic insulation of the bodyshell. Diesel engines were once notorious for their noise, but this has been greatly reduced. Vibration can arise from rattling doors or windows (in early cars, now thankfully eliminated), from the exhaust system, from the engine mounting, from unbalanced wheels, or from the road surface. The abatement of **n**oise and **v**ibration, together with the above-mentioned **h**arshness (or stiffness) of ride, constitutes the role of the 'NVH engineer'.

Obviously, ventilation and temperature control are also important. Passengers must be warmed in cold weather, cooled in hot weather, and given draughtless ventilation even when the windows are closed. This is achieved by the heating/air-conditioning system, which is usually installed behind the fascia panel with further ducts to the rear compartment and even, sometimes, to the side windows. There are also ducts (usually behind the rear side windows) that extract the stale air without allowing exhaust fumes to enter the passenger cabin.

Cost. As in all aspects of vehicle design, the competitive market demands that cost be kept to a minimum, but without jeopardizing the above objectives. This is one reason why some of the features of older cars such as chromium plating, two-colour bodies, quarter light windows and white-wall tyres have largely disappeared.

In summary, the above brief overview shows that the design of car platforms has become highly sophisticated and most automotive companies now offer a range of models to suit all tastes and pockets.

4.2.2 Body welding

The manufacture of a typical car involves three major stages: body construction, surface treatment (painting) and vehicle assembly. The galvanized steel panels that form the bodyshell are first welded

FIGURE 4.1

Automated car welding

(Image sourced from Wikipedia and available under the Creative Commons License)

together. Traditionally, this was done by hand and therefore was a labour-intensive process. Also, such welding depended on the skill of the welder and thereby did not necessarily give a uniform product. As cars move along the production line in a modern assembly plant, robots located on either side of the line advance on the body and make the welds, as shown in Figure 4.1. This is a highly automated process that gives a homogeneous and high-quality weld. The robots are synchronized to move in concert with the speed of the line. Although the capital investment in such technology is substantial, its introduction results in major savings in labour costs.

4.2.3 Surface treatment

After the component parts are welded together, the vehicle body is subjected to the following five-step process: pre-cleaning, phosphating (to inhibit corrosion), electrocoating of a primer coat of paint (electrostatic or electrophoretic deposition), rinsing (to remove non-adherent paint) and, finally, baking at 170–180 °C to cross-link and stabilize the polymer resin of the paint to yield a hard and reflective surface. The overall sequence of steps has to be repeated with a top coat of the desired colour. A finishing layer of lacquer may also be applied. Painting is a major part of car manufacture and represents a significant cost item. Much has been done in recent years to improve the quality of car-body paints and of the various chemical pre-treatments of the steel that help the paint to adhere to the metal. Research continues to develop a less-expensive process and to ensure a perfect end product.

The demand for cars with different colours gives rise to logistical problems. Inevitably, cars have to be painted in small batches of the same colour and then the equipment has to be cleaned before changing to another colour. Obviously, such a procedure delays the operation and increases its cost. One promising alternative approach is to use body panels made of polymer (*v.i.*) that are painted separately before assembling the vehicle. Only the metallic bodyshell would then have to be phosphated and primed. This practice would save on the capital cost of plant, reduce paint consumption, and speed up the overall process.

4.2.4 **Modern body materials**

In the 1920s and 1930s, some cars had bodies constructed of wood frames, typically ash, mounted on rigid steel chassis. The construction gave rise to some concern as the wood could splinter and prove hazardous in the event of a serious accident. Consequently, by the 1940s, most car bodies were made of pressed panels of mild steel. The design and production of the press tools is a specialist technology, and manufacturing the necessary dies takes a long time. For the largest or most intricate panels, such as the inner door skins or the floor-pan pressings, work on shaping the production press dies usually begins several years before the new model is scheduled to be released on the market.

In some respects, mild steel is one of the least-suitable materials for the construction of car bodies. As noted above, it is prone to rust, which is not only unsightly but also eats away the metal and so critically weakens the structure. Although lightweight pressings of aluminium alloy may be used for skin panels (e.g. doors) and are more corrosion-resistant than mild steel, they are not as strong, are much more difficult to weld, and are several times more expensive. Therefore, such pressings have not been widely used for bodyshells. Some manufacturers, however, are now employing aluminium alloy for body panels in their up-market cars to reduce weight and improve fuel consumption. In 2003, for instance, Jaguar introduced the third-generation Jaguar *XJ* with an aluminium body that made the car very light and thereby brought improvements in performance, agility and economy. More recently, the 2011 Audi *A6 Avant* has been made 70 kg lighter than its predecessor through the use of aluminium in place of steel.

The introduction of polymers, such as polyolefins (especially polypropylene), polyesters, polyurethanes and polyamides – all referred to generically as 'plastics' – has revolutionized the design of automotive components. Engineers have been given greater freedom to introduce innovative designs than was possible with pressed steel or light alloys. For components where strength is a prime requirement, the properties of such materials can be enhanced by the incorporation of second-phase fibres or pellets to form composites. Plastic panels are lighter than metal equivalents, while being of comparable strength; their application results in vehicles having greater fuel economy. Polymers are also less susceptible to dents and chipping by stones and, naturally, do not corrode.

The use of polymers for automotive components has presented many new challenges in their formation, attachment and painting. Thus, plastic panels are moulded rather than pressed, and are joined by adhesives rather than by welds. Painting by electrocoating is impracticable for an insulating plastic. Plastic items can be brittle and unyielding and, although the plastic itself is comparatively cheap, the tools needed to form it are more costly than those for stamping out pressed-steel or aluminium-alloy panels. For these reasons, the switch to plastics was initially confined to the manufacture of internal components of the cabin, which included fascias, steering wheels, roof and door linings, seats, and storage pockets. Nevertheless, in recent years, there has been an increase in the use of plastics for a variety of components external to the cabin shell, e.g. bumpers, door panels, radiator grills, wings, the bonnet/hood and the lid of the boot/trunk.

A particular attraction of body panels made from thermoplastic polyolefin is that, unlike metal alternatives, they can be recycled at the end of vehicle life. Technology has been developed to remove the surface paint and the bare panel is then mechanically shredded, followed by grinding and crushing to reduce the particle size. The plastic is next extruded and pelletized. New products made from this recycled material have properties identical to those made from the virgin resin. This procedure recovers a large quantity of plastic otherwise destined for landfill and reduces material costs for manufacturers.

4.3 Engines and transmissions

The function of the engine, of course, is to provide power by harnessing the energy stored in its fuel. As discussed in Chapter 1, petrol, steam and electricity were all in competition as power sources in the early days of the motor car but, due to the limitations of steam and electricity, the combination of petroleum (in the form of either petrol or diesel oil) with the internal-combustion engine (ICE) eventually emerged as the winner and has dominated the road transport scene ever since. Until the 1970s, the general pattern was to use petrol engines mainly in cars and diesel engines in commercial vehicles, especially the larger trucks and buses. From then onwards, however, a diesel-engine option was offered for passenger cars by major European companies such as Mercedes-Benz, Peugeot and Volkswagen. The technology was first deployed in taxis that cover high mileages, but later was taken up more generally as the engines were improved and made quieter. By contrast, in the USA and Japan diesel-powered passenger cars proved to be less popular and, until recently, were not offered by the local manufacturers. This geographic distinction has been driven by the higher cost of fuel in Europe coupled with the fact that the greater efficiency of the diesel engine results in better fuel economy.

In the coventional type of ICE – the piston engine – the combustion process produces hot, expanded gases that exert a force on the pistons in the cylinders. The linear movement of the pistons is converted into a rotary movement of the crankshaft which drives the wheels via a gearbox and differential (the 'transmission'). Other types of ICE may involve continuous combustion as in gas turbines, jet engines or rockets. Large gas turbines powered by natural gas are, for example, employed in electricity-generating plants. Combustion turbines are integral to jet engines, notably aircraft engines. They have also been used experimentally in a few cars, but have not enjoyed commercial success. Small auxiliary turbines, however, are commonly fitted to road-vehicle engines to boost performance by compressing air before it is admitted to the cylinders. Many European cars are now being fitted with these so-called 'turbochargers' (see Section 4.3.1.2), an example being the Ford *EcoBoost* engine.

4.3.1 Types of internal-combustion engine

Piston-driven ICEs are of two general types: spark-ignition (SI) engines and compression-ignition (CI) engines. Spark-ignition engines are fuelled by petrol (gasoline) and have been the customary power unit for cars and light goods vehicles. As their SI name indicates, a spark is needed to ignite the petrol–air mixture. By contrast, CI engines run on diesel, a heavier fraction of the petroleum distillate. The fuel–air mixture is compressed to a high pressure at which point it ignites spontaneously. This eliminates the need for a spark and the rather complex ignition systems that are employed in petrol engines.

A third type of ICE that has been used occasionally in cars is the rotary or Wankel engine. This has no cylinders, but a form of piston that rotates in a circular chamber. The rotary motion is then employed directly to turn the crankshaft. Although, in principle, this is a highly efficient configuration, in practice it has proved difficult to develop a rotary engine with satisfactory life and performance.

The principles of operation, performance characteristics and application of the different designs of engine are discussed in the following sections.

4.3.1.1 Two-stroke engines

The simplest type of ICE is the two-stroke (or two-cycle) petrol engine. This has a first 'compression stroke', during which the piston rises to compress the fuel–air mixture above and the resulting low pressure below draws more mixture from the crankcase through the inlet port. Spark ignition occurs as the piston approaches the top of its stroke ('top dead centre'). The latter action causes expansion of the product gases that drives the piston to the bottom of its travel – the combustion and expansion stroke, i.e. the 'power stroke'. As a result, the crankcase pressure increases and forces the mixture up through the transfer port. Thus there is one ignition event for every rotation of the crankshaft, in contrast to the four-stroke engine (*v.i.*), where ignition takes place only every second rotation. Consequently, the two-stroke engine has a higher power density which is roughly twice that of the four-stroke. The engine is also lighter because it is generally smaller.

The petrol–air mixture enters from the crankcase and flows into the cylinder at the completion of the power stroke immediately after the exhaust gases exit on the other side of the cylinder. The momentum of the crankshaft carries it past 'bottom dead centre' into the compression stroke. Thus the two-stroke engine must accomplish intake, combustion and exhaust in one back-and-forth movement of the piston. Since scavenging is incomplete and inefficient, the desired fuel mixture is difficult to obtain. Moreover, because the mixture enters through the crankcase, the same component cannot serve to hold the lubricant, as in a four-stroke engine (*v.i.*). Nevertheless, it is necessary to lubricate the crankshaft bearings, the piston connecting rod bearings and the cylinder wall. For this reason, lubricating oil is added to the petrol and this ensures that it is well distributed.

Most two-stroke engines are small single-cylinder engines that find their principal use in portable devices such as lawnmowers, hedge trimmers, verge strimmers and chain saws. Other applications include outboard motors for boats, jet skis and radio-controlled model planes. The engines have the required attributes of small size, reduced weight, high power output (for their size) and low cost. For certain applications (chain saws, for example), it is important that they can operate in any orientation. They are also sometimes used as motive power sources for mopeds, scooters and small motorcycles.

If two-stroke engines are simple in construction, lightweight, powerful and low cost, why are they rarely employed in cars and larger vehicles? The answer lies in their disadvantages, as follows:

- they wear faster and therefore have shorter service lives than four-stroke engines because the lubrication system is less effective;
- they consume considerable quantities of oil, about 30 g per litre of petrol, which, in a car, would translate to around a gallon of oil for every 1000 miles (1600 km) travelled;
- they produce an appreciable amount of atmospheric pollution, in part non-combusted oil droplets and smoke;
- they are inefficient and thus give poor performance in terms of miles per gallon (litres per 100 km);
- they are very noisy in operation.

For all these reasons, it is unlikely that two-stroke engines will ever be widely employed in four-wheeled vehicles, although some trucks in North America are fitted with large two-stroke diesel engines. Similar units are also operated in marine and industrial applications.

4.3.1.2 Four–stroke engines

Four-stroke automotive engines come in many different sizes and configurations. For instance, they can have any number of cylinders from one to sixteen – four, six or eight are the most common. Cars with three- and five-cylinder engines are also now appearing. Motorcycle engines may have one, two, four or even six separate cylinders. In a car engine, the cylinders may be bored vertically in the engine block (in-line configuration), or placed in a 'V' formation (see Figure 4.2), or horizontally opposed in a 'flat' arrangement which is often called the 'boxer' engine because the pistons move like boxers' gloves. Naturally, if one or other of the two last-mentioned configurations is selected, it is then necessary to have an even number of cylinders so that they are balanced, otherwise severe vibration would result. The virtues of a 'V' engine are that it is more compact than an equivalent in-line engine and weighs less. Engine size is especially important for a car with front-wheel drive, where the engine is aligned transversely and space is confined.

Until the 1960s, most British motorcycles had single cylinders, which were vertically mounted. For these to be smooth running, comparatively large flywheels were needed. By contrast, BMW developed motorcycles with twin-cylinder boxer engines, while Harley Davidson in the USA favoured V-twins. In the 1960s, Japanese manufacturers began to dominate the international scene; two of the leading companies were Honda and Yamaha. Since then, several European manufacturers have staged a comeback and have produced competitive machines. Today, many motorcycles are built with multi-cylinder engines positioned transversely across the frame of the machine.

American cars of the 1950s and 1960s typically had six or eight cylinders, with the latter being the routine choice for full-size saloons. Only European imports, which were generally smaller, had four cylinders. Even many sports cars had six or eight cylinders, e.g. the Chevrolet *Corvette*, the Ford *Mustang*, the Austin Healey *3000* and the Jaguar *E-type* – all now classic cars. During that period, petrol/gasoline was both cheap and plentiful so that fuel economy was less important than power. Since then, there have been major improvements in engine design that have led not only to much higher

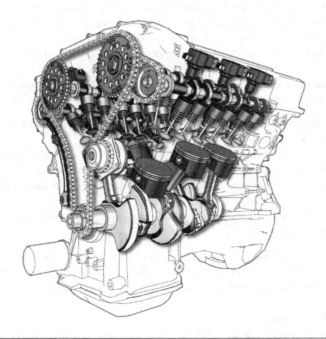

FIGURE 4.2

Modern six-cylinder 'V' engine

(Image sourced from Wikipedia and available under the Creative Commons License)

power-to-weight ratios but also to greater fuel efficiency. The result is that today many popular-sized cars have only four cylinders and yet produce the same power output as the six- or eight-cylinder models of yesteryear, but with far greater fuel economy. A recent example of this trend is provided by the new (2013) Opel/Vauxhall diesel engine that is to be fitted to the *Zafira Tourer* multi-purpose vehicle, which is a large family car. The 1.6-L four-cylinder engine has a lightweight aluminium block and is capable of delivering 134 bhp and 320 N m of torque (*v.i.*) while meeting the Euro VI emission regulations. It is claimed that the carbon dioxide liberated will be only 109 g km^{-1} and that the fuel efficiency will be 68 mpg (UK) (4.2 L per 100 km) on the mandated combined-cycle test.

Several models of car have been built with an odd number of cylinders. For instance, the German Auto-Union *DKW* car of the 1950s had a three-cylinder two-stroke engine and is regarded as a classic. Recently, it has been reported that Peugeot is planning to introduce a new range of three-cylinder four-stroke engines in its smaller cars, while Ford has announced that it is to fit a three-cylinder turbo-charged direct-injection petrol engine of 1.0-L capacity to its popular *Fiesta* and *Focus* marques. Clearly, with improvements in engine performance and concerns over high-priced fuel, the trend is towards smaller and more fuel-efficient engines. Five-cylinder engines have also become popular; Volvo, Audi and Ford are among the manufacturers that have offered such versions. The vehicles are seen as a compromise between four- and six-cylinder models in that they generate more power and torque than four-cylinder engines, while maintaining the smoothness and fuel economy of six-cylinder models. Also, unlike straight six-cylinder designs, the engines are sufficiently short to allow transverse mounting in the engine bay as required for front-wheel drive.

As mentioned above, the firing of each cylinder in a four-stroke engine takes place only every second revolution of the crankshaft, or every 720 degrees of rotation. The four strokes, or phases, of operation are shown in Figure 4.3 and consist of the following:

  (i) admission of the fuel–air mixture through the intake valves as the piston sinks in the cylinder (intake or induction stroke) – downwards piston movement;
 (ii) compression of the fuel–air mixture (compression stroke) – upwards piston movement;
(iii) spark ignition and expansion of the gases (power stroke) – downwards piston movement;
 (iv) burnt gases expelled through the exhaust valves (exhaust stroke) – upwards piston movement.

Each stroke corresponds to a 180-degree revolution of the crankshaft. In a four-cylinder engine, the power strokes are therefore sequential with no overlap. In a six-cylinder engine, firing occurs every 120 degrees and so there is 60 degrees of overlap and this improves the smoothness of the engine. By contrast, the cylinder fires every 90 degrees in an eight-cylinder engine to give even greater overlap and thereby extremely smooth running.

In a conventional four-stroke petrol engine, the mixture of fuel and air is made up in the carburettor and then passes to each cylinder in turn through the intake manifold and the intake valves in the cylinder head. Whereas simple engines have just one intake and one exhaust valve per cylinder, many car engines today have two or more of each since this makes for better distribution of the inflammable mixture within the cylinder and more ready exhaust of the burnt gases. In older designs of engine, the valves were opened and closed by vertical pushrods that were driven by the crankshaft. In modern engines, the cylinder head contains a camshaft with cams that operate rocker arms to open and close the valves. The timing of valve opening and closure is determined by the rotation of the crankshaft. In the case of overhead cams, a belt or chain coupling drives the camshaft at such a speed so as to synchronize the intake and outlet valve operations with the piston stroke. Many present-day cars have twin or dual

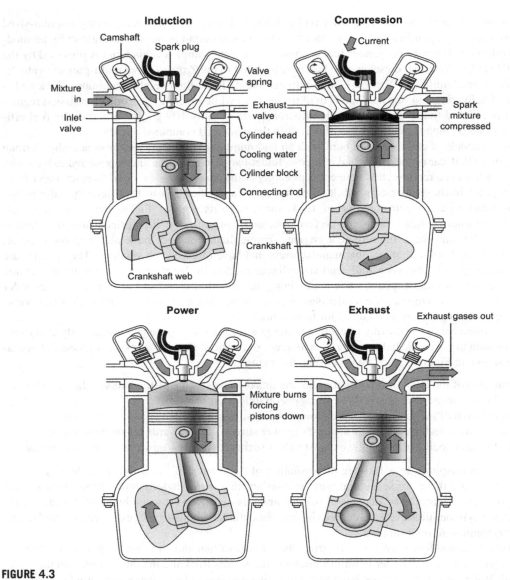

FIGURE 4.3

Construction and mode of operation of a four-stroke engine

camshafts that are mounted directly above the valves; one each to operate the inlet and exhaust valves. Since the cams now work directly on the valves, this arrangement eliminates the need for rocker arms; see Figure 4.4. Non-linear engines (i.e. 'V' or boxer engines) require two single or twin camshafts, one for each half of the engine.

(a)   **(b)**

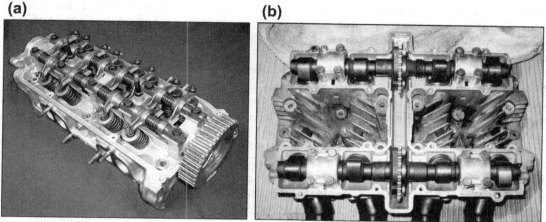

FIGURE 4.4

Overhead-camshaft four-cylinder engines. (a) Single overhead-camshaft engine with two inlet valves per cylinder (far side) and one exhaust valve (near side), (b) Twin overhead-camshaft engine with one inlet and one exhaust valve per cylinder

(Images sourced from Wikipedia and available under the Creative Common License)

Most engine blocks are constructed from cast iron with a head, also of cast iron, bolted on top of the block. In recent years, there has been a trend towards aluminium-alloy cylinder heads in order to save weight, and some vehicles now have aluminium engine blocks with cylinder liners of cast iron. The weight saving is important, particularly in racing cars, but aluminium engines are significantly more expensive than their iron counterparts. Due to a difference in expansion coefficients, an iron block with an aluminium head is more prone to leakage around the gasket seal than an engine with both parts made of the same metal.

The power output from the engine is determined by many factors and one of the most important is the quantity of air admitted at each intake stroke. In order to increase this quantity, the air can be pressurized before admission by means of a supercharger and/or turbocharger. There are two basic types of supercharger. The electrical version is simply an electrically driven fan that compresses the intake air. To be effective, significant power is required (1–3 kW) and this may involve the installation of larger batteries with matching alternators for recharging. By contrast, the mechanical supercharger is an air compressor that operates off the crankshaft with a belt drive. The disadvantage of the latter approach is that some shaft output power is lost in running the supercharger. In both cases, the compressed exhaust gases contain residual energy that is wasted.

A turbocharger is driven by the exhaust gases from the engine and powers a compressor on the same shaft to pressurize the incoming air. Obviously, the speed of the engine must be such that the exhaust gas has sufficient energy to operate the turbine, and this accounts for the well-known 'turbo lag' when the vehicle starts from stationary. Traditionally, turbochargers were fitted to high-power automotive engines in a quest for even greater acceleration and speed. In recent years, however, they have come to be employed routinely in more modest cars, both petrol and diesel models. With a turbocharger, it is possible to utilize a smaller (and cheaper) engine for the same power output so that fuel economy is enhanced and emissions are reduced. This is an important benefit when petroleum prices are high and when there is great concern over greenhouse gases. The use of a turbocharger together with an electric supercharger to make good the power sacrificed when a smaller-than-normal engine is used in pursuit of improved fuel economy is described in Section 5.2.5, Chapter 5.

On compression using a supercharger or turbine, air heats up and its density is reduced. This tends to defeat the objective of getting more oxygen into the cylinder and may cause overheating. Consequently, it is regular practice to fit an intercooler (a form of heat-exchanger) to extract heat from the compressed air and reject it to the atmosphere or to circulating water. The latter is preferable since gas–liquid heat-exchangers are more efficient than gas–gas alternatives.

The internal-combustion engine is an inherently inefficient source of power. A typical car engine is capable of transforming less than 20% of the energy of its fuel into usable power on the road; most of the remainder goes to waste in the form of heat or exhaust gases (see Figure 5.15, Chapter 5). The heat must be dissipated; otherwise, there would be rapid seizing up of the engine parts. All car engines are either air-cooled or water-cooled. Even very powerful engines can be kept cool by forced draughts of air (as the exceptionally fast Porsche *917* sports-racing cars have proved), and suitable ducting and efficient fans can maintain the temperature of any normal engine within acceptable limits. The two major problems with air-cooling, however, are keeping the engine reasonably quiet (a water jacket is a very efficient sound absorber), and heating and ventilating the passenger compartment (where the heat source is normally hot water tapped from the cooling circuits). These problems have been sufficient to persuade the overwhelming majority of manufacturers to use water-cooled engines, in which the combustion chambers and cylinder blocks are wrapped in water jackets that extract surplus heat. When the engine is running, the water continuously passes into and out of the engine on a closed circuit, and its temperature is kept in check by air passing through a cooling radiator, usually mounted in front of the engine. The water is circulated by a water pump driven by the engine. Additional cooling of the radiator, especially important when the car is standing motionless with the engine running, can be provided by a fan which is either belt-driven by the engine or powered by an electric motor.

The electrically-assisted turbocharger – the 'hybrid turbo' – consists of a high-speed turbine-generator combined with a high-speed, electric, air compressor. When the driver depresses the throttle, the hybrid turbo initially acts like an electric supercharger. The compressor motor is powered from the vehicle battery so that acceleration to full operating speed is achieved in less than 500 ms and thereby eliminates turbo lag. At high engine speeds, there is more energy generated by the turbine than is required by the compressor. Under these conditions, the excess energy can be used either to recharge the battery ready for the next acceleration phase or to power some of the auxiliary loads, such as an electric air-conditioning system. The hybrid turbo not only reduces turbo lag, but also results in up to 50% improvement in torque at low engine speeds.

The performance of an engine is defined in terms of its output power at a stated rotational speed (rpm) and also the torque it produces. The power output is measured in brake horsepower (bhp) or in kilowatts (kW) [1 bhp (UK) = 0.746 kW]. By no means all the power from the engine is available at the wheels as there are significant losses in the drive-train. The torque is the ability of an engine to rotate the crankshaft; it is a torsional moment, or couple, measured in Newton metres (N m) or foot pounds force (ft lbf) (1.356 N m = 1 ft lbf). The power output of an engine is related to its torque output through the speed of rotation of the crankshaft:

$$\text{Power (W)} = \text{torque (N m)} \times 2\pi \times \text{rotational speed (rps)}, \qquad (4.1)$$

or, in more practical automotive units:

$$\text{Power (kW)} = [\text{torque (N m)} \times 2\pi \times \text{rotational speed (rpm)}]/60\,000. \tag{4.2}$$

In traditional units, as employed in the USA:

$$\text{Power (bhp)} = [\text{torque (ft lbf)} \times 2\pi \times \text{rotational speed (rpm)}]/33\,000. \tag{4.3}$$

Note: some companies still quote engine power in terms of the PS unit – slightly less than 1 bhp (735 W) – although it was rendered obsolete by European Economic Community directives in 1992.

While it is true that very fast cars invariably have powerful engines, the horsepower rating of an engine is not the only guide to the performance of a given car. In practice, the torque rating often gives a better idea of the capabilities of an engine. The reason lies in the fact that, in different engines of equal horsepower, maximum torque may be available at quite different engine speeds, and this in turn determines the way in which the engine and, in particular, the gearbox must be used to obtain maximum performance. It is possible to obtain a good measure of an engine's road performance by examining graphs of its power and torque as a function of engine speed. For instance, if maximum torque is recorded at a relatively low engine speed and the curve is not too steep, then the engine will be a good performer at low speed so that the driver will not constantly have to use the lower gears in heavy traffic to keep the engine pulling smoothly. Automatic gearboxes are also best suited to high torque at relatively low engine speed. On the other hand, a steep torque curve is representative of a highly tuned engine that will deliver maximum performance (especially acceleration) by frequent use of the gearbox. Most sports-car engines are of this type. Examples of power and torque curves as a function of rotational speed for a Volvo with a common-rail diesel engine (*v.i.*) are shown in Figure 4.5. Peak power output is reached at 4000 rpm and the peak torque is at 2000–3000 rpm.

The torque available at the wheels is determined not only by the output from the engine but also by the gearing of the drive-train – more torque is available in low than in high gear. If an attempt is made to move a vehicle away in too high a gear, the engine stalls; there is simply insufficient torque to turn the transmission. Large American cars of the 1960s and 1970s with eight-cylinder engines had huge power outputs and therefore high torque at low rpm and that made it possible to drive at quite low speeds in a high gear with, consequently, less-frequent gear changing. The challenge today is to design small, economical engines that still provide high wheel torque at low rpm. This involves designing the gearbox appropriately and minimizing power losses in the drive-train.

It is predicted that small city cars of the future will be powered by highly efficient two-cylinder engines. A start in that direction has been made by Fiat with its 875-cc *Twinair* petrol engine. This is a two-cylinder turbocharged engine that has been fitted to the Fiat *500C* car. The engine weighs just 85 kg and produces 85 bhp (64 kW) with 145 N m (107 ft lbf) of torque. The carbon dioxide emissions are just 95 g km^{-1}, which is one of the lowest figures currently achievable for a four-seat car.

The Tata Motor Corporation of India has introduced an even smaller, low-specification, inexpensive car – the *Nano*. The vehicle is aimed at the Indian market and at other countries where people have the desire to move from small motorcycles and scooters to car ownership. The *Nano* has rear-wheel drive and a two-cylinder 624-cc aluminium engine, which develops just 35 bhp (26 kW) and 48 N m (35 ft lbf) of torque. This is seen as the 'people's car' for the developing world, just as the Ford *Model T* for the USA and the Volkswagen *Beetle* for Germany were people's cars in their day. The challenge in designing

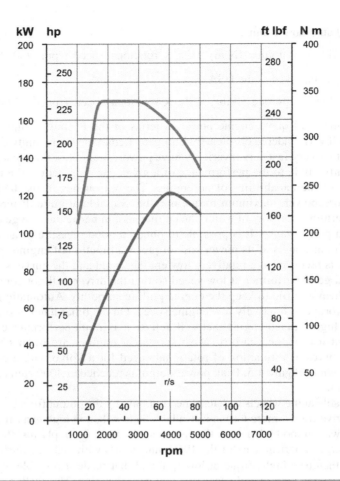

FIGURE 4.5

Power curve (lower) and torque curve (upper) as a function of engine speed

two-cylinder engines is to overcome problems of vibration and noise so as to ensure an acceptably smooth and pleasant ride. There are also difficulties with the bearings and with balancing rotational irregularities. These issues are being addressed and resolved.

In contrast to small, economical cars, manufacturers are also keen to improve the fuel economy and lower the emissions of large, luxury cars since this is a market sector where significant profits are made. One approach to enhancing fuel economy is 'cylinder de-activation', whereby an even number (usually two) of the engine's cylinders are closed down when the vehicle is cruising or descending a hill, i.e. when there is no call for maximum power. De-activation is accomplished by arranging that the inlet and exhaust valves of the cylinders remain closed temporarily so that no fresh fuel or air is admitted and, consequently, the burnt gases in the cylinder are simply compressed and expanded without delivering any power. The technique is employed, for example, in the new Audi *S7 Sportback,* which has a powerful eight-cylinder engine with four cylinders that shut down automatically when the car is cruising steadily or driving in town.

Another pathway to fuel economy in large cars has been proposed by Mercedes through the development of a hybrid of the petrol and the diesel engines known as the *Diesotto* – a name derived from diesel and Otto. This is a 1.8-L, high-compression, four-cylinder petrol engine that is equipped with a turbocharger. The spark plugs fire only when the driver accelerates hard; at other times, the fuel–air mixture ignites, as in a diesel, due to the generation of heat in the cylinder as a consequence of compression. The engine is said to provide a performance similar to that of a 3.5-L V6 petrol engine or a 3-L V6 turbo diesel, but with better fuel economy and reduced emissions. Further developmental work is required for this engine to become commercially viable. The aim is to fit it to the Mercedes *S Model* cars. Other manufacturers are working along similar lines.

Electric motors are noted for their high torque compared with that of internal-combustion engines. This is why the acceleration of electric trains from stations is more rapid than that of diesel trains. The same holds true for electric vehicles, which can often outpace internal-combustion-engined vehicles (ICEVs) away from the traffic lights – at least for the first few seconds. For example, it is claimed that the Tesla *Roadster* electric car (see Section 5.3.3.1, Chapter 5) can attain 60 mph (96 km h^{-1}) in 3.7 s from a standing start, a blistering acceleration.

4.3.1.3 Diesel engines

Diesel engines have long been employed in trucks, buses and heavy goods vehicles. In such applications, they soon developed a reputation as being noisy, clattering and polluting, with clouds of black smoke being emitted. For a long time, therefore, they were not even considered for use in automobiles. Over the years, however, major improvements have been made and modern small diesels are now employed extensively in cars, especially in Europe.

The degree of compression of the air in the cylinder of a diesel engine is greater than that for a petrol engine; see Box 4.1. Accordingly, the temperature of the air is raised considerably, to well over 500 °C. As the piston nears the end of its upward travel, a fine spray of fuel is injected from a nozzle near the top of the cylinder. Because the compressed air has become so hot, its mixture with the fuel ignites spontaneously without the need for a spark.

BOX 4.1 COMPRESSION RATIO

The compression ratio (CR) is defined as the ratio of the volume of the cylinder and its head space (including the pre-combustion chamber, if present) when the piston is at the bottom of its stroke to the volume of the head space when the piston is at the top of its travel ('top dead centre', tdc). Typically, petrol engines have a CR of 8–10, while diesel engines have a CR of 15–20. The CR of petrol engines is limited by the requirement that the fuel burns uniformly in the cylinder and does not ignite thermally prior to the spark (so-called 'engine knocking'). In a spark-ignition engine, the CR at which pre-ignition takes place is determined by the octane number of the petrol; see Box 4.2. High-octane fuel permits a high CR. Until about 30 years ago, lead tetraethyl was added to petrol as an anti-knock agent. This was phased out for environmental reasons and non-toxic additives are now sometimes used. Improvements in engine design over recent years have, however, led to satisfactory compression ratios with lower octane fuel.

For an engine where the cylinder pressure at the end of the exhaust stroke is 1 bar (10^5 Pa), the pressure at the point of ignition is approximately CR to the power 1.4. For example, if the CR is 10, then the pressure at top dead centre, P_{tdc}, is $10^{1.4} = 25.1$ bar.

A diesel engine can be adapted to run on almost any liquid fuel, but the most suitable and widely used diesel fuel is distilled from crude oil and is the next heavier fraction after kerosene. It is much less volatile than petrol. The fuel is delivered to each injector by a pump, and there is either one pump for each injector or else one main pump that supplies all the injectors in turn by means of a distributor valve. The pumps are of the reciprocating type, with spring-loaded plungers actuated by a camshaft driven by the engine. The accelerator control of the vehicle is connected to the pump mechanism, and alters the engine speed by varying the amount of fuel that is fed to the injectors. Spring-loaded needle-valves in the injectors are opened by the pressure of the shots of fuel delivered at the correct instant in the engine's cycle. The fuel is sprayed out through holes in the end of the injector and thereby broken up into a fine mist and distributed effectively around the combustion chamber.

When starting from cold, diesel engines may require the pre-heating of 'glow plugs' located in the cylinder head in order to initiate ignition. In some older diesels, which had comparatively low compression ratios, this procedure could take 20 to 30 s. The heat generated on compressing the fuel–air mixture in modern, high-compression, automotive diesel engines may be sufficient to eliminate the need for glow plugs. Nevertheless, the devices are still often fitted to aid starting in cold weather.

The quality of diesel fuel is described by its 'cetane number', which is a measure of the time delay between injecting fuel and the start of pressure rise as combustion takes place. Diesel engines run well with a cetane number between 40 and 55 – the higher the number, the shorter the time delay. High-performance engines run best with fuel that has a high cetane number. Diesel fuel may contain additives such as detergents to clean the fuel injectors and minimize carbon deposits.

Diesel engines have several advantages over spark-ignition engines for road transport. As well as their simpler ignition system, they are significantly more fuel-efficient (more miles per gallon, fewer litres per 100 km) and provide higher torque at lower rpm than petrol engines. Diesels also generally have a longer service life – 250 000 miles (402 300 km) is by no means unusual. For these reasons, taxis generally employ diesel engines. Diesel fuel is far less inflammable than petrol and therefore the engines are also safer in a serious road accident situation. All too often, major accidents on motorways lead to fatal fires when petrol ignites. Diesel fuel is cheaper to manufacture by simple distillation of petroleum and does not require the more sophisticated refinery operations (catalytic cracking, reforming, etc.) that are necessary to produce petrol.

On the negative side of the balance sheet, diesel engines are more expensive to manufacture than petrol engines because they have to compress their fuel–air mixture to about twice the pressure of that in a petrol engine and so demand a more robust structure. Also they develop less power than a petrol engine of comparable size. Diesel fuel has to be treated by hydrogenation in order to remove sulfur to the very low levels required by modern legislation. A further and serious limitation is that the fuel is unsuitable for operation at temperatures below about −15 °C, as it starts to solidify to a wax. In countries such as Canada where exceptionally low temperatures are encountered, diesel for trucks is replaced in winter by kerosene, which is a lighter distillate.

Heavy trucks invariably have direct-injection diesel engines, frequently turbocharged with intercoolers. The engines may be V-configuration with 10 or 12 cylinders that produce 300–600 bhp. Because of the large loads that have to be hauled, especially where the route includes steep hills, the transmission used in a heavy truck has more ratios than for a car, with as many as 16 gears.

4.3.1.4 Rotary engines

In a conventional engine, the reciprocating movement of the pistons has to be converted into a rotary movement in order to turn the crankshaft. This is achieved by means of connecting rods that join the pistons to the crankshaft. A simpler concept, in principle, would be to have an engine that operates in a rotating, rather than reciprocating, fashion. The first such engine was invented in 1957 by the German engineer Felix Wankel and, consequently, the generic technology is now often referred to as the Wankel engine.

In place of cylinders and pistons, a Wankel engine has a rotor that is essentially triangular in shape with convex surfaces; see Figure 4.6. This revolves in a housing which is machined such that its internal surface is a specified mathematical shape; namely, broadly elliptical. The apices of the rotor make contact with the housing at all times and thereby form three chambers, two of which are much larger than the third. The engine has exactly the same four strokes as a conventional piston engine. As the rotor moves around, it passes an intake port, takes in a fuel–air mixture, and moves further so that the gases are confined and then compressed as the chamber becomes constricted. At the point of

FIGURE 4.6

The Wankel engine

(Image sourced from Wikipedia and available under the Creative Commons License)

maximum compression, the spark plug(s) fire and the expanded gases force the rotor to move further around to the next phase where the chamber enlarges again. Finally, the burnt gases are expelled through the exhaust port.

The major attraction of the rotary engine is that it has so few moving parts. Gone are the pistons, connecting rods, camshaft, valves, valve springs, rockers, crankshaft and timing belt – in all, 30 to 50 moving parts – and these are replaced by just two or three rotors and an output shaft. This simplification makes for greater reliability and reduced manufacturing cost. Also, as the rotor spins continuously in one direction, rather than reciprocating, it produces a smoother ride. The power output is more uniform than in a piston engine and the rotor moves more slowly than do pistons, a feature that also contributes to the reliability of the Wankel engine. The downside of this engine is that it has a low compression ratio, which leads to poor thermodynamic efficiency and high fuel consumption. Also, there have been issues with the lifespan of the seals and with high fuel levels in the exhaust. Nevertheless, the attractive features are such that research continues into improving the rotary engine.

The Mazda car company in Japan has conducted by far the most development work, which was started in the 1960s. The Mazda *RX7* was the first rotary-engined car to reach commercialization and was on sale from 1978 to about 1995. This model was succeeded by the *RX8* sports car, which was withdrawn from the market in 2011. The engine of the *RX8*, trademarked the *Renesis*, had a nominal capacity of 1.3 L and was said to deliver 238 bhp (177 kW). This high power output from a compact, lightweight, reliable engine makes it ideal for use in racing cars, where fuel consumption and exhaust pollution are not prime considerations.

4.3.1.5 Gas turbines

The gas turbine is a power-generation source that has much greater overall efficiency than any of the foregoing engines. Accordingly, the technology is used to power aircraft and to generate electricity in gas-fired power stations.

Air enters the engine via a compressor and then passes through a rotating heat recuperator, where it is preheated by the exhaust gases prior to being mixed with fuel and burnt in the combustion chamber. The hot exhaust gases drive a power turbine (~ 90% of the output of which is used to propel the vehicle and ~ 10% to rotate the air compressor) and then return to the recuperator to preheat the incoming air.

The gas turbine is more efficient in a number of ways than conventional vehicle engines. First, like the Wankel engine, it operates rotationally and therefore is not required to convert linear motion to rotational motion. Second, the introduction of excess air ensures complete chemical reaction of the fuel and air mixture and thereby total consumption of the fuel. Third, this power source is more compact and lighter in weight than a piston engine and has about 80% fewer parts. Summing these facts together, it is evident that the gas turbine is a more economical source of power generation. Why then is the technology not used in the automotive industry? There are a number of reasons. The gas turbine, though an engineering marvel, has some limitations. Its efficiency is mainly dependent on the temperature that the turbine can withstand. The blades and the compressor have to work at very high temperatures, and they require the use of special alloys that must be machined to exceptionally fine tolerances. The price of the alloys and of the machine tools needed to shape them makes the engine uneconomic for use in cars. Further, the gas turbine can reach high thermal efficiencies only when operated at its rated speed. Thus when the unit is run at lower speeds, the thermal efficiency is greatly reduced. This feature would be a serious disadvantage in, for example, city traffic. In view of these limitations, it is understandable why the gas turbine has not been widely used in road transport. Nevertheless, a few prototype cars have been built.

The first car to be furnished with a gas-turbine engine was the Rover *JET1* in 1950. This innovation stemmed from wartime work that the Rover Company had undertaken in the development of jet engines for aircraft. The car had an inboard rear engine with a turbine that ran at up to 40 000 rpm and produced a maximum power of 100 bhp (75 kW) at 26 000 rpm. A later (1952) model delivered 230 bhp, which gave the car a maximum speed of 152 mph (245 km h^{-1}). Racing cars with gas-turbine engines were built and competed in the Le Mans race in 1963 and in 1965 before Rover discontinued its interest in the technology. Fiat experimented with a turbine-powered coupé in 1955, as did Renault the following year. Considerable experimental work has also been carried out on gas turbines for heavy goods vehicles, notably by British Leyland.

In the USA, the Chrysler car company spent many years of development that culminated in the production of their first gas-turbine car in significant numbers in 1963. A total of 55 vehicles were built and all but five were passed to the public for testing. Collectively, this fleet completed 1.1 million miles (1.8 million km) on the roads. The Chrysler turbine ran at up to 44 500 rpm and could operate on a wide range of liquid fuels. It generated 130 bhp (97 kW) of power and 425 ft lbf (576 N m) of torque. On account of the high rotational speed, considerably more torque was generated than in a typical diesel engine. The programme was ultimately abandoned when Chrysler ran into financial difficulties and most of the 55 cars were scrapped.

In summary, the major drawbacks with the gas-turbine car were its high fuel consumption and the high cost of both materials and production. Unless these problems can be overcome, it appears unlikely in the present climate that this engine technology will be resurrected as a competitive power source for road vehicles. There is, however, a developing interest in micro gas-turbines as a supplementary power source for extending the range of electric vehicles (see Section 5.3.2, Chapter 5).

4.3.2 Fuel supply

The traditional approach to fuelling a four-stroke petrol engine is by means of a carburettor, a device for mixing fuel droplets with air. The petrol is withdrawn from the storage tank by means of a pump, which may be either mechanically operated off the engine or electrically driven. The carburettor contains a float chamber in which the level of the stored petrol is controlled by means of a shut-off valve mounted on the float. Petrol is sucked from here into a second chamber where it is mixed with air and vaporized before passing into the inlet manifold, which, in turn, leads to the cylinder intake valves. An engine supplied with this system is said to be 'naturally aspirated'.

In recent years, the carburettor has been almost completely superseded by fuel injection, which is a quite different means of introducing fuel to the engine, namely, under pressure through injector nozzles. The technique applies to both petrol and diesel engines, although with significant differences. Petrol engines are usually equipped with indirect injection whereby the fuel is injected into the inlet manifold before entering the intake valve. Design of the injectors and optimization of the fuel–air mixture flow into the cylinder are both critical in ensuring good mixing and complete combustion. Computational fluid dynamics analysis has been used to model and optimize the flow while, in the laboratory, it has proved possible to observe and photograph the mixing in the cylinder by means of a laser flash shining through a transparent glass piston.

As discussed above, diesel engines operate at a much higher compression ratio than petrol engines. Because there is no spark, it is necessary for the temperature of the fuel–air mixture to reach a value where self-ignition takes place. Diesel fuel is injected at high pressure directly into the combustion chamber at the precise moment when the air has been compressed to its smallest volume. At such pressures, the air is hot enough to ignite the oil. In an alternative configuration, diesel is injected into a pre-chamber situated just above the cylinder head (indirect injection). Each configuration has its advantages and disadvantages. In particular, indirect injection takes place at a lower pressure (about 100 bar, 10 MPa) than direct injection (about 333 bar, 33 MPa). A modern version of direct injection for diesel engines is the 'common-rail' system. Instead of having individual fuel lines to each injection nozzle, there is a common feed pipe that operates at very high pressure (typically, 1000 bar, 100 MPa).

4.3.3 Engine ignition systems

The electric ignition system for the petrol engine was first devised in 1911 by Charles Kettering, who, as mentioned previously, also invented the starter motor. The principle is well known. A petrol engine needs a spark to ignite the fuel–air mixture in each of the cylinders. Ignition involves four basic and sequential functions: the provision of low-voltage electricity, amplification of the voltage to a high level, distribution of a high-voltage electrical current pulse to each of the combustion chambers and, finally, discharge in the form of sparks. These actions are carried out, respectively, by the generator, the induction coil, the distributor and the spark plugs, as follows.

(i) the generator in early vehicles was a hand-cranked magneto. After the invention of the battery-operated self-starter, a dynamo was used to produce direct current. Later, the dynamo was replaced by the more efficient alternator that delivers alternating current, which is then rectified;

(ii) the induction coil is an electrically simple component, essentially a transformer that induces a very high voltage in its secondary winding when the current through the primary winding is interrupted by the opening of contact-breaker points housed in the distributor;

(iii) the distributor directs the high voltage to the spark plugs;

(iv) the timing of the spark that ignites the fuel is critical to the efficient operation of a petrol engine. The objective is to ensure that the ignited gases attain maximum pressure in the cylinder to drive the piston down on the power stroke. The spark plug needs to fire shortly before the piston reaches top dead centre (tdc). This is because there is a finite short delay between the spark occurring and the build-up of maximum pressure, during which time the flame front propagates through the gases. As the engine speed increases, the spark has to take place progressively earlier before the piston reaches tdc (i.e. to be 'advanced') if maximum power, and therefore greatest efficiency, is to be obtained.

Under ideal conditions, the flame front propagates uniformly throughout the fuel–air mixture. If the spark is advanced too far, the mixture beyond the flame front can detonate spontaneously and explosively to cause a local shock wave – the phenomenon of 'engine knock'. The spark has to be delayed ('retarded') to eliminate knock. In cars built in the 1920s and 1930s, there was often provision for manually retarding the ignition timing to eliminate knock. Subsequently, this operation was done automatically. Modern engines may be equipped with a small piezoelectric microphone that detects the onset of knock and sends a signal to the electronic engine-management system, which, in turn, retards the ignition timing. Much research has been conducted into the design of cylinder heads and fuel admission so as to eliminate knock, to obtain maximum power output from the engine and to minimize pollutant emissions.

The overhead camshaft is driven by a belt from the crankshaft and the two components rotate in synchronization. The cams on the camshaft act on rockers that open and close the intake and exhaust valves at just the right instant. The rotor of the distributor, which controls the firing of the spark plugs, is also driven synchronously with the crankshaft. As the rotor turns, it opens and closes platinum contact-breaker points in the distributor and this action sends a short pulse of low-voltage (12 V) electricity to the primary of the induction coil. A high-voltage pulse is induced in the secondary of the coil and is sent, via the rotor arm, to the corresponding spark plug. The current then jumps the gap between the central electrode and the body of the plug to create a spark that ignites the fuel–air mixture. This ingenious invention has been in use in internal-combustion-engined vehicles for about 100 years. It has proved to be reliable, with the sole maintenance operation being a requirement to replace and reset the contact-breaker points periodically.

Since about 1980, electronic ignition has been introduced progressively. Instead of employing a distributor with mechanical contact-breaker points to set the time of sparking, the function is performed electronically by the computer that directs the engine-management system. Electronic ignition eliminates the maintenance required to clean and reset the points regularly, and also leads to smoother running. Several versions of the technique have been adopted. A recent design dispenses with the use of one high-voltage coil to serve all cylinders, and replaces it by a small coil above each sparking plug. The arrangement removes the need for high-voltage cables to each plug, which are a frequent source of trouble, and provides a pulse of more uniform voltage and duration, regardless of the engine speed. Almost all new petrol-engined cars are supplied with electronic ignition. Diesel engines, of course, have no need of this sophisticated ignition system since they have no spark plugs and rely on auto-ignition by compression.

In addition to the timing of the spark, the timing and duration of the valve opening are also critical for good engine performance and are determined by the profile of the cams on the camshaft as these control the valves. Traditionally, cam design is optimized for an average engine speed (rpm), but this results in reduced torque at low engine speeds and reduced power at high engine speeds. Numerous ingenious strategies, both all-mechanical and electromechanical, for controlling valve opening and varying its duration according to the engine speed have been proposed and patented. By employing such variable valve timing, it is possible to improve both the low-end torque and the high-end power output and thereby reduce fuel consumption. Variable valve timing is now favoured widely by many automotive companies.

4.3.4 Engine starting

All engines require a starter motor to turn them over before firing. In conventional vehicles, this is a straightforward, but powerful, direct-current electric motor. When the starter switch is activated by the driver, current flows to a solenoid attached to the starter motor. This current moves a lever into the solenoid that then causes a cogwheel of the motor to mesh with the teeth on the circumference of the flywheel. At the same time, an electrical contact is closed to allow a large current to flow and rotate the starter motor as well as the engaged flywheel. Typically, currents of hundreds of amperes are required to start the engine and are provided by the battery, which is generally a 12-V lead–acid module. The battery is recharged by the alternator–rectifier combination when the engine is running. Automotive batteries have improved enormously over the years and have far longer lives than formerly, even though they may be called upon to power many more functions. Although guarantees may be for two or three years, in practice batteries often operate for eight years or longer before failing. Moreover, modern car batteries no longer require periodic 'topping-up' with de-ionized water. Further information on the evolution of the lead–acid battery is given in Section 7.4, Chapter 7.

4.3.5 Power output

The power output of a spark-ignition engine and its thermodynamic efficiency are determined by the design of the engine, by the ratio of fuel to air, by the ignition timing and by the properties of the fuel used. As discussed above, the quantity of air in the cylinder can be boosted by means of supercharging or turbocharging. Tuning the engine is then largely a matter of optimizing the fuel-to-air ratio. The higher the compression ratio of the engine, the higher is its thermodynamic efficiency and therefore the higher is the power output. (Thermodynamic efficiency is the amount of work that the engine does in relation to the calorific value of the input fuel.) High-compression engines require 'high octane' fuel in order to avoid detonation outside the flame front; see Box 4.2.

Car engines are manufactured with a very wide range of power outputs to suit different customer requirements for speed and acceleration, or for economy of fuel usage. The smallest engines used in four-wheel cars had a capacity of just 500 cc, as in the original Fiat *500*. Today, the most economical engines employed in small saloons are around 900 cc, with an output of 80–90 bhp (60–67 kW) and carbon dioxide emissions of less than 100 g km^{-1}. At the opposite end of the scale, one of the largest engines fitted to expensive limited-edition cars is the Rolls Royce V12 of 6.8-L capacity with an output of 450 bhp (336 kW), while Mercedes has a 6.2-L V8 engine which produces 563 bhp and 479 ft lbf

---

BOX 4.2 OCTANE NUMBERS AND PRE-IGNITION

The octane number of a fuel is a measure of how likely it is to detonate outside the combustion front. The higher the octane number, the more the fuel can be compressed before engine knock is encountered.

The research octane number (RON) of a fuel is determined with a laboratory test engine in which the onset of knocking is observed. It is known that branched-chain paraffins have a higher octane number than their straight-chain analogues, so that one of the major operations of a petroleum refinery is to convert straight-chain hydrocarbons to branched-chain alternatives. Arbitrarily, pure iso-octane (2,2,4-trimethylpentane) is accorded a research octane number of 100, while the straight-chain paraffin *n*-heptane is given an octane number of zero. Hence, a petrol sample with the same anti-knock quality as that of a mixture containing 90% iso-octane and 10% *n*-heptane is said to have a RON of 90. Petrol is made up of a mixture of mostly branched-chain paraffins, with suitable additives, to give a RON in the range 90–100. Typically, 'regular' grade petrol in the UK will have an octane number of up to about 95, whereas 'premium' grade will lie in the range 95–100. The choice of which fuel to use depends on the engine. Originally, lead tetraethyl was added to delay the onset of self-ignition. When the extreme toxicity of organic lead became clearly established, it was replaced by less harmful additives such as methyl tertiary-butyl ether (typically known as MTBE).

Note: the research octane number, as determined in the laboratory, is not to be confused with the road octane number, which has a lower value and is more representative of real driving conditions. Confusingly, some countries use one and some the other at their pumps. In the USA and Canada, 'regular gas' has a road octane number of around 78–80.

(650 N m) of torque with carbon dioxide emissions of about 380 g km^{-1}. In the category of super-sports cars, the Bugatti *Veyron* with a power output of 1000 bhp (746 kW) is notable.

4.3.6 **Position of engine**

The design of an engine, especially with regard to the number and layout of its cylinders, is at least partly determined by where, within the bodywork, the engine is to be located. In most cars, the engine is at the front, which appears to offer designers the most scope for making the best use of the total space available. Some cars (e.g. the well-known VW *Beetle*), however, had their engines mounted at the back. A few two-seat sports cars are 'mid-engined', with the unit behind the seats but ahead of the rear wheels. This placement represents a rather wasteful use of the total space, but designers believe it improves the high-speed handling of a car.

Traditionally, engines located at the front of the car, under the bonnet, were positioned longitudinally in the direction of travel. The driveshaft (also called the propeller shaft) emerged from the rear of the engine and drove the flywheel, which smoothed the output of a reciprocating engine. After the gearbox, the shaft passed under the car to the differential mounted in line with the rear wheels Power was then transmitted to the rear wheels via the differential and two half-shafts ('rear-wheel drive' configuration). The advantage of this layout is that there is generally sufficient space under the bonnet to accommodate quite large engines (six or eight in-line cylinders) with good access for working on the engine. The downside is the intrusion of the driveshaft into the passenger cabin and the lack of traction if the driver inadvertently reverses on to soft or muddy ground.

In the 1950s, the idea arose of front-wheel drive. This notion involves mounting the engine laterally across the car. The first full-production car with a transverse engine was the highly successful British Motor Corporation (BMC) *Mini*, which was the brainchild of the famous car designer Sir Alex Issigonis; an example of the car is shown in Figure 4.7(a). The design brief for the BMC *Mini* was to accommodate four or five passengers in a small car that was 10 ft (3.0 m) in length. These specifications could only be met by mounting the engine transversely. The concept gave rise to a host of technical issues that included the development of a new 'hydrolastic' suspension, and the sourcing of 10-in (25-cm)-diameter wheels and tyres to fit. Without a drive-train under the cabin, the car was of limited height as well as length. This was a revolutionary and iconic design of car, which was so successful that it remained in production for over 40 years, from 1959 to 2000. In the latter year, the company was taken over by BMW, who now produce the new model *Mini*, which is significantly larger in all dimensions and is classified as a compact car.

The Fiat *128*, introduced in 1969, was the next production car to have a transverse engine. Soon, other automotive companies took up the technology and today most major manufacturers produce models with front-wheel drive. Much earlier, some car companies had decided to offer models with rear engines, in which, effectively, the engine and luggage compartments changed place. These small cars were popular before the transversely mounted engine with front-wheel drive was developed as the configuration made it possible to have rear-wheel drive without the passage of the driveshaft through the passenger cabin. The engine was often air-cooled as there was little space to accommodate a radiator system. Fiat manufactured the popular Fiat *500* in the form of three models (successively 479, 499 and 594 cc capacity, which produced 13, 17 and 23 bhp, respectively) that weighed only 500 kg; see Figure 4.7(b); the cars were available from 1957 to 1975. As mentioned above, the most famous of these rear-engine cars was the Volkswagen *Beetle*, shown in Figure 4.7(c), which was introduced in

FIGURE 4.7

Classic small cars: (a) BMC *Mini*; (b) Fiat *500*; (c) Volkswagen *Beetle*

(All images sourced from Wikipedia and available under the Creative Commons License)

Germany in 1938 as the 'people's car'. Production of the *Beetle* continued for 65 years until 2003, which makes it the longest-running model ever; over 21 million were produced and sold throughout the world.

4.3.7 Current developments in engine technology

A great deal of research and development is being conducted on improved engines and transmissions. One promising concept is the 'homogeneous charge compression ignition' (HCCI) engine. This is the broad principle behind the new Mercedes *Diesotto* engine, described earlier. Fuel (petrol and/or diesel) is well-mixed with air and then compressed to the point of auto-ignition. A lean fuel–air mixture is employed, which permits a high compression ratio (> 15), i.e. like a diesel, and gives a good fuel efficiency (low litres per 100 km). The HCCI engines run comparatively cool and burn cleanly, so that no soot is formed and very small amounts of nitrogen oxides (NO_x) are released. The levels of unburnt fuel and carbon monoxide are rather higher than with a conventional petrol engine, so a catalytic converter is still required. The difficulties encountered with HCCI engines lie in the area of controlling the combustion process. With SI engines, the instant of ignition is regulated by the spark, which is at precisely the right point in the cycle. Similarly, with CI engines the ignition is timed by the instant of diesel injection. In the HCCI engine, however, there is no direct initiator of combustion. Auto-ignition depends upon the composition of the fuel–air mixture and its temperature. To control these factors, it is necessary to use feedback from the engine to a management system. If the cylinder pressure is allowed to rise too high, then the engine may be damaged. Work is continuing to perfect HCCI technology given that it offers the advantages of higher fuel efficiency, lower emissions of soot and NO_x, and the ability to burn a range of liquid fuels.

Other engine developments include exhaust gas recycling (EGR), cylinder pressure sensing, variable valve lift, active fuel management, stop–start technology and regenerative braking (see Section 5.2, Chapter 5).

4.3.8 Transmission

The drive-train (or transmission) transfers power from the engine to the wheels, via the gearbox, the differential and the half-shafts. The crankshaft of the engine rotates at a much faster rate than the wheels, particularly on start-up or when travelling slowly in traffic. For this reason, it is necessary to have a gearbox[2] that reduces the rotational speed and raises the torque available. The lower the gear used, the larger is the gear ratio and the more the torque that is transmitted to the wheels. When the vehicle encounters a hill, its speed will drop and the engine will falter and stall unless more torque can be made available. A slow-running engine cannot provide sufficient torque for climbing hills or for starting from rest. Selecting a lower gear enables the engine to run faster in relation to the road wheels and thus multiplies the torque.

Early cars had just three or four forward gears and a reverse gear. Most four-cylinder cars today have five or six forward gears, with a growing trend towards six in order to economize on fuel when driving on long, straight highways at high speed. Powerful V6 and V8 engines provide high torque at low rpm and may not require so many gears. Conversely, some larger vehicles (trucks, buses) and cross-country vehicles have an extra ultra-low gear for negotiating steep inclines, or for coping with wet or muddy ground where the traction is poor. Interestingly, as hybrid cars move towards greater degrees of electrification (see Chapter 5) the requirement for multiple gears may decline because electric motors are noted for high torque.

Gearboxes are of two types – manual and automatic. Manual gearboxes require a clutch to disengage the engine from the drive while the gears are being changed. For a smooth gear change, it is necessary to adjust the engine speed to the new gear speed. In early cars, this involved 'double-declutching', i.e. when shifting down, the driver gave the accelerator a 'blip' between clutch movements so as to adjust the engine speed. Later, this action became automatic through the development of a mechanical 'synchromesh' system in the gearbox that automatically selects the correct forward gear ratio. There is further fixed gear reduction in the differential in order to bring the shaft speed down to that of the wheels. The differential also allows the inner wheel of the powered axle to turn more slowly than the outer wheel when the vehicle rounds a bend.

Automatic gearboxes change gear without driver intervention to match the prevailing torque demand. The technology uses a fluid 'torque converter' instead of a clutch to manage the connection between the engine and the onward transmission and has a 'sun-and-planet' arrangement of wheels to provide a set of fixed-ratio gears. The fluid employed to transmit the power is a refined petroleum-based product. The torque converter allows the vehicle to remain in gear while stationary with the engine

---

[2] In N. America, the gearbox is often referred to as 'the transmission', whereas it is actually only one component of the transmission, which additionally includes the driveshaft, differential and half-shafts.

idling. This avoids the necessity of juggling a clutch and gear lever when travelling slowly and when stationary in urban traffic. There is, however, some energy loss in the torque converter and so vehicles fitted with automatic transmission normally give rather poorer fuel consumption than those with manual transmission. Also, as the torque converter is not 100% effective, there is a tendency for the vehicle to creep forward when it is stationary with the gear selector set in the 'drive' position and it may therefore be necessary to apply the brake.

Between the USA and Europe, there is a distinct difference in customer preference for the type of transmission. Almost all saloon cars in the USA have automatic transmission and it is often necessary to place a special order if manual transmission is desired. By contrast, in Europe a high proportion of cars are supplied with manual transmission. In part, this is a cultural difference; many Europeans seem to prefer to exercise choice over which gear is to be employed, particularly on narrow, winding and steep roads, whereas Americans seem perfectly content to allow the car to make the decision and enjoy the simpler driving mode.

A number of alternatives to manual transmission and automatic transmission have been devised. One is dual- (or double-) clutch transmission (DCT), in which there are two independent clutches housed in a single unit. One clutch controls even-numbered gears (two, four and six), whereas the other controls the odd-numbered gears (one, three, five and reverse). This permits gears to be changed without interrupting the torque to the wheels. Many of the major vehicle manufacturers are starting to equip their prestige cars with DCT, although there is a cost penalty associated with this added complication.

Another approach is that of continuously variable transmission (CVT), which provides step-less gear shifting between maximum and minimum values. The idea of a CVT was said to have originated with Leonardo da Vinci in 1490, but the first European patent application was made by Daimler and Benz in 1886. A CVT drive permits the engine to run at its most efficient number of rpm for a range of output vehicle speeds and so conserve fuel without any input from the driver. Several different principles and engineering designs are employed in CVTs. The most popular is the variable-diameter pulley drive; see Figure 4.8. Two conical pulleys are split into halves perpendicular to their axis of rotation. A flexible belt, preferably a steel band, joins the two pulleys and sits in the groove between the two halves. This belt is V-shaped to mesh with the surfaces of the pulleys. The gear ratio is changed by moving the two halves of one pulley closer together and the two halves of the other further apart. This action causes the belt to ride higher up the first pulley and lower down on the other, so changing the effective diameter of the pulleys and therefore the gear ratio. Other types of CVT employ (i) rotating discs and roller bearings to transmit power between the discs, or (ii) a hydraulic fluid to transmit power.

Continuously variable transmissions are widely employed in small-scale agricultural machinery, snowmobiles and some motor scooters. Japanese car makers – notably Subaru, Nissan, Honda and Toyota – have been pioneers in the introduction of the technology into cars. Some of the systems employ belt-driven pulleys, some use cones and roller bearings, and others are electronically controlled. Clearly, CVTs are becoming increasingly common as a technology for reducing fuel consumption. In 2010, for example, over 10% of light-duty vehicles sold in the USA incorporated these transmissions.

Parallel hybrid electric vehicles (see Section 5.2, Chapter 5) need to transmit power and torque from an internal-combustion engine and an electric motor, and to vary the respective contributions of these different power sources to the overall output. This requirement poses a quite different problem from that of standard vehicles with a single power source. Accordingly, power-sharing transmissions (PSTs)

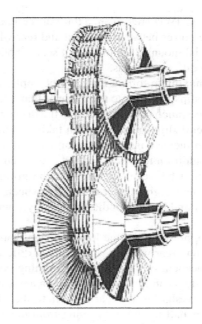

FIGURE 4.8

Schematic illustration of continuously variable transmission (belt driven)

(Image sourced from Wikipedia and available under the Creative Commons License)

have been devised for hybrid vehicles and are used in the Toyota and Lexus hybrids. Development work continues as new hybrid vehicle concepts evolve.

Motorcycles have traditionally had a chain transmission to the rear wheel, but some manufacturers have employed a shaft drive. The latter is more rigid and does not stretch like a chain. Harley Davidson motorcycles are equipped with belt drive, a third option.

4.3.9 Two-wheel and four-wheel drives

When a vehicle is turning, the driven wheels need to rotate at different speeds – the wheel that is on the outside of the curve rotates faster than that on the inside. The differential distributes torque (angular force) evenly, but angular velocity (turning speed) unevenly, such that the average for the two half-shafts equals that of the input driveshaft. This operation is effected by a clever mechanical device in the differential that involves inter-meshing bevel gear wheels with teeth cut in a spiral form.

Vehicles that are intended for use across rough or muddy fields or tracks, or other slippery conditions such as snow or ice, are equipped with four-wheel drive (4WD or 4 × 4). This facility may be either a permanent 4WD (also known as all-wheel drive, AWD), or part-time 4WD that can be engaged in a two-wheel drive (2WD) vehicle in situations where the ground is such as to require maximum traction to avoid wheel slippage. Typical vehicles in the 4WD category are Jeeps, Land Rovers and many sports utility vehicles (SUVs). There are several different designs of 4WD but, broadly, each powered pair of wheels requires a differential to distribute power and torque between the left and right sides.

When all four wheels of a part-time 4WD are driven, a third differential (or transfer case) mounted on the driveshaft serves to distribute power between the front and rear wheels. By contrast, a permanent 4WD may have a single differential mounted centrally that serves both pairs of wheels. This arrangement allows each pair to rotate at different speeds.

'Traction' is defined as the maximum torque that may be applied to the wheels before slippage occurs. Immediately one wheel slips, the torque available at that wheel drops to almost zero. Longitudinal slippage results from too much power applied to the drive wheels, as for example on a wet or muddy road. By contrast, lateral slippage stems from taking a sharp curve too rapidly. The former action is often no more than inconvenient (except on ice), while the latter can be deadly. The factors that control slippage, in addition to power applied and speed, are the weight over the driven axle and the coefficient of friction between the tyre and the ground. The former factor favours front-wheel drive where the engine weight is over the driven wheels; the latter factor is determined by the nature of the ground itself and by the condition of the tyre with respect to tread pattern and wear (*v.i.*).

When driving around a curve, not only are the left- and right-hand wheels revolving at different speeds, but so also are the front and rear wheels, which follow different trajectories. The role of the transfer case in 4WD is to take account of this when apportioning power between the front and rear axles so that a part-time 4WD system can function properly on any surface. When the transfer case is locked, thereby tying the front driveshaft to the rear driveshaft, the wheels spin at the same rate and each pair of drive wheels receives half the torque coming from the engine. This situation inevitably results in tyre slippage when going round a bend on a dry, flat surface and consequently leads to excessive tyre wear. For this reason, the use of the locked mode of part-time 4WDs should be confined to low-speed travel on slippery or muddy surfaces. The transfer case may also contain an extra set of low gears for use on steep hills. With a maximum speed of about 5 mph (8 km h^{-1}) when these gears are engaged, a vast torque is applied to the wheels and this enables the vehicle to crawl up very steep gradients. In an all-wheel drive (permanent 4WD) vehicle, the use of only one central differential and the fact that the axles may rotate at different speeds means that driving on normal roads does not present any operational difficulties.

4.4 Suspension, steering, brakes

In the 19th century, horse-drawn carriages and stage coaches were fitted with wooden or steel leaf springs to absorb shocks and to provide a measure of comfort to the passengers when travelling over poorly made roads. This principle was adapted and refined for use on automobiles. The smoothness of the 'ride' of a modern vehicle is determined by its springs, by its shock absorbers, and by the struts that link the body and the wheels – collectively, referred to as 'the suspension'. The purpose of springs is to absorb impacts from irregularities in the road surface or from bumpy terrain. Shock absorbers serve to dampen oscillations in the springs as they bounce back from being compressed. The overall role of the suspension is to maximize the friction between the tyres and the road, to maintain steering stability with good handling, and to ensure the comfort of the passengers.

Designing suspension systems is difficult because the topography of any road – and therefore the nature of the job tackled by the suspension – alters constantly and unpredictably. For this reason, not only must each road wheel be efficiently suspended but also the entire system has to be carefully matched from one side of the vehicle to the other, as well as from the front wheels to the rear wheels. Only if the designer can achieve these features will the car prove to be stable at all speeds and over different types of road surface.

4.4.1 Springs

Early cars had steel leaf springs, made up of layers of mild steel strapped together. Some trucks still have leaf springs, as shown in Figure 4.9(a). On the other hand, most cars have coil springs, see Figure 4.9(b), which are capable of greater compression than leaf springs. Coil springs, however, vary greatly in their stiffness, i.e. their ease of compression. Naturally, their mechanical behaviour is determined by the properties of the metal alloy used to form the coil – namely, its gauge and intrinsic shear modulus – as well as by the diameter of the coil itself. Strong springs, which compress with difficulty, are fitted to heavy vehicles such as trucks. When the truck is fully laden, the springs are able to support the total vehicle weight over rough ground, but when unladen the weight is insufficient to compress the spring fully and the ride is bouncy. Conversely, vehicles such as luxury cars, buses and taxis that are not intended to go over rough ground have comparatively weak springs in order to give a soft ride. Some cars that are available as 'sports models' have stiffer suspensions than their family-oriented counterparts.

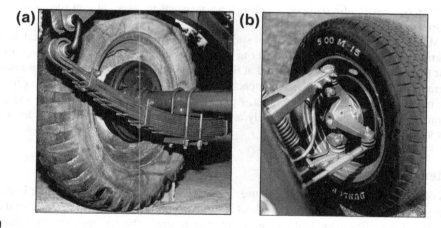

FIGURE 4.9

Vehicle springing: (a) leaf springs under a truck; (b) coil spring

(Images sourced from Wikipedia and available under the Creative Commons License)

4.4.2 Shock absorbers

Vehicle springs absorb energy but are not good at releasing it; rather, they oscillate until finally the energy is dissipated as heat. Shock absorbers dampen oscillation of the springs by taking up their kinetic energy, although they also serve to smooth the ride to some degree. The shock absorber consists of a piston within a cylinder that contains a hydraulic liquid or, sometimes, air. The energy of the spring is converted to heat in the viscous hydraulic fluid when it flows through a small orifice. In the alternative version, the heat is expelled as hot air through a valve and replaced by cold air. The shock absorber can be selected to suit the vehicle and the terrain. In some advanced designs, it is even possible for the device to be tuned remotely by the driver, or automatically by a computer, to accommodate variations in the terrain during a journey and to match the loading of the vehicle.

There are several other advanced types of shock absorber based on electro-rheological fluids or magneto-rheological fluids which use, respectively, an electrical field or a magnetic field to change the viscosity of the hydraulic fluid. It is likely that one or other of these will be more widely adopted in the years ahead so as to give the smoothest possible ride. Modern heavy trucks generally have air suspensions with hydraulic dampers fitted for safe handling and an even ride.

4.4.3 Struts

The wheel is attached to the body of the vehicle by means of metal struts. Two configurations are widely adopted – namely, the double-wishbone strut and the MacPherson strut; these are shown schematically in Figure 4.10(a) and Figure 4.10(b), respectively. Each version has its advantages and limitations. Double-wishbone struts were once common on cars, but are now mostly used on trucks and larger vehicles. The system has two sets of transverse links that connect the top and bottom extremities of individual wheel carriers to the structure. The suspension spring is usually a steel coil clamped between the body structure of the car at one end and one of the wishbone members at the other. In addition, there will be a telescopic hydraulic damper which is very resistant to large, abrupt movements but quite compliant to small, slow ones.

In the MacPherson strut system, the coil spring and the damper tube are combined in one near-vertical member fixed to the stub-axle carrier at the bottom and, by a flexible mounting, to the body structure at the top; the system is completed at the bottom end by a lower wishbone to resist braking and acceleration forces. This provides a compact unit, well-suited to front-wheel drive cars, in which the wheels move independently of each other. Apart from being cheap to manufacture and install, the MacPherson strut spreads shock loads widely around the structure and thereby provides a quiet and refined suspension system.

4.4.4 Steering

Two types of vehicle steering are in general use. The venerable rack-and-pinion system is the simplest and most popular for cars. As the steering wheel is turned, a pinion attached to the base of the steering column moves along a linear-toothed rack to which it is meshed. The arrangement converts the rotary movement of the wheel to a horizontal movement along the transverse axis of the vehicle. The rack is attached at each end to tie rods which transmit the movement to the wheels. This method of steering is positive and provides rapid feedback to the driver. A second steering mechanism, the so-called recirculating-ball system, is used on some heavy trucks and SUVs because of its robustness and greater mechanical advantage. The latter attribute makes turning the wheel easier but, with the advent of power steering, this action is now less of an issue and many heavy vehicles are now adopting rack-and-pinion steering.

Cars with front-wheel drive have much of their weight over the front wheels and therefore it becomes more difficult to turn the steering wheel. This factor, together with the use of wider tyres on SUVs and larger cars, compounds the difficulty to the point that unassisted steering would be almost

(a)

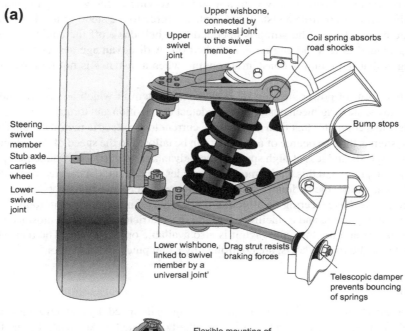

Upper wishbone, connected by universal joint to the swivel member

Upper swivel joint

Coil spring absorbs road shocks

Steering swivel member

Bump stops

Stub axle carries wheel

Lower swivel joint

Lower wishbone, linked to swivel member by a universal joint'

Drag strut resists braking forces

Telescopic damper prevents bouncing of springs

(b)

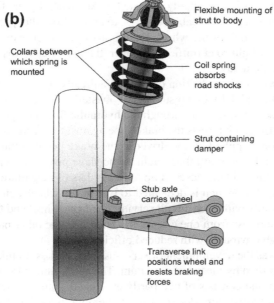

Flexible mounting of strut to body

Collars between which spring is mounted

Coil spring absorbs road shocks

Strut containing damper

Stub axle carries wheel

Transverse link positions wheel and resists braking forces

FIGURE 4.10

Two common forms of suspension: (a) double-wishbone; (b) MacPherson strut impossible for many people. Thus most vehicles are today equipped with power steering, a system that is normally operated hydraulically with mineral oil serving as the working medium. A double-acting hydraulic cylinder controls a piston that applies a force to the steering mechanism to augment the effort made by the driver. The pump is operated by a belt-drive off the engine and therefore no assistance is provided unless the engine is switched on. A disadvantage associated with hydraulic power steering is that the pump is constantly running when assistance is not required and this consumes fuel.

An alternative form of power steering is an electrical system in which an electric motor provides assistive torque to the steering mechanism. Sensors detect the position and torque of the steering column and feed the data to a computer that then controls the current to the electric motor. A major advantage of the electrical system is that the degree of assistance can be tailored to the speed of the vehicle, with more assistance at low speeds and less at high speed. Other advantages are: (i) a hydraulic pump, a belt drive and hoses are not required and this leaves more space under the bonnet; (ii) the electrical system is more fuel efficient in that, unlike the hydraulic system, it only uses energy when it is operating.

Some vehicles now have steering on all four wheels. This arrangement was introduced primarily to permit a tighter turning circle and to facilitate parking in a restricted space. It is most useful for trucks, heavy goods vehicles and tractors, although it is also available on a few cars. The rear wheels, which cannot turn as far as the front wheels, are controlled by a computer and actuators.

4.4.5 Brakes

The role of brakes is to bring a vehicle to a halt, or to reduce its speed, by converting its kinetic energy to heat through friction. Early road vehicles had manually operated brakes whereby the force exerted on the brake pedal was transferred mechanically, via cables or rods, to brake shoes that pressed against the inner surface of a rotating drum in the hub of the wheel. Such brakes, however, were barely adequate for low-speed vehicles in an era of light road traffic. Moreover, the use of asbestos fibres in the brake shoes would be totally unacceptable today.

The next stage in brake evolution was the introduction of the hydraulically operated drum system. In this design, a rod attached to the brake pedal drives a piston into a 'master' cylinder that is filled with hydraulic oil. The resulting pressure rise is transmitted through the hydraulic fluid in pipes to a 'slave' cylinder at each road wheel, the piston of which thrusts the brake shoe against the inside of the rotating brake drum; see Figure 4.11(a). The force applied by the driver to the brake pedal is thus magnified, first by leverage at the master piston, and then multiplied again at the slave pistons by virtue of their smaller diameter; the combined effect can yield increases of up to two orders of magnitude. Invariably, there are two separate brake circuits, each of which is dedicated to two of the wheels, either the front and rear pairs separately, or split diagonally with each circuit serving one front wheel and the opposing rear wheel. In the event of hydraulic fluid loss from one circuit, the brakes on the other pair of wheels will bring the vehicle to a stop, although obviously with reduced efficiency.

The third stage of brake evolution was the development of servo-assisted brakes. In this design, the force exerted on the brake pedal is amplified by means of a vacuum. The servo-unit is located between the brake pedal and the master cylinder and consists of two partly evacuated chambers separated by a moveable diaphragm or piston. The partial vacuum derives from the engine inlet manifold or, in the case of fuel injection, from a separate vacuum pump. When the brake pedal is depressed, air is admitted to one chamber and this pushes the diaphragm or piston into the other chamber. The amount of servo

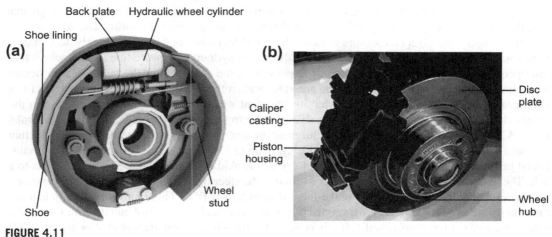

FIGURE 4.11

Braking systems: (a) drum; (b) disc

(Images sourced from Wikipedia and available under the Creative Commons License)

assistance increases with the physical force used on the pedal. Again, a connecting arm drives the piston of the oil-filled master cylinder that transmits force hydraulically to the brake shoes. Thus, the vacuum servo mechanism magnifies the applied force. Since the servo operation relies on the engine to create the vacuum, it is inoperative when the engine is not running, for example when the vehicle is being towed or coasting downhill. Servo-assisted drum brakes were universal until the advent of disc brakes and are still fitted to some vehicles, particularly to the rear wheels.

The fourth stage of brake evolution was the invention of the disc brake; see Figure 4.11(b). Instead of brake shoes rubbing against the inside of a rotating drum, two brake pads are clamped on either side of a revolving cast-iron or steel disc which is mounted on the axle or inside the wheel. The pads are contained in a brake caliper and, again, a hydraulic system is used to augment the effort applied by the driver. This new form of brake was invented in England early in the 20th century and was used for a while on Lanchester cars, but at first achieved little success. It was not until the mid-1950s that the technical problems were overcome by the Dunlop Company and disc brakes came into automotive use on British cars such as Jaguar and Triumph models. The Jaguar *C-Type* was the first racing sports car to adopt disc brakes and, with them, the vehicle won the 24 Hours Le Mans race in 1953. For a long while, most cars had disc brakes fitted only to the front wheels, with drum brakes at the rear. The first production cars to have disc brakes on all four wheels were the 1954 Austin Healey *100S* and the Jensen *541*. In the early 1960s, disc brakes were widely adopted by German and US car companies. Disc brakes are much more effective than drum brakes, especially in wet weather, and are now standard equipment on most cars. They are also fitted on heavy trucks to meet stopping distance regulations in Europe.

Disc brakes are also fitted to most modern motorcycles. In this application, holes are drilled in the large discs to promote ventilation and cooling. Even mountain bicycles are now fitted with disc brakes.

Since 2007, all new passenger cars sold in Europe have to be equipped with an anti-lock braking system (ABS). This is a device to prevent one or more wheels from 'locking up' and skidding on a slippery surface when the forward momentum keeps the car moving at a speed significantly greater than that at which the tyres are spinning. The difference in speed is the 'slip ratio'. The ABS system has four components: (i) speed sensors at each wheel to detect when one wheel suddenly decelerates, as it would immediately before lock-up; (ii) a valve in the hydraulic pressure line to each wheel that can open and close rapidly; (iii) a pump to re-pressurize the system after the valve to a particular wheel has opened and thereby lowered the pressure; and (iv) an electronic control unit (ECU), i.e. a microprocessor. When a sensor detects that its attendant wheel is suddenly slowing relative to the others, the ECU opens the corresponding valve in order to reduce the pressure and hence the brake force. Almost instantaneously, the valve closes again and the pump restores the pressure, so that maximum braking is applied. By repeating this cycle up to 15 times per second, the braking is maintained just below the level that would cause the wheel to skid until the vehicle is safely brought to a halt. The action of ABS makes it far easier to control the steering and to maintain a straight course.

An adjunct to ABS is electronic brake-force distribution (EBD). This innovation uses much the same hardware as an ABS, but is programmed differently. An electronic control unit determines the slip ratio of each of the tyres individually. If it is noticed that the wheels are in danger of slipping, the EBD system modulates the amount of hydraulic fluid going to each wheel brake and so applies an appropriate amount of force to that brake to stop the wheel from skidding. This maximizes stopping power while maintaining driver control. A related system is electronic stability control (ESC), also sometimes called electronic stability programme (ESP), which prevents slippage when braking hard and turning at the same time. Without an ESC facility, the vehicle may over-steer or under-steer with consequent loss of control. An extra yaw sensor (a gyroscopic device that measures angular velocity around a vertical axis) detects the direction of the front wheel, i.e. the rotation of the vehicle, as it turns and compares the measurement with the angle of the steering wheel. If the car begins to under-steer, the inner rear brake is activated to increase the rotation of the vehicle. If the car begins to over-steer, the outer rear brake is activated to decrease the rotation. The torque created around the vehicle's vertical axis opposes the skid and brings the vehicle back in line with the driver's intended direction. Most new cars are now being equipped with this technology which, it is claimed, prevents many serious and fatal accidents.

When a vehicle's brakes are applied, the kinetic energy of motion is normally dissipated as heat. It is now possible, however, to recover some of this energy as electricity through the use of an electrical generator as a brake. This is useful for both hybrid electric and pure-electric vehicles where the recovered energy is fed to the traction battery and subsequently used to power the vehicle or for other electrical operations. This is known as 'regenerative braking' and is discussed further in Section 5.2.1, Chapter 5.

Heavy trucks have lots of momentum and therefore need to have particularly effective brakes to bring them to a halt. Air brakes are generally employed. An air compressor is powered by the engine and the compressed air is stored in a tank mounted on the chassis. When the vehicle is stationary, the brakes are clamped on and can only be released by air pressure. Many heavy trucks also utilize engine compression to retard the vehicle when descending a hill so as to save wear and tear on the brakes.

4.5 Exhaust systems and emissions

A car exhaust system has two main functions. First, it takes hot waste gases from the engine to a point where they can be released into the atmosphere without danger to the occupants of the vehicle. Second, it reduces the noise made when used gases are expelled by the engine. The latter function is achieved by means of a silencer that contains baffle plates.

The gases produced in an engine expand with great force and are released into the exhaust system under pressure. Each time the gases pass into the exhaust manifold, a shock wave is created. Since the shock waves occur at the rate of several thousand a minute, the noise from cars would be socially unacceptable if it were not moderated. The extent of such a nuisance is sometimes experienced with motorcycles or with cars that have broken exhaust systems. Initially, the shock wave travels at supersonic speed but, after a short distance down the exhaust pipe, it slows down somewhat to below the speed of sound. By the time exhaust gases leave the silencer, they have expanded sufficiently for their pressure to have fallen to near that of the outside air and most of the noise has been curbed.

Concern over exhaust gases (or 'emissions') in the atmosphere first arose in Los Angeles on account of the 'smog' that was caused by a particular set of geographical and climatic conditions. Later, there were fresh concerns over the emission of toxic lead compounds (derived from lead tetraethyl in petrol); sulfur dioxide, which causes acid rain (from sulfur compounds in the fuel); carbon monoxide (a poisonous gas); and, finally, carbon dioxide as a major contributor to climate change (see Section 2.7, Chapter 2). The problems of lead and sulfur were overcome by modifying the fuel composition. The release of carbon monoxide, NO_x and unburnt fuel was addressed through the development of catalytic converters for treating the exhaust from petrol engines.

4.5.1 Exhaust catalyst technology: petrol engines

In the late 1960s, the Johnson Matthey Company in the UK was developing platinum-group catalysts to extract gaseous pollutants from industrial effluents; for example, NO_x from the tail gas of nitric acid plants. It was found that the NO_x could be chemically reduced by methane (CH_4) over such a catalyst and that rhodium was particularly effective for this reaction. The company therefore directed its research effort to address the new problem of developing catalysts for the treatment of exhaust gases from petrol engines. Initially, two types of catalyst were employed. The engine was run slightly rich to suit the first catalyst, platinum–rhodium supported on a ceramic matrix, which under such fuel conditions reduced NO_x to nitrogen. Air was then introduced and a second catalyst (platinum–palladium) was employed to effect the oxidation of any unburnt hydrocarbons or carbon monoxide to carbon dioxide. It was essential that the first catalyst did not further reduce the nitrogen to ammonia (NH_3) as this would be oxidized back to NO_x over the second catalyst. These catalytic converters were introduced around 1975 and were fitted to cars in the USA and Japan. The design consisted of a ceramic honeycomb structure contained in a metallic housing, with the catalyst deposited on the internal walls of the honeycomb. The exhaust gases pass through the honeycomb where the catalyst promotes the oxidation of the pollutants.

Subsequent research at Johnson Matthey led to the development of the so-called 'three-way catalyst' to control simultaneously all three pollutants – hydrocarbons, carbon monoxide and NO_x. This major advance was facilitated by increasingly sophisticated fuel-management systems such as direct fuel injection. The problems to be overcome included not only pollution control under steady-state operation, but also the development of a catalyst that could achieve rapid heat-up on engine starting and maintain good stability for the life of the vehicle. Moreover, exhaust catalysts are rapidly poisoned by lead and this, in addition to the health hazard, was a further reason for the demise of lead additives in petrol. Today, more than 50 million cars per annum are fitted with exhaust catalysts. Meanwhile, to meet the ever-increasing demands of stringent environmental legislation, the research on catalysts is continuing.

4.5.2 Exhaust treatment: diesel engines

Diesel engines present more of a challenge. The combustion atmosphere is rich in both nitrogen and oxygen, and since diesels run hotter than petrol engines more NO_x is produced in the exhaust. The problem is being tackled, particularly for heavy vehicles, through the introduction of a system whereby aqueous urea is added to the exhaust gases; the additive reduces the NO_x to nitrogen over a catalyst. New catalyst technology is being evolved to meet the challenge of future strict legislation for heavy-duty diesel emissions. Already considerable progress has been made and the exhaust from urban buses, for example, is now far cleaner than 20 years ago. The environmental performance of light diesel engines, as used in cars and vans, has also been greatly improved, partly as a result of direct fuel injection.

Diesel engines can also emit copious quantities of black smoke (soot) that consists of finely-divided particles of carbon. This is the most serious pollutant from diesel engines and a cause of much environmental concern. To date, the most practical way to control the release of soot has been to trap the particles in a fine ceramic filter. The device must satisfy a number of requirements, which include straightforward and low-cost mass production, ability to sequester particles over a wide range of sizes (0.2–150 μm), durability and resistance to high temperatures – all without imposing a high back-pressure on the engine. To meet the overall specification, ceramic honeycombs are employed and are similar to the catalytic converters used on petrol engines. Johnson Matthey again has pioneered this field through the development of 'continuously regenerating trap' (CRT®) technology, which not only collects and burns the soot but also removes carbon monoxide and residual hydrocarbons. The appliance differs from the catalytic converter in that alternate channels are blocked off at opposite ends so that the exhaust gases are forced through the porous walls of the structure. The pores are sufficiently fine to capture the smallest of particles. Cordierite, a naturally occurring aluminosilicate, is the favoured ceramic, although silicon carbide has also been used. Modern diesel particulate filters (DPFs) can achieve particle-removal efficiencies that exceed 95%. Since 2009, legislation has required that all new diesel cars sold in the European Union must be equipped with DPFs.

Filters have to be regenerated periodically to burn off the deposited carbon and thereby ensure that they remain functional. The procedure may be performed either passively or actively. In the passive mode, the DPF is located as close as possible to the exhaust manifold where the gases are hot. At low engine speeds, soot is deposited on the filter, but at high speeds the exhaust gases are at their hottest and they serve to remove the carbon. This is possible because the exit gases from a diesel engine contain considerably more oxygen than those from a petrol engine. Sometimes, a precious metal catalyst is incorporated in the ceramic element to facilitate the combustion process. The active regeneration mode is used when geometrical constraints in the engine bay do not permit installation of the filter close to the exhaust manifold. One approach to active management is to increase the temperature of the exit gases periodically, either by injecting extra fuel after the exhaust stroke or by having a small fuel burner placed after the turbocharger. Another method is to introduce a catalytic fuel additive that reduces the temperature of soot combustion from above 600 °C to 350–450 °C. Sensors are fitted to measure the back-pressure and/or the temperature and then feed the data to the engine management computer. When the filter eventually becomes blocked and requires regeneration, the computer activates the addition of the catalyst to the fuel from a small auxiliary container. Alternatively, in a less automated version, there is a fascia light to warn the driver that the DPF needs attention. The consequences of failing to take the necessary remedial action can be severe – they range from high fuel consumption to damage of the DPF unit by overheating, or even a fire – and hence call for an expensive repair.

The clean-up of engine exhaust leaves carbon dioxide as the main environmental challenge that faces the automotive industry. Whereas the situation would obviously be ameliorated by the development of more efficient engines and vehicles to minimize fuel consumption, a complete solution requires the substitution of electricity or hydrogen for petroleum as the prime source of motive power and, even then, only if these two 'fuels' are derived from non-carbon origins, i.e. nuclear or renewable energy. Advances in electrically-propelled vehicles and the prospects for hydrogen-powered vehicles are treated in Chapter 5 and Chapter 8, respectively.

4.5.3 Exhaust systems

Exhaust systems comprise the exhaust manifold from the engine, the catalytic converter (for petrol engines), the DPF (for diesel engines), the silencer/muffler, and the tailpipe. Exhaust manifolds and silencers are normally made of mild-steel tubing and sheet, which corrode progressively under the action of hot gases, including water vapour. The silencer is generally the first part of the exhaust system to corrode through as this is towards the rear where it is cooler and where acid, formed from reaction of NO_x with steam, condenses. As soon as a hole develops in the silencer, the noise level increases dramatically. Corrosion of mild steel is inevitable and, conventionally, exhaust systems have been treated as 'consumables', along with fuel, oil, batteries and tyres. The life of a silencer is prolonged if the mild steel is aluminized. Some cars have been fitted with stainless-steel exhaust systems, but these are expensive and can still fail through breakdown of internal welds in the silencer. Exhausts now last much longer than formerly since sulfur has been removed from the fuel and the bulk of the NO_x from the exhaust.

4.6 Other key components

4.6.1 Wheels

Traditionally, road wheels are made of steel and comprise a stamped inner part welded to a rolled outer rim. The key advantages of steel wheels are a low manufacturing cost and high strength. Usually, the wheels last the life of the vehicle without much attention. The main disadvantages lie in their considerable weight and the amount of kinetic energy that they store when in motion; this energy has to be absorbed by the brakes.

Alloy wheels are lighter than steel ones and will therefore accelerate faster, stop more rapidly, and reduce the load on the suspension and the steering linkages. The wheels, which were first produced by casting a magnesium alloy, had the opposite properties to those of steel – namely, low mass, but high cost and relatively poor strength. Moreover, magnesium wheels were prone to corrosion and had one other serious limitation – in the event of a major puncture in which the tyre was ripped at speed, there was a chance that the wheel, as well as the tyre, would catch fire. Once magnesium ignites it is extremely difficult to extinguish. Thus magnesium wheels were used primarily for racing and sports cars before the casting and forging of aluminium alloys was perfected.

Early attempts at casting wheels from aluminium alloy were not successful as the resulting products were too brittle. Refinements in both alloy composition and casting technique in the 1960s overcame this problem and soon aluminium-alloy wheels went into mass production. They are now fitted to many cars, either as standard or as an optional extra. The attraction of alloy wheels is in part cosmetic, with many different designs possible, but technically their advantages lie in light weight and good thermal conductivity. The latter property facilitates the dissipation of energy in braking and thereby helps to prevent brakes from overheating, a factor which is important in high-speed vehicles. Disadvantages are a tendency to galvanic corrosion, which can result in disfigurement and in leakage of air from the tyre, as well as difficulty of repair when a wheel is bent out of shape. A minority of alloy wheels are forged rather than cast. These are lighter and stronger, but more expensive.

4.6.2 Tyres

Early tyres were made of solid rubber. In 1839, Charles Goodyear discovered that treating molten natural rubber latex with sulfur improved both its elasticity and its strength. This process is known as 'vulcanization'. Initially, the new material was used to form solid 'cushioning' tyres for horse-drawn carriages in place of the steel bands that, until that time, had been wrapped around wooden wheels made by the wheelwrights of old. The 'pneumatic' tyre was patented by R.W. Thompson in 1845 and took the form of several narrow inflated tubes packed around the wheel inside a leather casing. Unfortunately, the design proved to be impractical and it was not until 1888 that John Dunlop, a Scottish veterinary surgeon working in Belfast, produced the first pneumatic tyre to become commercially successful (see Section 1.1, Chapter 1). The tyre, which was designed for bicycles, used a single inner tube and was manufactured by the Dunlop Rubber Company from 1889 onwards. Two years later, the brothers Michelin in France developed a detachable pneumatic tyre that consisted of a separate tube with an outer cover bolted on to the wheel rim. It is remarkable that these three pioneers gave their names to three of the world's great tyre companies of the 20th century – Goodyear in the USA, Dunlop in the UK and Michelin in France.

Tyre development continued apace and, in 1915, the Palmer Tire Company of Detroit demonstrated the use of rubberized cord fabric to make 'cord tyres'. The cords were laid parallel to each other and incorporated into rubber sheets. These sheets were cut on a bias and placed on top of each other to build up the tyre casing. The resulting product became known as the 'cross-ply' tyre and it found regular application in automobiles until the 1970s. Steel cords, which were much stronger than the original fibre cords, were incorporated in the cross-ply tyres of heavy trucks from 1937 onwards and are still in use today.

Cross-ply tyres were not known for their longevity and a major breakthrough came in 1948 when Michelin introduced the 'radial tyre'. As the name suggests, steel or fabric bands are wrapped radially across the casing, rather than circumferentially. These tyres have sidewalls that are considerably more flexible than those of cross-plies and therefore offer much longer life. Whereas European and Japanese automotive companies soon adopted the radial design, US manufacturers did not follow suit until the 1970s. It is estimated that, worldwide, over one billion (10^9) tyres are now produced annually, of which the majority are radials.

The design of tyres to meet specific applications is a sophisticated business. There are two main components – namely, the casing that provides the strength to enable the tyre to retain its shape throughout its life, and the tread that provides the traction. Casings now make an airtight seal with the wheel rim and therefore it is no longer necessary for car tyres to have inner tubes, although heavy trucks and buses often still use them as an added safety precaution. Three important factors are: (i) grip on the road surface, (ii) rolling resistance, which contributes to fuel consumption, and (iii) rate of wear – all of which are determined largely by the chemical composition and structure of the tread. To an extent, these factors are mutually incompatible and compromises have to be made within a so-called 'performance triangle'. Tyre behaviour in wet weather is a further key consideration and this is mostly a function of the geometric design of the tread. Channels are formed in the tread to allow water to run away and so prevent the vehicle from 'aquaplaning'. In this connection, racing cars are fitted with different designs of tyres according to whether the conditions are dry or wet.

Modern tyres are designed to be as robust and hard-wearing as possible, consistent with combining a comfortable and safe ride with good manoeuvrability of the vehicle. The tyres are formed from natural or synthetic rubbers (with additives, including notably carbon black), fabric and wires. The nature and quantity of synthetic rubbers and other chemicals employed determine the degree to which the tread is either soft or hard. Soft rubber gives shorter stopping distances as the tread deforms more readily and provides greater contact with the road. On the other hand, soft tyres wear faster and have higher rolling-resistance. Drivers in countries with cold climates often change tyres between summer and winter. The soft winter tyres provide more traction on ice or snow. During summer months, a harder tyre is preferable as it is more durable.

In the UK, the legal lower limit for tread depth is 1.6 mm and it is mandatory for a tyre to be changed when it wears down to this level. A marker is incorporated in the material to indicate when this tread depth is reached. Tyres replaced at this stage, though no longer considered safe, still constitute a large amount of material and society faces a challenge in deciding how best to dispose of them. Whereas some tyres are shredded and incorporated into road surfaces, the majority have traditionally been sent to land-fill or burnt on open dumps. Given that the last two practices are not environmentally acceptable today, research into efficient and benign methods of tyre recycling is now in progress. One promising approach is pyrolysis (heating to a high temperature in a kiln in the absence of air), which produces combustible gas, a liquid fuel, steel and a carbonaceous char from which more carbon black can be made.

4.6.3 Lights

The first road vehicles had acetylene lamps or paraffin/kerosene lamps for use at night, but these were of limited utility, and were soon replaced by electric lighting when tungsten filament lamps became available. Initially, these lamps operated on a 6-V supply of direct current from the car battery. This power source serves many functions besides lighting, among which two of the most important are the supply of power to the starter motor and to the ignition system. As engines grew larger, a 6-V system was found to be inadequate and, by the 1940s, 12 V was adopted almost universally. On the other hand, 6-V batteries remained the preferred choice for many motorcycles.

The early incandescent headlamps were large, especially on luxury cars, but not particularly effective. The dazzling of on-coming drivers was a problem and so two-filament bulbs, one for high beam and one for low beam, were introduced along with a dip switch to enable the driver to select between beams. Early dip switches were foot-operated but later they became more conveniently located on a stalk attached to the steering column. Today, dipped-beam headlamps must meet regulations that ensure that the vision of on-coming drivers is not dangerously impaired.

In some countries, notably in Scandinavia, Canada and in some American States, it is a requirement or practice to drive with so-called 'daylight running' lights that are illuminated at all times, even in sunlight. Certain car manufacturers fit these as standard and arrange that they are switched on automatically when the engine is started.

Early vehicles usually had just one small red rear lamp, designed as much to illuminate the rear number plate as to warn drivers approaching from behind that there was a vehicle ahead. Even trucks had totally inadequate rear lighting – a complete contrast to the situation today, where often trucks are emblazoned with lights. Ever since the 1950s, it has been accepted that all vehicles should have at least two red rear lamps, one on each side of the body. There are also red brake lights (stop lights) of higher power that are often incorporated within the same housing as the rear lamps. Regulations require a minimum contrast in intensity so that brake lights are not confused with rear lights. Until quite recently, incandescent filament lamps were employed for both purposes but, with the advent of light-emitting diodes (LEDs) as an alternative to filament lamps, it has become standard practice to fit a third red LED brake light, which is mounted centrally and placed considerably higher than the other brake lights. This is thought to reduce the incidence of rear-end collisions, in part because LEDs light up fractionally faster (by about 0.2 s) than filament lamps when the brake pedal is pressed. Many cars also have a powerful fog lamp at the rear with which the driver can warn following drivers when visibility is low. These lamps are also red but they are often single lamps on the offside of the car and thus should not be confused with the configuration of normal rear or brake lights. Unfortunately, in Europe, it is common to find twin rear fog lamps and therefore it is just possible that they may distract attention from the brake lights.

Most new production vehicles are equipped with twin white lamps for assistance when reversing (reversing or back-up lights), which are activated immediately the driver selects reverse gear. Trucks, additionally, have an audible warning when they are reversing to alert pedestrians, and this feature is increasingly becoming commonplace with cars. There is also a requirement to have a white light (or two) for illumination of the rear number plate.

At first, road vehicles were not furnished with direction indicators (turn signals) and it was therefore the responsibility of the driver to give hand signals. In the UK, small retractable semaphore arms ('trafficators') were introduced in the 1930s to indicate the driver's intention to turn the vehicle; these devices operated with a low-wattage bulb and often were barely visible. Bright flashing direction indicators were first introduced in the USA in the 1940s, with red lights at the rear. Later, amber-coloured indicators became almost universal, except in North America, where rear indicators with red lights are still permitted. Regulations often specify the rate of flashing and also require a fascia warning light, together with an audible clicking sound, to alert the driver to the fact that the signals are operating. The signals are self-cancelling after the turn has been completed. Vehicles must also be equipped with hazard warning lamps, although this requirement is normally satisfied by all the direction indicators acting together when the driver presses the hazard lamp button.

In summary, the modern European passenger car has at the rear a minimum of six red lights (two rear, three brake, one fog), two white lights (reversing), one or two white number-plate lights, and two amber lights (indicators). The industry has come a long way since the days of the solitary acetylene headlamp!

The tungsten–halogen headlamp bulb (also known as 'quartz halogen') provided a major breakthrough in vehicle lighting. First introduced into Europe in 1962, it is now widely employed. A small quantity of a halogen, generally iodine, is introduced into the lamp. The use of a quartz or hard glass enclosure permits the lamp to be operated at a higher temperature than a traditional bulb and consequently it gives a greater output of light. If a conventional lamp were operated at this temperature, it would soon blacken due to the evaporation of filament tungsten and its deposition on the glass envelope. The halogen serves to remove this deposit chemically and to re-deposit it on the hot filament, thereby enhancing luminous intensity and sustaining longer life.

A later development was the xenon high-intensity discharge lamp, used for headlamps on many new cars. The most recent lighting innovation – the LED – is starting to be utilized in headlamps as well as for the high-level brake light discussed earlier. Buses and trucks, in particular, are equipped with such lighting. The attraction of LEDs, in addition to their fast rise time when switched on, is their high efficiency (i.e. lumens emitted per watt).

Undoubtedly, there are still further advances to be made in vehicle lighting. One proposal is to employ fibre optics to distribute light to various locations around the vehicle from a single central source. This strategy should result in reduced manufacturing costs and less drain on the battery. Another option is variable-intensity direction indicator lighting that is enhanced by day so as to improve visibility and reduced at night to avoid glare.

4.7 Safety

As mentioned earlier, the monocoque construction of the modern car is intrinsically safer than a vehicle with a rigid chassis. This is due, at least in part, to the incorporation of a 'crumple zone' which collapses in a collision and absorbs some of the energy of impact without transferring it directly to the vehicle occupants.

Seat belts, to restrain vehicle occupants from violent movement during a crash, were first introduced by the Nash Motors Company in the USA in 1950 and soon taken up by other car manufacturers. Initially, these safety belts were confined to the front seats and were simply lap belts, as are used in aircraft and some buses, but their limitations were soon realized and full shoulder harnesses, with three-point fixing, were adopted. Today, most cars have shoulder-type seat belts for all occupants. Modern belts are retractable and cleverly designed so that they may be reeled out if pulled gently, thus allowing some freedom of movement to the seat occupant, but they are instantly locked in position if jerked suddenly, as in a collision. Seat belts, together with laminated-glass windscreens, are known to have saved thousands of lives and in many countries the wearing of belts is now compulsory. The world's first legislation was put in place in 1970 in the state of Victoria, Australia, whereby the wearing of a seat belt became mandatory for drivers and front-seat passengers.

Other safety features to be introduced in the past few decades include recessed steering wheels, reinforced roofs, and roll bars for convertibles that have no rigid roof. Head restraints fitted to all seats are designed to limit whiplash injuries to the neck in rear-end collisions. An important safety feature on some cars is the provision of rear doors that can be arranged to open freely from the outside while remaining locked on the inside. This prevents child passengers from opening the doors when the car is moving. Nowadays, virtually all side doors are hinged at the front, so that they tend to stay closed when the car is in forward motion.

The inflatable airbag was first employed in the 1980s to protect the driver from being impaled on the steering column in a front-end collision. The bag is neatly stowed in the centre of the steering wheel. In the event of a sudden deceleration, as in a crash, the airbag inflates violently, in a fraction of a second, in front of the driver's chest and face, and thereby should reduce the risk of serious injury. Later, a second unit was installed in the fascia to prevent the front-seat passenger hitting his/her head against the windscreen. Subsequently, further airbags have been provided in the sides of the front seats to provide protection during a side-impact collision. In some cars, a similar facility has now been included for the benefit of rear-seat passengers. The latest innovations are curtain airbags for side-impacts, deployed from the roof, and even airbags incorporated in the seatbelts. Airbags are only intended to inflate in moderate-to-severe crashes and thus they supplement seatbelts, which provide basic protection in lesser collisions.

The mode of operation of airbags is of interest. A small accelerometer records the initiation of the collision. The instrument contains a mechanical component that moves to change the capacitive element of an electronic circuit and this perturbation is detected by the electronics on an integrated chip. An electrical current is immediately sent to a pyrotechnic device that, in turn, actuates the airbag. The device rapidly heats a chemical compound that decomposes explosively with liberation of nitrogen gas to inflate the bag. The overall time for this operation is 60–80 ms from the first moment of vehicle contact.

Recently, the emphasis in vehicle safety has moved from mitigating the consequences of an accident to preventing the accident from happening at all. Developments in electronic braking control, tyres and in improved lighting (*v.s.*) all contribute to this goal.

4.8 Accessories

4.8.1 Driving aids

Some driving aids, such as rear windows with electrical heating wires incorporated in them for demisting/defrosting to maintain rear-view visibility during adverse weather conditions, are quite basic and almost universal. External driving mirrors, originally sited on the wings of the car, are now invariably attached to the door frames. They may also have heating elements for de-frosting. Most cars have a sensor to measure the external temperature and a display to inform the driver, with an 'ice alert' when the temperature approaches freezing.

The driver's seat may be adjusted, not just back and forth to suit leg length, but also up and down to accommodate torso height. The back of the seat is designed to provide lumbar support (vital for drivers with back problems) and its rake may be adjusted to suit preference.

A great convenience for long-distance driving on motorways is cruise control, which permits the driver to set the speed at which he/she wishes to travel and then remove his/her foot from the accelerator pedal. Immediately the brake pedal is touched, the cruise control is disabled and the car reverts to manual control. Fuel economy is helped by the introduction of the 'eco-meter', which gives a digital display of instantaneous fuel consumption in real time. This encourages the driver to adjust the mode of driving to conserve fuel.

A recent innovation is the introduction of parking sensors (proximity sensors) which provide audible information on the distance between the car and another vehicle (or solid object), either in front or behind. The sensors usually operate by means of ultrasonics. The rear sensor is activated when the driver selects reverse gear; the front sensor is activated manually via a switch. The car emits an ultrasonic beam that is reflected from the solid object back to a receiver. Since the speed of sound is known, the transit time can be used to calculate a distance measurement that, in turn, is converted to an audible signal, the pulse-frequency of which rises as the distance between the two objects decreases. Proximity sensors are particularly useful in tight parking situations. There is sometimes also a rear-view camera so that the driver can actually see the low-level situation behind before reversing. These sensors have saved many minor collisions.

Another feature available on some cars is a detector that gives audible warning if the driver strays from the lane in which he/she is driving. This is an important 'wake-up call' for drivers who are becoming drowsy.

Many vehicles are now equipped with intruder alarms although they are not universally regarded as a blessing because of the incidence of false alarms triggered by high winds or other innocent causes. Another 'driving aid', but one which is not legal in all jurisdictions, gives advance warning of speed cameras installed on the roadside, or even of mobile police speed traps.

One of the most remarkable driving aids is satellite navigation ('sat nav'), which was introduced around 2000. The functioning of the device is totally dependent on the ability to determine accurately the position of an object on the earth's surface with respect to satellites in known orbits. There are several international satellite systems that can be employed; the oldest and most widely used is the United States Global Positioning System (GPS). This consists of 24 satellites that orbit the earth at a height of ~ 20 000 km (12 500 miles). Each satellite transmits a characteristic signal from which its precise position can be determined together with the time at which the signal was sent. From the timing delay in receiving the signal, the sat nav calculates its exact distance from each of several satellites. By a process of triangulation, it is then possible to compute the location of the car on earth. The result is superimposed on a detailed ground map to show where the vehicle is situated, with an accuracy level of within about 50 m. A driver travelling from point A to destination B can simply input this information to the sat nav, which then calculates the desired route (most direct or quickest) and displays this on an electronic map as the journey proceeds. Optionally, a voice-over issues oral instructions during the approach to junctions or turns. The sat nav system is particularly valuable for journeys in unfamiliar territory at night or in adverse weather conditions.

4.8.2 Convenience

Minor convenience items that are often found in modern cars include: lockable glove boxes and other storage facilities, lockable fuel filler caps, cup-holders for drinks, courtesy lights activated by unlocking and locking doors, instrument lights that can be dimmed or brightened, map-reading lights, etc. Electrically powered front windows are now a standard feature and most cars also have them in the rear. Where the rear windows are powered, the driver has over-riding control. There is a separate electric motor for each window sited in the door panels.

The central computer in a modern car fulfils many functions. As well as its essential role in engine management, braking, etc., it also provides useful information on vehicle performance. This includes calculating and displaying the fuel consumption, both instantaneously and integrated over a journey, and this allows adjustment of the driving mode to maximize economy.

Central locking is much appreciated and may be either key-operated on one door, or remotely activated via a radio frequency signal from the key fob, which contains a battery and a small radio transmitter. All doors are locked simultaneously, including that of the boot. Another option is a proximity device carried by the driver that automatically unlocks or locks the door as he/she approaches or leaves the vehicle. Another useful feature is the facility to lock all the doors from inside when travelling in urban or lawless areas if there is the possibility of intrusion. Automatic garage doors that are activated to open by a wireless beam as the car approaches are also welcome, especially in wet or freezing weather.

Hatchback cars have a wide-opening rear door, which allows ease of storage of goods in the boot. Folding away the rear seats to create a much larger platform for freight is of further benefit. Even better is the ability to lower horizontally either two-thirds or one-third of the seats, thereby making for flexibility in carrying goods and passengers together.

4.8.3 Comfort

The most important item in this category, after seating, is climate control. The earliest cars had no provision for heating. Later, in the 1940s, cars had primitive cabin heating with little temperature control that was supplied by simply directing hot water from the engine through a radiator in the footwell and distributing the heat by means of a blower. Modern cars have air-conditioning (climate control) that is much more sophisticated and provides both heating and cooling, as well as humidity control. The heated/cooled/dehumidified air is distributed throughout the cabin by blowers to multiple vents that can be opened or shut. Heated air derives, as before, from the circulating engine coolant, while cooled air is produced by a dedicated refrigerator that is usually belt-driven by the engine. The driver has control over the cabin temperature as well as the ability to direct the air stream to specific vents, such as the windscreen for demisting internally or defrosting externally. There are often air vents for passengers in the rear seats. The most up-to-date systems use temperature sensors to monitor the cabin air and the car's computer automatically controls the temperature at a selected set point.

Another comfort feature is the provision of electrically-heated seats. These are particularly appreciated on first setting out on a cold day before the engine coolant has warmed up. Some cars also have seats that are pre-programmed to adjust to the optimum position for each of the regular drivers.

4.8.4 Entertainment and communications

Cars have long been fitted with analogue radios with two or four speakers, which were often located in the door panels and/or behind the rear seats. The radios were frequently supplemented by provision for playing tapes. Nowadays, however, tape systems have overwhelmingly been replaced by compact disc (CD) players, which can be built into the instrument panel or may be located in the boot of the car with provision for changing the disc and track remotely.

Radio reception can be erratic as the car travels around the country and moves from one radio zone to another. For this reason, car FM radios are equipped with the Radio Data System (RDS), which automatically seeks out and then focuses on the nearest antenna broadcasting a particular station to ensure best reception. Provision is also made for interrupting radio programmes briefly to receive important reports of weather conditions, road accidents and zones of high traffic density. Analogue radios are now largely being replaced by digital radios. Many new cars are fitted with satellite radios so that signal fade is no longer a problem.

Mobile (cell) phones have become ubiquitous. In many countries, it is an offence to use a hand-held phone while driving, but hands-free phones may be permitted. Thus, cars now incorporate provision for hands-free phones that includes automatic dialing, a microphone and a loudspeaker. There may also be a special receptacle in the central console between the driver and front passenger to hold the equipment.

4.9 The future for internal-combustion-engined vehicles

In summary, this chapter has outlined the many advances in automotive technology that have been made in recent years. The developments encompass improvements in vehicle design, engine efficiency, fuel economy, environmental clean-up, and safety, as well as comfort and convenience for both the driver and the passengers. The reliability of vehicles is now far greater than it was 20 or 30 years ago, despite their increased sophistication and complexity. In-built computer systems are at the centre of all aspects of vehicle operation. Thirty years ago, it was generally possible for car owners with mechanical and electrical competence to diagnose and repair faults on their own vehicles. This is no longer possible. Specialized garage equipment is required to determine faults, the resolution of which, more often than not, involves replacement rather than repair. On the other hand, the improved reliability of vehicles means that visits to the repair shop are far less frequent than formerly.

The focus of current vehicle development, under the stimulus of environmental legislation, is on improving engine and transmission efficiency in order to reduce greenhouse gas emissions, fuel consumption and pollution. Significant progress has been made, and is still being made in this direction. In the USA, the average sales-weighted fuel economy of new light-duty vehicles (cars and light goods vans) rose from 13.1 miles per US gallon (18 L per 100 km) in 1974 to 22.4 mpg (10.5 L per 100 km) in 2009. Advances in fuel efficiency during the past couple of decades have been achieved through the introduction of multi-valve cylinders, sometimes with variable valve timing, direct-injection engines, continuously variable transmissions, cylinder deactivation, turbochargers and hybrid electric vehicles. Meanwhile, advances in pollution control have been made through catalytic converters for petrol engines and the control of NO_x and particulate emissions from diesel engines. At best, however, as the world's road transport fleet grows inexorably, these developments are only an interim solution to the long-term global problems of petroleum depletion and carbon dioxide emissions. The next engineering phase, now under way, is that of electrifying the power-train, as discussed in Chapter 5.

The development of ICEVs over the past several decades has been impressive in all aspects, but there is some concern that the introduction of ever more sophisticated technology to improve vehicle performance, limit emissions and fuel consumption, and enhance safety may result in unacceptably high costs for the motorist. To satisfy the potentially huge markets in Asia and Africa, the production of a relatively simple and inexpensive car (such as the Tata *Nano*) may represent the way forward.

Another area that must be addressed is the harmonization of national regulations for the construction of vehicles. This is important in a global industry. To quote the International Organization of Motor Vehicle Manufacturers (www.oica.net):

Automakers are faced with a wide variety of different regulations in different countries, often aimed at achieving the same purpose, but differing for historical reasons. Harmonizing these regulations world-wide offers savings in technical resources, which can be applied elsewhere to produce better, cleaner, safer vehicles and it offers the possibility of reducing production complexity, resulting in lower costs and prices and a wider choice of vehicles available to all consumers.

Other areas where there is scope for harmonization across the world are: (i) drive cycles (urban, extra-urban and combined) for determining fuel consumption and (ii) vehicle emission standards together with test methods for measuring emissions. These are often key issues for customers who need to be confident in the performance specifications advertised for vehicles. In many countries, emission levels determine taxes charged and fuel consumption has a direct impact on the cost of driving. The responsibility for developing 'global technical regulations' lies with the expert working groups of the United Nations. Among the areas that are currently being addressed are safety, braking, controls and displays, and emissions.

What is the future for conventional road transport? In the near term, it seems certain that ICEVs will continue to play an important role in the vehicle mix. The dual trend is likely to be towards cars with improved fuel economy and with increasing use of alternative fuels (see Chapter 3). At the same time, the automobile industry has to meet the regulations for fuel economy and emissions and to satisfy the customer as regards design, performance and cost.

The proliferation of 'gadgets' based on electronics and computers or micro-processors, both for vehicle control and for accessories, is likely to continue, particularly if there are further basic advances in automation, in lighting and in displays. The many electrical motors and other power-hungry devices now incorporated in cars are making increasing demands on the battery. High currents dictate harnesses of thicker-gauge copper wiring that are expensive and heavy. For this reason there has been interest in moving to a higher-voltage system. At present, almost all vehicles have 12-V batteries and 14-V alternators for charging, but a move to 36-V or 48-V batteries is currently under investigation in Europe, and a German standard has been prepared to cover this possibility. The introduction of 'stop–start' technology to save fuel makes different, and conflicting, demands on the battery (see Section 5.2.1, Chapter 5), while the advent of lithium-ion batteries (with 3-V cells) in the marketplace is a further complicating factor.

With all these activities ongoing – technical advances in vehicle design and engineering, manufacturers striving to meet government targets for reduced fuel consumption and emissions, new global technical regulations being introduced, the development of autonomous cars – a challenging future is foreseen for ICEVs.

The global population of registered road vehicles exceeded one billion in 2010 and the era of 'two billion cars' has already been foreseen. It is doubtful whether world resources can sustain this number of road vehicles if they are all ICEVs. Engineering can go only so far in improving efficiency and reducing fuel consumption and emissions. Beyond that, a radical change in lifestyles, in societal infrastructures and in road transport modes will be needed. These topics are reviewed in the final chapter of this book.

Progressive Electrification of Road Vehicles

5

5.1 Electricity to the rescue

For most of the 20th century, road vehicles were powered almost exclusively by internal-combustion engines (ICEs) that were fuelled by petroleum products. More recently, concerns have arisen over the future availability of these fuels and the emissions that they produce. It was anticipated that fuel supply could be interrupted by political events, such as took place in the Middle East during 1973–1974, or could be curtailed as demand soars and world oil reserves are depleted. Already, the price of crude oil has escalated from a few US dollars a barrel in the 1970s to around 100 US dollars today. These and other 'drivers for change' are described in Chapter 2.

With such uneasiness escalating and becoming universal, there has arisen considerable enthusiasm for developing alternative means of vehicle propulsion that will reduce, or better still eliminate, the present overwhelming dependence on oil. Although, at quite an early stage (1989), the Audi company gave some thought to the introduction of hybrid electric vehicles (HEVs) as a possible solution, the main effort across the industry was initially devoted to vehicles with only an electric motor powered by onboard energy from a rechargeable battery that was replenished from the mains supply – so-called 'battery electric vehicles' (BEVs). Since the ICE was eliminated, such vehicles offered an ideal technology for coping with the problems of decreasing oil availability, urban pollution and greenhouse gas emissions.

The battery was clearly the key to improving the performance of electric road vehicles to a level that would be commercially viable. Obviously, the driving range between battery recharges is determined by the amount of electrical energy that is stored on board. For example, depending on the size and weight of the vehicle, about 20 to 30 kWh would sustain travel for around 100 miles (160 km). High power would be equally essential to deliver acceptable acceleration and hill-climbing ability. Moreover, since there would be limitations on the current that could be passed around the vehicle, a high-voltage multi-cell battery would have to be employed. Given that potential purchasers of electric vehicles were accustomed to re-fuelling their ICE cars within about two minutes, a lengthy time for recharging the battery (hours) after driving only 100 miles would prove very unattractive. Lead–acid and nickel–cadmium batteries (both with liquid electrolyte) were employed by most developers during the earliest exploration of BEVs for widespread public use. Unfortunately, however, 25-kWh batteries of either of these chemistries occupied large volumes and thereby limited the space available for passengers and luggage. Moreover, the batteries were heavy so that a significant fraction of the stored energy was wastefully expended in their conveyance – a self-defeating outcome. A further problem was the restricted temperature range over which the batteries could operate satisfactorily; they were not ideally suited to either very hot or very cold climates. For BEVs to become attractive to a discerning public, their batteries must also be safe and long-lasting. Lead–acid and nickel–cadmium batteries both present minimal safety hazard but their service life in BEVs is viewed as less than ideal.

A large effort was launched to find a better battery system – namely, one that would provide appreciably more energy with reduced penalties in terms of weight and volume, be capable of very rapid recharge, and have an adequate service life. The periodic table was scoured and new battery chemistries were conceived, notably: nickel–metal-hydride, sodium–sulfur, sodium–nickel-chloride and lithium-ion (see Chapter 7). Even though the search continues today, there appears to be little prospect in the near term of finding a battery system that can provide a driving range between charges of much more than 150 miles (240 km), withstand rapid recharging for a satisfactory life and be manufactured at a competitive cost.

The efforts of vehicle designers to reduce fuel consumption and emissions can, in general, only succeed if the options that they offer amount to a value proposition that is agreeable to the purchaser. There is a trade-off between capital cost and running costs (including fuel) that each prospective buyer has to evaluate individually. Studies have indicated that the majority of automobile owners will be prepared to accept a vehicle with improved fuel economy, but only if the purchase price is competitive with that of a conventional vehicle. One exception to the maxim that 'cost is everything' has been in the area of public 'transit' buses. For instance, some authorities have been willing to subsidize up to 90% of initial cost so that fleet managers are able to opt for vehicles with expensive batteries without exceeding their budgets. In all other cases, the normal rule of the automotive industry prevails and manufacturers will strive to take every opportunity to reduce cost.

Given the limitations of BEVs and doubts about developing an entirely new and affordable battery with satisfactory performance and safety, attention turned to hybrid vehicles. In principle, any vehicle that has more than one power source might be classified as a 'hybrid'. This term therefore encompasses 'all-electric' vehicles that employ two different types of battery (one for stored energy, the other to meet peak power demands), or a fuel cell and a battery, or a battery and a supercapacitor. The designation also includes electric vehicles in which the main battery-powered drive is supplemented in peak-load conditions by taking energy stored in a flywheel or in a hydraulic accumulator. To date, however, the only hybrid type that has been seriously developed for road transport is one in which a battery-powered motor is combined with a heat engine to provide high-efficiency drive without 'range anxiety'. Thus the industry has returned to the concept of the HEV, which had been first conceived in Germany by Porsche in 1901 and, as mentioned above, was re-launched by Audi in 1989.

Initially, there was some doubt over the practicability of HEVs, on the grounds that their complexity would lead to unacceptable costs. In the end, however, sales of the Honda *Insight* and especially those of the Toyota *Prius*, which were the first HEVs to enter the market, far exceeded the expectations of many commentators. Credit for the success of such a radical change must be given to the vision of these two Japanese companies and to their engineers who produced designs with less complexity than had been feared and therefore avoided serious increases in manufacturing costs. One particularly elegant development was the integrated starter–generator (ISG, sometimes referred to as the integrated starter–alternator). This device combines two electrical machines into one unit that is used both to start the ICE and to channel energy recovered from braking back into the battery. The latter function is commonly referred to as 'regenerative braking' (*v.i.*).

Hybrids do not eliminate emissions but they do reduce them by making use of the hydrocarbon fuel more efficiently and, crucially, the vehicles do not suffer from the range limitations that beset BEVs. Hybrids are also in tune with the passage of emissions legislation because there is a full range of designs from the simplest form, which produces the least reduction in emissions, through to the most comprehensive, which yields the most benefit. Thus it is possible to introduce vehicles with increasing degrees of sophistication and accompanying cost/benefit improvements, as and when required.

The three general categories of HEV power-train that have been considered – series, parallel and power-split (or mixed) hybrids – are shown schematically in Figure 5.1 and described further in Section 5.2.2.1 below. The incorporation of an ISG compensates to some extent for the added complexity which results from an increased degree of electrical function in hybrid designs. In the parallel hybrid, the ICE is the main propulsion unit and the electric motor operates in parallel to provide up to 30 kW of electrical boost power in addition to the ISG function. By contrast, the series hybrid employs an electric motor to deliver all of the motive power, up to 50 kW, with a small ICE to recharge the battery, when required, and so extend the vehicle range. The power-split hybrid is a blend of series and parallel arrangements as employed in the Toyota *Prius*.

This chapter covers the several types of advanced vehicle that are intended, through intelligent use of electrification, to improve fuel economy and to decrease gaseous emissions. The various options are examined in order of increased electrical function, namely:

* stop–start vehicles, SSVs (often, incorrectly, known as 'micro-hybrids');
* mild/medium hybrids and full hybrids (defined below);
* plug-in hybrids, PHEVs;
* extended-range electric vehicles, E-REVs;
* battery electric vehicles, BEVs.

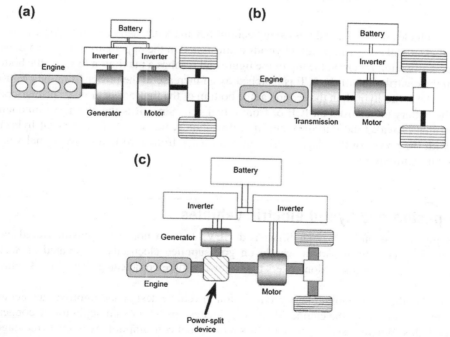

FIGURE 5.1

Alternative configurations of HEV power-train: (a) series hybrid; (b) parallel hybrid; (c) power-split hybrid. The components that are labelled 'inverter' have to operate bi-directionally, i.e. (i) as inverters, during the drive mode; (ii) as controlled rectifiers during the recovery of regenerative braking energy. Although the units provide bi-directional functionality, they are often just called 'inverters' in the technical literature

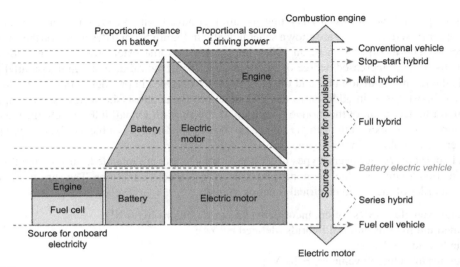

FIGURE 5.2

Types and configurations of hybrid vehicles

Fuel cell vehicles (FCVs) are addressed in Sections 8.8 and 8.9, Chapter 8. The progressive increase in the use of electrical energy in parallel hybrids is illustrated in the upper part of Figure 5.2, where the degree of electrification is zero at the top of the figure and reaches a maximum towards the bottom. The decrease in the contribution of the ICE is matched by an increase in the power from the electric motor and battery. The series hybrid, in the section at the bottom of the figure, can draw its primary power for charging the battery either from an ICE or from a fuel cell. Since it is the electric component of the power-train that provides the improvement in fuel economy, the so-called 'degree of hybridization' increases with the power of the electric drive that is used – from zero in a conventional vehicle to a maximum with a full hybrid.

5.2 Stop–start and hybrid electric vehicles

The following discussion deals with SSVs and HEVs that do not use mains-delivered electricity. Electric vehicles that rely in part or wholly on grid-supplied electricity are treated in Section 5.3. Consequently, the present discussion covers the following three broad categories of road vehicle:

- SSVs involve the least revolutionary modification to vehicle design and improve fuel economy simply by shutting down the engine when the vehicle stops – for example, in traffic congestion or at traffic lights. Whereas such a strategy does reduce fuel consumption by a few percentage points, the new operating schedule requires far more engine-start events – a duty that has an adverse impact on the battery life, and poses challenges for the continued powering of electrical components that must remain active when the vehicle is stationary. Some vehicles that operate with this stop–start system may also make use of the added facility for channelling energy captured by regenerative braking back into the battery via an ISG. At present, the batteries in SSVs are rated at 12 V but, in future, modules of 48 V may be used. The stop–start principle was initially introduced in cars with manual transmission but some automatics are now equipped with the system. To date, the majority of SSVs have been produced and sold in Europe.

- Medium hybrids (sometimes referred to as 'mild hybrids') draw some power from the electric motor to assist the ICE during vehicle launch and for overtaking. They also recover more energy from regenerative braking than is possible in SSVs. All of this additional capability requires higher levels of power in the electrical system and this, in turn, calls for battery packs of higher voltage (generally in the range 100 to 150 V) if currents through the vehicle wiring are to be kept within reasonable bounds.
- In full hybrids, the capability to provide a modest amount of all-electric driving range is added to the functions that are available in medium hybrids. Since such operation demands even higher power levels, full hybrids generally employ battery packs with voltages above 200 V.

5.2.1 Stop–start vehicles

Although medium and full hybrids were the first HEVs to reach commercialization, sales of such vehicles have been rapidly overtaken by those of SSVs that, as mentioned above, have principally taken place in Europe. Within five years, the number of SSVs purchased had reached almost three million, largely because automotive manufacturers see this type of vehicle as the least expensive way to achieve the fleet-wide emissions targets that will come into force in the European Union in 2015 with heavy penalties for non-compliance; see Section 2.5.1, Chapter 2. The power system of a representative SSV is depicted in Figure 5.3. The starter motor and the alternator are shown separately but, as discussed earlier, they may be combined into a single ISG unit, which is either belt-driven (B-ISG) or crankshaft-mounted (C-ISG) when regenerative braking is to be used. The latter arrangement is capable of handling higher power levels than a belt-driven device.

To achieve reductions in both emissions and fuel consumption, SSVs are designed to draw on one or more of four distinct functions, as follows.

- *Stop–start.* This function switches off the ICE automatically when the car becomes stationary. For example, in the BMW *EfficientDynamics* system, shown in Figure 5.3, the action is triggered by the driver stopping the car, engaging neutral and releasing the clutch. The engine restarts when the driver depresses the clutch once more.

  The decreased use of the engine manifestly saves fuel and lowers emissions. Nevertheless, for the system to be acceptable to the driver, the high number of extra engine starts must be accomplished reliably, with traction power available immediately after switch-on, and without excessive noise or vibration. The restart brings with it a temporary increase in fuel consumption that reduces the benefit of the engine-off period, but there is a net improvement in terms of fuel economy provided that the off period lasts for 20 seconds or more.

  During the time that the vehicle is stationary, all of the onboard electrical devices are provided with power from the battery. Obviously, the engine restarts that follow the automatic stops also result in a significantly increased number of high-rate load events. Together, these features represent a far more demanding performance requirement than that for conventional automotive batteries, i.e. an ability to withstand a greater number of cycles and deeper cycling, as well as a

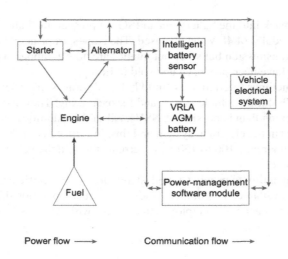

Power flow ⟶   **Communication flow** ⟶

FIGURE 5.3

Schematic of the BMW SSV electrical power system (VRLA, valve-regulated lead–acid; AGM, absorptive glass mat)

(Adapted from S. Schaek, A.O. Stoermer and E. Hockgeiger, 2009, J. Power Sources, 190: 173–183, Elsevier copyright)

high charge-acceptance capability to enable the recovery of a desirable state-of-charge (SoC) after the end of each stationary period.

Tests performed under the New European Drive Cycle (NEDC) have demonstrated that SSVs without regenerative braking operate with the lead–acid battery at a SoC of 90% or above and exhibit an average fuel saving of around 6% in comparison with vehicles that employ none of the SSV features. Clearly, the actual savings depend on the level of traffic congestion and generally will be greater in city driving than on the motorway.

- *Regenerative braking ('regen').* When brakes are applied to car wheels, a considerable amount of energy is wasted as heat on the friction surfaces of the components. Some of this mechanical energy can be transformed to electrical energy and stored in the battery by allowing the moment of inertia of the vehicle to drive the rotation of the alternator. The current from the alternator is controlled to a suitable output voltage. At times when the current exceeds the demand of the electrical consumers, there is surplus electrical energy that can raise the SoC of the battery. Given that the charge-acceptance is very low near top-of-charge, batteries in systems that are designed to incorporate regenerative braking are intentionally operated at a partial state-of-charge (PSoC), often within a window of 60 to 80%. Onboard electronic systems are set to maintain the SoC within the desired range. If the top of this range is exceeded, then the regenerative braking system is temporarily disabled.

Most SSVs employ lead–acid batteries as the least expensive option, even when regenerative braking is deployed. The use of lead–acid under PSoC conditions causes a problem in that lead sulfate (the product of the discharge reaction; see Section 7.4.1, Chapter 7) tends to accumulate on the negative plate. This so-called 'sulfation' leads to loss of power and early battery failure. The BMW *EfficientDynamics* system ameliorates this debilitating effect by periodically providing the battery with a 'refreshing charge' in the form of an excursion to top-of-charge and back down to the operational PSoC. The amount of benefit that can be gleaned from regenerative braking depends on the power level that is achievable with the electromechanical components that are available. For SSVs that have a typical generator capacity of up to 4 kW and operate with a 12-V battery, the amount of regenerative-braking energy that can be recovered is limited. The higher battery voltages in medium and full hybrids bring an advantage in this respect. Vehicles that are able to recover a good proportion of the energy are reported to achieve fuel economies of up to 12% as measured under the NEDC.

- *Passive boost.* In this function, the alternator is de-energized while the car is accelerating. Consequently, more of the engine's power becomes available for propulsion. On the other hand, an additional demand is placed on the battery, which must now supply the load requirements of the electrical system while the alternator is 'out of commission'. Such operation exposes the battery to a greater level of high-rate cycling.
- *Charge-voltage control.* This feature is implemented to minimize the amount of energy needed for recharging the battery. Overcharging (and the inefficiency that it introduces) is avoided and the battery is not charged at times when the efficiency of the engine is low, such as during idling.

5.2.1.1 Battery issues

The SSV functions listed above clearly place greater demands on batteries. In addition to the traditional starting–lighting–ignition (SLI) duties, the batteries must fulfil other requirements, namely: withstand more and deeper discharges due to stop phases and, where implemented, passive boost; operate at a lower SoC to allow for successful regeneration of braking energy; sustain very high in-rush currents. At present, SSVs still incorporate 12-V lead–acid batteries, but these differ in detail as determined by which one of the above-mentioned economy-boosting functions is installed.

In the simplest version of an SSV, i.e. one that employs only the stop–start function, it is still possible that flooded lead–acid batteries may be used (see Section 7.4, Chapter 7), but their life in duty will be short unless design modifications are incorporated. So-called 'enhanced flooded batteries' (EFBs) have been proposed that, in comparison with standard flooded counterparts, can achieve somewhat higher cycle-lives and are better able to withstand the problems associated with the development of local changes in acid strength within the electrolyte (a phenomenon known as 'acid-stratification'). The improvements have been achieved by means of more robust 'active masses' (i.e. the chemicals involved in the charge–discharge reactions) and by reducing the mobility of the electrolyte.

Enhanced flooded batteries also offer the advantages of minimal increased cost and an ability to withstand the elevated temperature of the vehicle engine compartment. To date, however, valve-regulated lead–acid (VRLA) batteries which employ absorptive glass mat (AGM) separators have been preferred when the SSV also incorporates regenerative braking; see Section 7.4, Chapter 7. Although these are more expensive than flooded batteries, the VRLA design offers the considerable advantage that it does not suffer from acid stratification to any significant degree.

Projections for the anticipated sales worldwide of emerging vehicle technologies are given in Figure 5.4. The analysis indicates that SSVs will dominate for the next several years, whereas the sum total of HEVs will account for only a small fraction of the market.

5.2.2 Medium and full hybrids

The ICE is notoriously inefficient. Maximum performance is obtained only within a very small area of its operational regime, and even there the efficiency is rather poor, i.e. around 40%, as shown in Figure 5.5. Moreover, this optimum condition is met only rarely and the torque available at the wheels

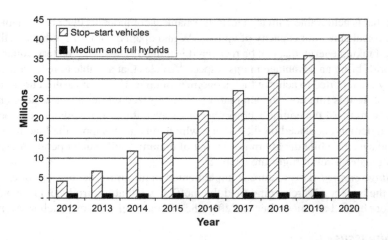

FIGURE 5.4

Projected rate of penetration of the automotive market by emerging vehicle technologies: total sales expected for SSVs (hatched columns) and for medium and full hybrids (solid columns)

(With permission from Pike Research)

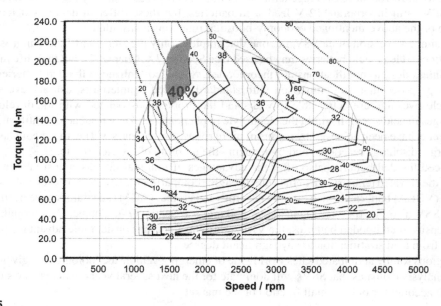

FIGURE 5.5

Typical efficiency map of an internal-combustion engine (a turbo direct-injection diesel) with contour lines of constant efficiency overlaid with lines of constant power in kilowatts (the smooth curves). The small shaded area in the top left-hand quarter of the map represents the region within which efficiency reaches its maximum value as power is increased

(Image from H Kabza in Encyclopedia of Electrochemical Power Sources, Volume 1, page 249, Elsevier copyright)

when starting off is quite limited. By contrast, an electric motor matches the requirements for traction very well. It provides maximum torque at zero speed and consumes no energy during standstill. Furthermore, such a machine can be driven in the reverse sense by an applied force, to act as a generator that converts mechanical energy on the shaft into electrical energy for regenerative braking. The major disadvantage of the electric motor is that it requires electrical energy to be provided on board the vehicle. The supply of sufficient electrical energy to allow for a significant contribution to vehicle propulsion brings disadvantages in terms of volume and weight, irrespective of whether the source of electrical energy is a battery or a fuel cell.

The individual merits of the ICE and the electric motor are combined in medium and full hybrids. The electric motor can be engaged to provide traction at times when the ICE would be operating in its least efficient manner, such as pulling away from standstill. The ICE can then be operated in a region closer to where it is more efficient (shaded area in Figure 5.5) and thereby the range limitation of the BEV is effectively avoided. Vehicle systems that deploy both a heat engine and an electric motor for motive power are more complex and somewhat more expensive than conventional power-trains, but bring with them the benefits of better fuel economy and greater reductions in emissions than are achievable with simple SSVs.

Both medium hybrids, such as the Honda *Insight,* and full hybrids, such as the Toyota *Prius,* are characterized by the ability of the electric motor to support the ICE (so-called 'power-assist') at times when additional power is required for propulsion, pulling away from a stationary position, and overtaking. The electrical power levels required by these types of hybrid dictate that powerful C-ISG systems must be used and that the battery voltage must be far greater than 12 V. The Ah capacity of the batteries used in power-assist hybrids is generally low (< 7 Ah) so that the amount of stored energy is similar to that stored in the battery of an SSV, 1–2 kWh; see Table 5.1. As pointed out earlier, the higher power levels allow a greater fraction of the available regenerative braking energy to be recovered. These two hybrids offer superior fuel economy compared with the simple SSV but do attract a greater cost, which is largely due to the use of a more expensive battery pack (i.e. including battery management).

Full hybrids, such as the Toyota *Prius,* have sufficient electrical power to enable the vehicle to have a small all-electric range. The *Prius* drives off with the aid of the 27-kW electric motor and the 1.8-L 73-kW ICE takes over only when the speed reaches 10–15 mph (16–24 km h^{-1}). In this way, the inefficient use of the ICE at very low speeds is circumvented and additional fuel economy is achieved, but with the imposition of a price increment. The *Mk III* version of the *Prius*, introduced in 2009, has an electric water pump, which represents a further step in the progressive use of electric components in place of belt-driven ones and adds to the improvement in fuel economy.

The expansion of electrical functionality in road vehicles that are designed to cut back both fuel consumption and the emission of combustion products leads inevitably to an increase in the amount of energy that must be supplied by the battery. In part, the additional energy can be provided by the selection of a battery with larger energy capacity but such a strategy introduces weight and cost penalties. The increase in battery size can be kept to a minimum if maximum use is made of regenerative braking energy, which, as previously observed, leads to the adoption of high-voltage batteries and consequent upgrading of electrical networks. It has generally been assumed that the maximum safe direct-current voltage for automotive use is 60 V, although some authorities now question whether this should not be 56 V or even 54 V. Special precautions must be adopted to prevent unsafe access to the high voltages that are necessary in HEVs (and BEVs). The debate over the optimum voltage for an SSV is ongoing but, for the first generation of such vehicles, it appears likely that any departure from the 12-V system

Table 5.1 Stop–Start Vehicles and Hybrid Vehicles with Little/No Facility for Electric-Only Travel

Attributes	Stop–Start Vehicles (Low Voltage)			Medium Hybrids	Full Hybrids
Battery voltage (V)	12–48			100–150	200+
Capacity (Ah)	90–25			7–6	6–5
Stored energy (kWh)	1.08–1.20			0.7–0.9	1.0–1.2
Discharge power (kW)	>5			>15	> 35
Stop–start	✓	✓	✓	✓	✓
Regenerative braking		✓	✓	✓	✓
Supercharger (downsize and boost)			✓		
Power-assist				✓	✓
Limited all-electric range					✓
Energy saved (%)[a]	6–8	12	15–25	8–20	15–40
Supplementary cost (US$)[a]	0–600	1000	1500	2250–4200	4200–7000
Cost-to-benefit ratio for 1% reduction in carbon dioxide emissions (US$)	~100	~80	60–100	~200	~200

[a]*Compared with an equivalent internal-combustion-engined vehicle that is performing an identical driving cycle.*

will be to 48 V. This step will provide a margin to accommodate overcharge before the hazard level is reached.

5.2.2.1 Hybrid configuration

As disclosed above, medium or full hybrids can be conceived in one or other of several broad structural configurations, namely: series hybrids, parallel hybrids and power-split (or mixed) hybrids.

The series hybrid has a purely electric traction system in which the electric motor drives the wheels and can recover kinetic energy during braking events; see Figure 5.1(a). A primary energy converter (either an ICE driving a generator, or a fuel cell) provides electricity, which is converted from energy stored chemically as gasoline or as hydrogen, to the electric motor (and the battery). An advantage of the series hybrid scheme is that no gearbox is needed. If the amount of electrical energy stored in the battery is large, then the ICE–generator system can run smoothly at the average power demanded with high efficiency to recharge the battery continuously. The 'extended-range electric vehicle' (E-REV), which is described in more detail below, is an example of a series hybrid. The electric traction drive may consist of either a single central electric motor that powers the driveshaft or an arrangement of two or more hub motors at the wheels.

The traction force in the parallel hybrid is supplied by a combination of the mechanical power from the ICE and the mechanical power output of the electric motor; see Figure 5.1(b). The system may be either on the same axle, which amounts to 'torque addition', or on separate axles, which constitutes 'power addition'. An advantage of the parallel configuration is that power from the ICE is transferred almost directly to the wheels so that the overall losses of the chain of components in the series configuration are avoided. The parallel hybrid concept can be implemented in a variety of ways. One approach would be to have a conventional drive from the ICE to the front axle and to have the electric drive applied to the rear axle. Power from the two drives can be coordinated through a high-level management system.

The power-split hybrid, the third category of hybrid configuration, employs a combination of features taken from the series and parallel configurations. The power flow is divided in a variable proportion between the mechanical and the electrical paths. The system should be capable of very high fuel efficiency but is more complex than the series or parallel designs and has more components. For example, the Toyota *Prius* power-split scheme, as shown in Figure 5.1(c), makes use of a planetary gear to share the engine power between a direct mechanical path to the wheels and an electric route. The latter passes, in turn, through a first electrical machine (a starter–generator), a first inverter, a second inverter and a second electrical machine (which doubles as a power motor and a generator for regenerative braking). The Toyota version of power-split technology, known as *Hybrid Synergy Drive*, is equivalent to a continuously variable transmission, but the gear ratio is changed by a variable electrical input to the power-train, as determined by the engine-management system. This arrangement permits the engine to operate at its most efficient all the time.

5.2.2.2 Battery issues

The demand for very frequent discharge and charge, due to the use of electrical energy in the power-assist mode, together with the recovery of regenerative braking energy, should be favourable for a healthy battery life because only small changes in the SoC are involved in most of these events. In practice, however, the consequent mode of cycling is detrimental to battery longevity in that it calls for very high power and hence high currents. Further, given that medium and full hybrids employ multi-cell packs, imbalances in cell SoC must be prevented and the temperature of the whole assembly must be maintained within specified limits if satisfactory battery performance is to be achieved. These crucially important requirements are met with a battery-management system (BMS).

One possible means of reducing the wear on the battery that is caused by high-current episodes is to employ a parallel electrochemical system in which a supercapacitor copes with the high-power charge/discharge pulses and the battery is only called upon to supply the energy requirements of the vehicle. The extra volume of the supercapacitor and the additional cost are disadvantages of this scheme as it stands, but an elegant solution may be at hand. At least in the case of lead–acid chemistry, it has been shown that the capacitor function can be built into the battery by making one of the plates perform a hybrid function – namely, charging and discharging both in a galvanic manner and as a capacitor. This device is known as the UltraBattery™ (see Section 7.10, Chapter 7) and could, in principle, be deployed to other battery chemistries to enable their performance to be augmented by the addition of a capacitive function.

5.2.2.3 Electric machines

It is important that the electric drive in an HEV should be highly efficient and maintenance-free. Conventional direct-current (DC) machines, where the wear on brushes that are operated at high specific power can be problematical, are avoided. Consequently, alternating-current (AC) synchronous or asynchronous rotating-field machines are preferred in medium and full hybrids.

In a permanent-magnet synchronous motor, the rotating magnetic field of the stator imposes a torque on the magnetic field of the rotor and thereby the speed of the rotor matches that of the rotating magnetic field of the stator. The magnetic flux of the rotor is built up by means of permanent magnets that are made from alloys of transition metals (e.g. iron, cobalt) and rare-earth elements (e.g. neodymium, samarium). Since these materials are expensive, and because synchronous motors are complicated to manufacture and to control, such machines are more costly than the alternative asynchronous machines.

On the other hand, the magnets offer the merits of a rather wide air-gap (more than 1.0 mm) between the rotor and the stator, which, in turn, allows not only good tolerance for wear in the crankshaft bearings but also very high torque at zero speed.

The functioning principle of AC induction or asynchronous machines is akin to that of a transformer – namely, the rotating field in the stator induces currents in the rotor (acting like the secondary of the transformer) that then interact with the rotating magnetic field to cause rotational motion. This inductive process does not provide the direct link between the rotations of the two parts of the machine that is the characteristic of the synchronous motor. Instead, the amount of movement of the rotor depends on the electrical load. The air-gap between the rotor and the stator has to be rather narrow (of the order of 0.5 mm) and this could pose a problem when asynchronous motors are used as ISGs in concert with diesel engines. Wear of the crankshaft bearings can reach a point where the small air-gap is insufficient to avoid fouling between the moving parts.

5.2.3 Power electronics

The flow of energy between the battery pack and/or supercapacitor bank and the electric motor(s) is managed by a power electronics unit. This component has to operate bi-directionally, namely: (i) as an inverter, during the drive mode, to convert electrical energy from the DC form supplied by the storage facility to the AC that is required by the motor; (ii) as a controlled rectifier during the recovery of regenerative braking energy to convert AC to DC for charging the battery and/or supercapacitor. Although the unit provides bi-directional functionality, it is often just called an 'inverter' in the technical literature. When the operating voltage of the power electronics does not exceed around 100 V, metal-oxide-semiconductor field-effect transistors (MOSFETs) act as the electronic switches. For systems running at higher voltages, insulated-gate bipolar transistors (IGBTs) are employed. High levels of electromagnetic interference can be encountered in the vicinity of the power electronics of HEVs (and BEVs) due to the high switching frequency of the inverters and the substantial currents involved. Consequently, precautions have to be taken to prevent any disturbance to the communications or control systems of the vehicle.

5.2.4 Energy characteristics and economics of stop–start vehicles and hybrids

The extent to which fuel savings can be made through the upgrading of internal-combustion-engined vehicles (ICEVs) to the SSV and HEV designs depends largely on the nature of the driving profiles to which they are exposed. It is clear that the savings will be greatest for travel in congested urban traffic where there is a high proportion of stop-and-go driving. The benefit is also determined by the performance of the base-line vehicle before hybridization. For example, a 10% improvement in the fuel economy of a large 'sports utility vehicle' (SUV) will save more fuel per mile/km driven than a 10% improvement for a small car that already achieves 50 mpg (80 km per gallon).

Some indication of the operational characteristics of SSVs and medium and full hybrids, together with their mean energy savings and costs, is given above in Table 5.1. An important point to note is that the amount of energy stored in the batteries in all three types of vehicle is virtually the same at 1–2 kWh. This feature emphasizes a clear distinction between these vehicles, with relatively small batteries, on the one hand, and the group of plug-in hybrid, range-extender and battery electric vehicles, with rather large batteries, on the other. The latter group, which can be driven for substantial ranges in an all-electric mode, are described in Section 5.3 below.

With respect to the economics of running a hybrid, two important factors to consider are: (i) the overall savings potential, as compared with the additional manufacturing cost of the vehicle upgrade; (ii) the payback time. The final row of data in Table 5.1 gives some idea of the cost-to-benefit ratio but, of course, the payback time depends on the current price of fuel and the distance driven over a given period.

There is a considerable difference in the amount of battery energy throughput for the different classes of hybrid. In terms of discharged energy, a 12-V SSV that is undertaking two stop–start cycles (0.3 Ah per cycle at an average voltage of 11 V) per km will use 6.6 Wh km^{-1}. A minimum power-assist vehicle (medium hybrid) is expected to require 25 Wh each time it engages in a power–assist event and this demand can amount to an average of about 50 Wh km^{-1}. For a maximum power-assist vehicle (full hybrid), the energy throughput will be closer to 100 Wh km^{-1}.

The development of the markets for each of the types of vehicle listed in Table 5.1 is proceeding at different rates from country to country throughout the world. Diesel engines are not favoured in automobiles in the USA and Japan, where concerns over their emissions of particulates have proved a major barrier. Consequently, efforts to achieve fuel economy in these two countries have initially focused on medium and full hybrids. By contrast, this is not the preference in Europe. Because of the benefit provided by electrical power-assist, medium and full hybrids are able to make use of smaller engines in pursuit of fuel economies in comparison with non-hybrid equivalents. The top speed of each of these two hybrid variants, however, is determined by the internal-combustion engine and thereby introduces a performance limitation that is unpopular in Europe.

Legislative initiatives in Europe (see Chapter 2) call for each manufacturer to reduce carbon dioxide emissions from its entire fleet. There are a number of strategies for achieving the target reductions. The four favoured approaches, which may be used alone or in combination, are: stop–start, hybridization, switch to diesel, or 'downsize and boost' (which is described in the following section). The respective technologies vary considerably in cost, as well as in effectiveness according to the driving schedule. The fuel economies that can be achieved by stop–start and by hybridization are mainly confined to urban driving situations, with little benefit appearing on the highway. By comparison, the fuel economies obtained by the other two alternative technologies are operative in all types of driving. Stop–start vehicles, with or without downsize and boost, offer the best cost-to-benefit ratio among all the vehicle concepts that retain batteries of small capacity (1–2 kWh) as long as they are able to reach the absolute emissions levels that are mandated.

The latest addition to the range of systems that are aimed at reducing carbon dioxide emissions to the lowest possible level, while retaining the facility of a large, luxury family car, is the diesel–electric hybrid. This combines the known efficiency of the diesel engine with that of a full electrical hybrid. The approach has been adopted by Citroën in its *DS5 Hybrid 4*, which was introduced in 2012 and combines a 163-bhp diesel engine that drives the front wheels with a 37-bhp electric motor that powers the rear wheels. It is claimed that, despite its size, the car emits only 99 g CO_2 per km. The diesel–electric hybrid appears to be a useful way forward for large vehicles and an extension to smaller models may be anticipated, provided that the cost can be kept within reasonable bounds.

5.2.5 Downsize and boost

The higher cost premiums associated with medium hybrids, and even more so with full hybrids, are an incentive for car manufacturers to seek alternative means of achieving fuel economy for their basic models. As mentioned in the previous section, a switch from a petrol engine to a diesel can provide a substantial improvement, but diesel cars have so far found little favour in either the USA or Japan. One approach that is being actively pursued by most of the manufacturers is to increase the fuel economy of the conventional vehicle by reducing the size of the engine and making good the consequent loss of power by the addition of a turbocharger; see Section 4.3.1.2, Chapter 4. This, then, is not a hybrid vehicle but its competitor.

Although it may be an inappropriate name for a green technology, the turbocharger raises fuel efficiency substantially and is now included in around 75% of new cars in Europe. Whereas the mention of turbochargers might have once conjured up images of loud and powerful engines, the technology has, in recent years, become a tool of choice for cutting carbon emissions.

The process of 'downsize and boost' can be taken further by applying the strategy to a stop–start vehicle and adding an electric supercharger, which allows the use of an even smaller engine without loss of performance. (The different characteristics of a turbocharger and a supercharger are discussed in Section 4.3.1.2, Chapter 4.) In this way, reductions in emissions can be achieved to rival those of medium and full hybrids, but at a lower cost. Two demonstration programmes have served to illustrate the effectiveness of this concept, as follows.

The first programme involved a VW *Passat* with a 1.4-L engine and the addition of a B-ISG and a variable-torque electric supercharger. The modified vehicle was labelled as the *LC Super Hybrid* (actually a misnomer as the vehicle is not a hybrid). The power and acceleration performance of the modified vehicle are superior to those of the original vehicle and, more significantly, are comparable with those of a 2-L Volvo, as shown by the data listed in Table 5.2. Furthermore, the emissions performance of the *LC Super Hybrid* is significantly better than that of either of the other two vehicles, i.e. by 19% and 26%, respectively. In January 2013 the team that produced the *LC Super Hybrid* was honoured with the Low Carbon Champions award by the UK Low Carbon Vehicle Partnership, a public–private enterprise that exists to accelerate a sustainable shift towards lower-carbon vehicles and fuels. Ironically, it is the inclusion of a carbon component in the lead–acid battery that allows the vehicle to recover sufficient braking energy to power the electric supercharger on which the clean credentials of the vehicle depend. (Further details of the effect of carbon are provided in Section 7.4.4.2, Chapter 7.)

In a separate demonstration, members of the team that mounted a project labelled 'Hyboost' – Ford, Valeo, Controlled Power Technologies (CPT) Ltd, the Advanced Lead–Acid Battery Consortium

Table 5.2 Emissions and Performance (New European Drive Cycle) Achieved by Engine Downsizing Coupled with the Introduction of a Belt-Driven Integrated Starter–Generator and an Electric Supercharger

Features	Unconverted VW Passat	VW Passat LC Super Hybrid	Volvo S40
Engine volume (L)	1.4	1.4	2.0
Power (hp)	122	142	145
Acceleration, 0–100 km h^{-1} (time in s)	11.1	8.7	9.5
Fuel (mpg, imperial)	45.6	50.5	37.2
L per 100 km	6.2	5.6	7.5
Carbon dioxide (g km^{-1})	140	130	176

(ALABC) and Imperial College, London – retrofitted a 2-L Ford *Focus* with a radically downsized (1.0-L) three-cylinder gasoline engine. Driving performance was then restored by turbocharging the engine and adding a CPT electric supercharger, as shown in Figure 5.6(a). Stop–start functionality was provided through the use of the Valeo Starter Alternator Reversible System (StARS) and energy recovery was assisted with the aid of a 220-F supercapacitor. Due largely to the rather enterprising reduction in engine size, the decrease in carbon dioxide emissions reached over 40% and, as in the *LC Super Hybrid* project, this was achieved with no loss in performance in comparison with the base vehicle. The respective contributions to the emissions reduction from the different component changes were as follows:

- the unmodified 2-L Ford *Focus* base vehicle provided 107 kW and produced 169 g CO_2 per km;
- downsizing to a 1-L boosted direct-injection engine provided 105 kW and reduced carbon dioxide emissions by 25%;
- adding stop–start and 6 kW of regenerative braking energy lowered carbon dioxide emissions by a further 10%;
- by optimizing the management of the gas flows in the system through the addition of cooled exhaust gas recirculation and by revising the turbo match with the electric supercharger, an additional 6% reduction in carbon dioxide emissions was achieved;
- overall, the HyBoost vehicle released 99.7 g CO_2 per km.

The marked reduction in carbon dioxide emissions is, of course, a direct consequence of an equivalent reduction in fuel consumption. The findings of these two expositions of the 'downsize-and-boost' principle suggest that fuel economy and emissions benefits equivalent to those offered by hybrids with expensive high-voltage systems may become available at considerably lower cost with 12-V (or perhaps 48-V) systems that are fitted with appropriate accessory equipment. Some indication of the cost advantage of such an approach is given in the third column under the 'Stop–Start Vehicles' heading of Table 5.1.

The legislated targets for carbon dioxide emissions that are in prospect for 2015 and beyond would be far easier to reach for small vehicles than for large ones. Large cars with large engines inevitably

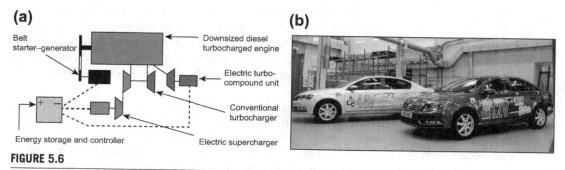

(a)

Belt starter–generator

Downsized diesel turbocharged engine

Electric turbo-compound unit

Conventional turbocharger

Energy storage and controller

Electric supercharger

(b)

FIGURE 5.6

(a) Schematic of the 'Hyboost' system; (b) 12-V and 48-V downsize-and-boost vehicles

(Figures reproduced with permission from the Advanced Lead–Acid starter–generator Battery Consortium (ALABC))

consume more fuel and produce more carbon dioxide than small ones. A downsize-and-boost strategy for the high-end sector will have to eliminate more emissions than for small vehicles and will require a more effective fuel economy scheme than is achievable with a 12-V battery system. Serious consideration is therefore now being given to a new electrical system operating at 48 V that will allow the recovery of more braking energy without the need to breach the electric shock hazard threshold that arises at around 60 V. Demonstrator downsize-and-boost vehicles based on 12-V and 48-V systems, respectively, are shown in Figure 5.6(b).

5.2.6 Hybrid electric buses and trucks

Commercial vehicles in urban traffic situations are a very promising field of application for HEV deployment. The most widespread use to date has been in urban bus services, which involve frequent acceleration and deceleration owing to both scheduled stops and traffic constraints. Each instance of acceleration requires substantial propulsion energy, part of which is wasted through braking during deceleration of buses with conventional drive-trains. As with some other forms of HEV, a good fraction of the regenerative braking energy can be stored in the battery of a hybrid bus and then used to assist with subsequent propulsion events. Thus it is possible for the ICE to be downsized so that it runs at a higher fraction of its maximum load and thereby achieves greater fuel efficiency. Another advantage is that the ICE may be switched off and replaced by the electrical system at times when power demand is very low. Avoiding engine operation at zero or light load in this way brings additional improvement in fuel economy. Hybrid electric buses have been in regular use for more than a decade and their effectiveness in reducing both fuel consumption and carbon dioxide emissions is well established. The National Renewable Energy Laboratory (NREL) in the USA conducted a study of the New York City Transit fleet from September 2004 to May 2005. A comparison was made of the performance of three types of bus; namely diesel, compressed natural gas (CNG) and diesel–electric hybrid. The buses were each being driven at a rate of about 30 000 miles (48 000 km) per year and for 15 hours per day. The fuel economy of the diesel–electric hybrids was found to be about 35% and 100% better than that for conventional diesel buses and CNG-fuelled buses, respectively.

Hybrid buses may have either a series- or parallel hybrid architecture. The series configuration is deemed best for vehicles that travel at low average speed and have frequent stops, as for urban buses in heavy traffic. A powerful motor–generator and a battery that can accept charge at a high rate are essential if much of the regenerative braking energy that is available in series hybrid buses is to be recovered. By contrast, parallel hybrids are best for buses on rural routes where stops are less frequent and where the diesel engine is the prime source of motive power. The parallel system has lower losses as there is no round-trip efficiency loss through the electric systems as with a series hybrid.

Parallel hybrid electric buses can be designed as either mild or full hybrids depending on how much of the total traction is provided by the electric power source. In the mild-hybrid version, only a small fraction of the driving power is supplied by electric energy and such vehicles generally do not have an electric-only driving mode. The electrical energy generated is used mainly for accessory loads such as air-conditioning, hydraulics and lights. By contrast, the total traction power for full hybrid buses can be provided either by the ICE or by the electric system. By employing the electric-only mode of propulsion while the bus is in the most densely populated streets of a city, hybrid electric buses offer a valuable spin-off in terms of a reduction in air borne pollutants, such as particulate matter and nitrogen oxides. Although the full hybrid design is more costly than the mild-hybrid counterpart, the majority of the parallel hybrid electric buses launched to date are of the former type.

Early use of inexpensive lead–acid batteries of inappropriate design in hybrid electric buses resulted in short cycle-life and a need for the application of a periodic 'conditioning process' to balance the SoC of each cell. Despite the arrival of newer forms of lead–acid, such as the UltraBattery™ (see Section 7.10, Chapter 7), that are able to operate for a long life without conditioning, almost all buses now use nickel–metal-hydride or lithium-ion batteries with the cost penalty covered by subsidies. Studies have shown that the purchase price of hybrids is 60–80% greater than for conventional diesel buses. A further consideration is that the life of present battery technologies is unlikely to match that of a bus. The latter may be as much as 20 years so that battery replacement may well become an additional financial burden.

Despite the aforementioned concerns, fleets of hybrid buses have been launched in many parts of the world, including London, Oxford and Manchester in the UK, but here, as elsewhere, there has been a need for government subsidy. The British-made buses are mostly series hybrids, in which a small diesel engine coupled to a generator produces electricity to drive the electric motor that powers the wheels. The system design is illustrated schematically in Figure 5.7, together with a photograph of a London hybrid double-deck bus. Considerable experience with these buses has shown that fuel consumption and emissions of carbon dioxide are up to 40% lower than with a conventional diesel bus, i.e. in broad agreement with the NREL study in New York (*v.s.*).

Transport for London, the local government body responsible for most aspects of the transport system in Greater London, has instituted a programme to place hybrid buses on many of its routes; well over 200 are now in regular service. The programme is experimental in that there is a mixture of single- and double-deck buses. These vehicles, which have been produced by various manufacturers, are mostly series hybrids, and a variety of battery types are employed. Given the low speed of travel and frequent stops in London traffic, it is envisaged that series hybrids will prove to be the more suitable with respect to minimizing fuel consumption and emissions.

Diesel–electric hybrid trucks are produced by some of the world's major manufacturers of heavy-duty vehicles. The trucks are parallel hybrids that employ the same technology and drive-trains as are used in parallel hybrid buses. The operation of a parallel hybrid is shown schematically in Figure 5.8, and consists of four stages, namely: (i) moving off using electric drive only; (ii) acceleration using both the diesel engine and the electric motor; (iii) cruising with the diesel engine only; (iv) slowing down with energy from regenerative braking being fed back into the battery. To date, it seems that hybrid electric buses have proved more popular than hybrid trucks, probably because transit buses are urban whereas trucks are inter-urban.

5.2.7 Hybrid trolleybus concepts

Traditional trolleybuses are powered from the mains by means of overhead cables. Although they have proved perfectly satisfactory in many cities around the globe, there are certain disadvantages associated with this method of electric traction. For example, the infrastructure required is extensive

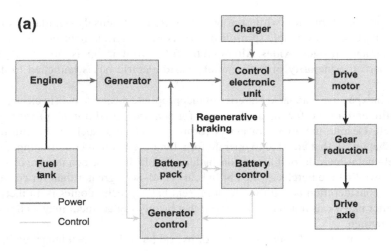

FIGURE 5.7

UK hybrid bus: (a) series hybrid configuration; (b) on duty in London

(Image b sourced from Wikipedia, and available under the Creative Commons License)

and a trolleybus cannot overtake another that is stationary except at places where there is a loop in the power-supply cables. It has been recognized that the introduction of an auxiliary traction battery to form a mains–battery hybrid would permit a limited degree of autonomous operation and also provide electricity for lighting, heating and other onboard services at times when there is disconnection or failure of the mains network. As an added advantage, a hybrid trolleybus on reaching the end of the line in a city suburb would then be able to tour the neighbourhood and thereby collect and

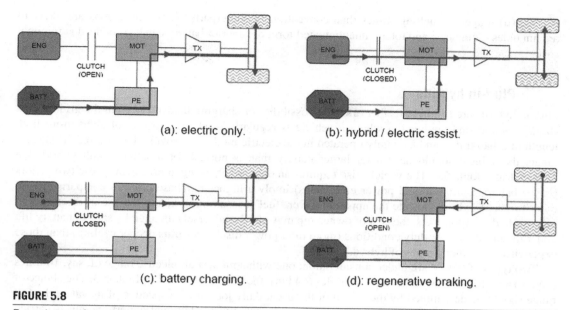

(a): electric only.   (b): hybrid / electric assist.

(c): battery charging.   (d): regenerative braking.

FIGURE 5.8

Four phases of operation of a parallel hybrid drive-train

(Image sourced from Wikipedia and available under the Creative Commons License)

deliver passengers nearer their homes. Nonetheless, mains–battery hybrid trolleybuses have attracted little attention, possibly because of the additional cost of providing the storage battery.

During the 1980s and 1990s, 'Duo-buses' were operated in Essen and other German cities. These were trolleybuses fitted with diesel engines to propel the vehicle when away from the overhead power supply. The Duo-bus is a straightforward approach to reducing pollution in city centres, by using trolleybuses rather than conventional buses, while at the same time having flexibility to operate around the suburbs. The downside to the Duo-bus is the expense of having two independent propulsion units – one electric and one diesel.

5.3 Electric vehicles with batteries charged from the mains

Plug-in hybrids (PHEVs), extended-range electric vehicles (E-REVs) and battery electric vehicles (BEVs) obtain at least part of their electrical energy from an external source, namely the grid, and therefore differ from the HEVs described above in Section 5.2. Although the three classes of vehicle all appear to offer the prospect of very large fuel economies or the elimination of petroleum fuels altogether, they generally make use of much larger batteries and this leads to a substantially higher cost per vehicle.

To date, battery electric cars have not found popular appeal for a variety of reasons. In addition to the cost penalty, the vehicles have the two key disadvantages of: (i) shorter driving ranges, i.e. up to 150 miles (240 km) at best, before battery recharging, compared with the > 500 mile (> 800 km) ranges of

ICEVs; (ii) longer 'refuelling' times than conventional counterparts. These limitations are likely to remain unless some new and totally unanticipated form of superior battery can be produced at minimal cost.

5.3.1 Plug-in hybrids

Plug-in hybrids are full hybrids that offer the possibility of charging the battery from the grid. Such a strategy would find application where a vehicle is regularly used for journeys of short-to-medium length in urban traffic and is mainly operated in the electric mode. To provide the required all-electric range, the vehicle should have a far larger battery than is normal for a standard full hybrid; see Table 5.1 and Table 5.3. The vehicle also requires an onboard charging interface and these two factors (larger battery and additional power electronics) imply a higher purchase price in comparison with other forms of hybrid. Despite the apparent lower 'fuel' consumption due to the use of electricity in place of hydrocarbon fuels, the cost of ownership may also be adversely affected by shorter battery life under the more arduous duty conditions (more high-power discharges **plus** deeper cycling) than those experienced by other classes of hybrid.

Two types of PHEV are under consideration: one with a modest all-electric range of, say, 10 miles (16 km) and the other with, perhaps, 40 miles (64 km); the details are listed in Table 5.3. The choice of range should be determined by the length of the usual daily journey. Irrespective of its value, the full all-electric range should be used as often as possible, otherwise additional battery weight is carried around constantly for no purpose. Every 10 miles (16 km) of all-electric range requires an onboard storage of 3–4 kWh, as determined by the size of the vehicle.

Plug-in hybrids can operate in two distinct modes: the 'charge-sustaining' mode or the 'charge-depleting' mode. In the former, the battery is maintained at a constant SoC with the amount of energy discharged during power-assist operations being balanced by charging from the generator and from regenerative braking. In the charge-depleting mode, however, the battery is discharged more than it is charged during the course of a journey and the shortfall is made good by drawing electricity from the mains supply. Not only must the battery in a PHEV store sufficient energy to furnish the required

Table 5.3 Comparison of Typical Battery Characteristics for PHEVs, E-REVs and BEVs

Parameters	Plug-in Hybrid 10 Mile	Plug-in Hybrid 40 Mile	Extended-Range Electric Vehicle	All-Electric Vehicle
Battery size (kWh)	4	12	10–20	20–30
Lithium-ion weight (kg)	49	77	143–286	286–429
Lead–acid weight (kg)	113	387	333–666	666–999
Lithium-ion volume (L)	10.6	36.3	66.7–133.4	133.4–200.1
Lead–acid volume (L)	49	166	143–286	286–429
Lithium-ion cost (US$)	1700	5800	5000–10 000	10 000–15 000
Lead-acid cost (US$)	510	1740	1500–3000	3000–4500

Data for PHEV are for high power. Data for E-REV and BEV are for high energy.
Weight, volume and costs are estimates for a fully-managed pack.

all-electric range, i.e. have a deep-cycle facility, but it must also provide the high-power performance (especially for charging) that is demanded by all hybrids. This is an exacting regime for the battery but the electrical energy that is drawn from the external source does reduce the amount of petroleum that is used by the vehicle and thus produces an apparently substantial improvement in fuel economy.

The main difference of the PHEV from the standard full hybrid is the size of the battery, which, in the charge-depleting mode, is designed to support a much greater all-electric range. The ICE remains the main source of motive power, however, and full performance is only achieved through the ICE and the electric motor operating together. In the all-electric mode, the acceleration and the top speed are reduced in comparison with the hybrid power mode. The attributes of the charge-sustaining mode of operation are different. Since the battery is maintained at a near-constant SoC by the engine (as for a conventional automotive battery) the demands made upon it are less severe.

Plug-in hybrid cars are now being introduced to the market. The General Motors' Chevrolet *Volt*, which is labelled the *Ampera* in Europe, is a series PHEV equipped with a 16-kWh battery pack that gives an all-electric range of 40 miles (64 km); an example of the car is shown Figure 5.9. The *Volt* has been on sale in the USA from the end of 2010 at a price of around US$40 000. It has a 1.4-L ICE to charge the 175-kg battery pack and to power the 112-kW electric motor that drives the front wheels. Full charging from the mains takes about three hours at 220 V and eight hours at 110 V.

Elsewhere, Volvo is to sell its *V60* model as a parallel hybrid PHEV; it is said to be the first PHEV to be equipped with a diesel engine. The five-cylinder 2.4-L engine produces 215 bhp (160 kW) to drive the front wheels, while the rear wheels are powered by a 70-bhp (52-kW) electric motor. The car has a range of around 30 miles (48 km) in pure-electric mode. In 2012, the Toyota *Prius* was introduced as a PHEV equipped with a 4.4-kWh lithium-ion battery that delivered an all-electric range of 11 miles (18 km). The three examples – the *Volt*, the *V60* and the *Prius* – illustrate the many possible options for

FIGURE 5.9

Chevrolet *Volt* plug-in hybrid electric vehicle

(Image taken from Wikipedia and available under the Creative Commons License)

PHEVs in terms of power-train architecture, the type and size of engine employed, the sizes of the batteries and electric motors utilized, and whether the twin power units drive the same two wheels or whether separate four-wheel drive is chosen. As always, the designer is striving for optimum performance at minimal cost and has many variables to juggle.

Plug-in hybrids are inevitably expensive to produce in small numbers and, even with government subsidies, sales are unlikely to be large initially. Nevertheless, it is only by 'going electric' that the major automotive manufacturers will be able to meet the future emissions targets that are being set by authorities around the world.

5.3.2 Extended-range electric vehicles

The extended-range electric vehicle (E-REV) is effectively an all-electric vehicle, with all the motive power provided by an electric motor, but with a small ICE present to generate additional electric power. Alternatively, it may be viewed as a series hybrid with a much larger battery, namely, 10–20 kWh; see Table 5.3. When the battery is discharged to a specified level, the ICE is switched on to run a generator that, in turn, supplies power to the electric motor and/or recharges the battery. With this arrangement, the range limitation that is inherent in a BEV can be overcome. For moderate distances, E-REVs can operate in full-electric mode and are then as clean and energy-efficient as BEVs (unlike parallel hybrids and other series hybrids with their smaller batteries and very limited electric range). For longer distances, E-REVs utilize the ICE to keep the battery charged, but consume noticeably less fuel than conventional ICEVs for the following two reasons:

(i) The engine of an E-REV is significantly smaller than that of a conventional ICEV – it only needs to meet average power demands because peak power is delivered by the battery pack. The engine of an ICEV, on the other hand, must also cover peak-power surges, e.g. accelerations.

(ii) The engine of an E-REV operates at a constant, highly efficient, rotation speed; whereas that of an ICEV often runs at low or high rotation speeds during which, in both situations, its efficiency is low.

The different modes of E-REV operation are shown schematically in Figure 5.10. The vehicle begins its journey with the battery SoC close to 100%. All the vehicle power is provided by the electric motor, which draws energy only from the battery, and there are no local exhaust emissions. The battery is partly recharged with each regenerative braking event. When the battery is depleted to a pre-ordained SoC – marked in Figure 5.10 at three levels of increasing severity, viz., green, orange and red – the vehicle switches to extended-range mode. While the vehicle is operating in this mode the ICE is switched on as and when necessary to keep the battery within the SoC range marked by the green and red dashed lines. After the journey, the battery SoC is returned to 100% with power taken from the grid.

A future possibility would be to replace the piston engine with a micro gas-turbine as the range extender. Jaguar has produced the *C-X75* hybrid concept car, which is an E-REV with two small gas turbines (each 35 kg) to charge the battery (15-kWh lithium-ion). Four 145-kW electric motors, one at each of the wheels, can drive the 1350-kg vehicle up to 205 mph (330 km h^{-1}) with a total torque of 1600 N m. The *C-X75* has an electric-only range of 70 miles (113 km), and a 60-L fuel tank.

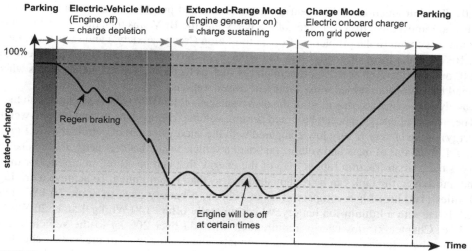

FIGURE 5.10

Battery state-of-charge in an extended-range electric vehicle (E-REV) when driven sequentially in electric-vehicle mode and extended-range mode before being charged from the grid while stationary

(Image from H. Mettlach in Encyclopedia of Electrochemical Power Sources, Volume 1, page 289, Elsevier copyright)

5.3.3 Battery electric vehicles

Battery electric vehicles are the least complicated of the technologies that are under consideration for road transport, with considerably fewer component parts than a conventional ICEV – no ignition system and no gearbox, for example. At first glance, BEVs also appear to offer the ultimate solution in the pursuit of freedom from fossil fuels and the elimination of environmental pollution. This ideal asset, however, would only be valid if all of the electricity used to propel the vehicles were to be derived from sources that do not involve the combustion of hydrocarbon fuels; see Section 5.5.2 below. At the time of writing, there are very few areas of the world where this utopian situation exists. In Europe, for example, the primary energy mix is such that a compact-class BEV with a lithium-ion battery will be responsible for 87 g CO_2 per km driven as a result of the greenhouse gas emissions at the power stations where the electricity is generated. Consequently, the conclusion might be drawn that there is no great advantage to the BEV from the viewpoint of carbon dioxide emissions, although the advantages of avoiding local pollution and depletion of petroleum stocks still obtain. The widespread uptake of BEVs and PHEVs would not be expected to lead to a great increase in the demand for electricity-generating plant, provided that the majority of the vehicle recharging could take place at off-peak times (e.g. overnight). There would simply be an increase in the proportion of total electricity generated that could be produced by base-load plant.

5.3.3.1 Battery electric cars and light-duty vehicles

At the beginning of the 20th century, there were more electric cars on the road than there were vehicles powered by ICEs. For a short time, there was competition between BEVs and ICEVs, each limited by its characteristic disadvantages – BEVs were short of range and long on recharge time, whereas ICEVs required considerable muscular effort for hand-crank starting and were less reliable, breaking down frequently. The introduction of the electric-starter system and progressive engineering improvements that followed removed the disadvantages of the ICEV and the BEV was then quickly eclipsed because no solution was found to the problems of range limitation ('range anxiety') and recharge time. Nevertheless, BEVs in the form of goods vans and other light-duty vehicles continued to be employed in Europe throughout the 20th century, often in places such as factories, airports and hospitals where their silence and freedom from fumes were important considerations.

It was always recognized that most of the disadvantages of the BEV were associated with the battery itself. The amount of energy that can be stored per unit volume (energy density) and per unit weight (specific energy) of the battery is very low compared with the energy content of petroleum; see Figure 7.20, Chapter 7. Limitations of mass and volume on board a vehicle mean that the energy that can be stored in a battery is but a small fraction (about 1%) of the energy that can be stored in a petrol or diesel fuel tank and, moreover, the tank can be replenished in less than two minutes. To achieve a driving range of 100 miles (160 km) for a compact class of vehicle, a battery pack of some 30 kWh has to be installed. Even with a lithium-ion battery, which typically offers 140 Wh kg^{-1} and 320 Wh L^{-1} (see Figure 7.16, Chapter 7), this energy requirement will add over 200 kg to the vehicle weight and occupy twice the volume of a hydrocarbon fuel tank. The lifetime cost of the battery should also be factored into considerations of the value proposition for BEVs. Nickel–metal-hydride batteries appear capable of lasting the life of the vehicle in which they are installed. Nevertheless, there is a widespread expectation that BEVs of the future will make use of lithium-ion batteries, which have greater specific energy and thus will provide longer driving range. The price of lithium-ion packs that are fully instrumented to cope with safety issues has yet to fall below US$500 per kWh so that the batteries for a compact class of BEV are almost as expensive as a complete ICEV of equivalent size. It is essential, therefore, that lithium-ion batteries designed for BEVs should not have to be replaced during the life of a vehicle.

The Nissan *Leaf*, shown in Figure 5.11(a), is a five-door hatchback BEV with a 24-kWh air-cooled laminated lithium-ion battery, an 80-kW synchronous AC motor, an onboard 3.3-kW charger and a regenerative-braking facility. The driving range is 100 miles (160 km), as measured on the New European Drive Cycle. The battery costs roughly half the total selling price of the vehicle. Full recharging can be accomplished in eight hours at 240 V and, optionally, 80% charging can be achieved from public quick-charge stations in 30 minutes. By the end of 2013 the *Leaf* was already the best-selling electric car ever to be produced, with world sales rapidly approaching 100000. The US Environmental Protection Agency (EPA) has rated the *Leaf* to have a fuel economy of 99 miles (158 km) per US gallon gasoline equivalent (2.4 L per 100 km) for a combination of urban and highway driving. This performance relates to a charging time of seven hours on 240 V and a driving range of 73 miles (117 km), under varying driving conditions and climate controls. The battery pack is expected to retain 70 to 80% of its capacity after 10 years but its actual lifespan depends on how often fast charging (480 V DC) is used and also on driving patterns and environmental factors. The battery is guaranteed by Nissan for eight years or 100000 miles (160000 km). In addition to the main battery, the *Leaf* also has an auxiliary 12-V lead–acid battery that provides power to the car computer systems and accessories such as the audio system, supplemental restraint systems, headlights and windscreen wipers.

The rapidly rising strength of the Chinese automotive industry is exemplified by the launch in London of a fleet of 50 *BYD e6* minicabs (see Figure 5.11(b)) that have been manufactured by BYD Auto Co. Ltd ('Build Your Dreams') in Shenzhen. These vehicles are reported to offer a driving range between charges of over 180 miles (290 km) and are powered by a lithium–iron-phosphate battery (see Section 7.6.1, Chapter 7) with a life of more than 4000 cycles.

FIGURE 5.11

Recent electric cars: (a) Nissan *Leaf*; (b) *BYD e6* (c) *G-Wiz*; (d) Fiat *Panda Elettrica*

(Images sourced from Wikipedia and available under the Creative Commons License)

Some manufacturers are testing the market with less-expensive batteries. Lead–acid batteries are available at around US$150 per kWh, so that a 30-kWh unit will contribute US$4500 to the initial purchase of a vehicle and could be replaced at least twice within the lifetime cost for lithium-ion. It is more likely that the lead–acid option will be taken up for cases where shorter daily range, perhaps 40 miles (64 km), is required, as is the case for many city dwellers and commuters. For instance, analysis has shown that three-quarters of daily journeys in the USA are of less than 50 miles.

The *REVAi* is a much smaller BEV that was manufactured in Bangalore and has been sold in the UK as the *G-Wiz*; Figure 5.11(c) shows an example on charge in a London street. The compact car has a 9.6-kWh lead–acid battery pack and a range of 48 miles (77 km). It weighs 665 kg (including 270 kg of batteries) and has a 13-kW AC motor. Recharge time is reported as eight hours. In 2009, an updated version with lithium-ion batteries was released that reduced the mass by 100 kg and the recharge time to six hours, while increasing the range to 75 miles (120 km) – but also nearly doubling the price!

Other BEV manufacturers have experimented with the use of high-temperature ZEBRA (sodium–metal-chloride) batteries in small cars; see Section 7.7, Chapter 7. These included the Fiat *Panda Elettrica*, shown in Figure 5.11(d), and the Norwegian *THINK City* car. The attraction of this battery is that it is virtually maintenance-free; has a proven deep-cycle life; and can be used in any climate, hot or cold, as it operates at 250–350 °C inside a thermally managed container. The characteristics of this battery, however, make it more suited to use in fleet vehicles such as buses rather than private cars.

Not all BEV development is driven by the needs of economy alone. The Tesla *Roadster*, built in California, was a high-priced (US$109 000 in the USA), high-performance BEV that was produced for a limited period between 2008 and 2012. It had a three-phase 215-kW induction motor driving through a single-speed 8.27-to-1 gearbox, and a 53-kWh lithium-ion battery pack that weighed 450 kg. The vehicle mass was 1235 kg, to which the motor contributed only 52 kg and delivered 400 N m torque up to 6000 rpm. The battery-to-wheel efficiency was 88%. Vehicle performance was listed as: a driving range of > 200 miles (320 km), 0 to 60 mph (96 km h^{-1}) acceleration in 3.7 s and a top speed of 100 mph (160 km h^{-1}). In Australia, a Tesla *Roadster* that was participating in the 2009 Global Green Challenge – an evolution of the prestigious World Solar Challenge (see Section 5.4 below) and designed to showcase the latest advances in hybrid, electric, solar, low-emission and alternative energy vehicles – travelled 311 miles (501 km) down the Stuart Highway to break the world record distance for driving an electric car on a single charge. A new model, the Tesla *Model S*, has been developed with a low-profile 'skateboard' chassis that also accommodates the battery pack.

Toyota had stood back from BEV development while enjoying the success of its hybrid *Prius*. In May 2010, however, it announced that it would invest US$50 million in US-based Tesla to develop jointly a new low-priced BEV, which was basically a Toyota *RAV4* with a Tesla power-train. The list price for the new model was US$49 800 but after government incentives and tax breaks it was reduced to around US$40 000. It remains to be seen whether the Tesla technology will have widespread appeal when it is given the marketing momentum that Toyota is able to apply.

For a BEV to be competitive, not only the daily driving range but also the time to achieve a full battery charge must be acceptable. If charging is to be undertaken in a home with, for instance, a 220–240-V supply, then a 30-kWh battery will require around 10 h to recharge fully. This time could be reduced to 10–15 min at public fast-charging stations provided that the battery is capable of accommodating a fast charge – batteries designed for electric vehicles are optimized for energy rather than power. On the other hand, such chargers carry large currents and require the use of high-precision electronic components and algorithms; they are therefore expensive devices. If the time taken for battery replenishment is to equal that involved in refilling a gasoline tank (around a minute), then a charger of almost 2 MW would be needed and the battery electrolyte would boil – obviously, a totally impracticable proposition.

Over decades, there has been an enormous effort to improve battery performance through exploration of the characteristics of different chemistries and through the development of sophisticated materials. The search for electrode couples that might prove superior in terms of higher specific energy, higher specific power and longer life has ranged over much of the periodic table and, as observed earlier, it seems unlikely that any dramatic breakthrough with an entirely new and practical technology will be made. Meanwhile, steady improvements will undoubtedly be made in established systems, especially those based on lithium. Electric cars have been viewed as a possible means of road transport for over a century and, as shown in Table 5.4, their performance has improved over that time, but not in a manner that could be regarded as a quantum leap.

Despite the clear and persistent shortcomings of BEVs, the significant numbers now appearing on the streets of major cities demonstrate that there are some applications where the benefits of local freedom from pollution and noise prevail. Although initial purchase prices are likely to remain high, there are compensating benefits; for example, the energy costs (for electricity rather than fossil fuels) are low, the energy efficiency is higher than for an ICEV, the vehicle has a long life and the stop–start performance in traffic is very good. Some advantages and disadvantages of BEVs are summarized in Table 5.5.

Consideration of the attributes of BEVs reaffirms the conclusion that they are best suited to short daily journeys in a temperate climate, especially in an urban environment and where the driver has access to private overnight charging. Commuting to and from work is an ideal application, as is door-to-door delivery. Numerous post offices in Germany, Austria, Australia, the USA and elsewhere have experimented with pilot fleets of BEVs for mail delivery, but so far these have not proved entirely suitable or economic. A significant problem has been the cold winter climate in many countries. With further technical developments, reduced costs brought about by mass production and with the ever-rising price of petroleum fuels, this conclusion may yet change.

Two conditions are necessary if the barriers to the large-scale introduction of BEVs are to be overcome. First, petroleum fuels need to be in short supply or, at very least, considerably more expensive than electricity. This condition is closer to being met in Europe than in the USA, where gasoline is still relatively inexpensive. Second, both national and local governments must actively encourage the uptake of BEVs through reduced taxes, through regulations that limit the use of ICEVs in towns and cities, and through active steps such as preferential parking opportunities for BEVs with the installation of recharging facilities in public car parks.

Although some authorities are offering to waive tax revenues for 'zero-emission' vehicles, the success of the BEV as a part of future road transport schemes will only become clear if it does not rely on tax benefits. As the price of petrol and diesel fuels rises, the economic case for vehicles that do not depend on either fuel becomes stronger. There is certainly a degree of enthusiasm for BEVs among manufacturers, governments and local authorities, as well as dedicated environmentalists. Public recharging facilities are being set up in parts of the UK, in European cities and across the USA; see Section 6.4, Chapter 6. The extent to which such enthusiasm will translate into actual sales remains to be seen.

An international Electric Vehicles Initiative was launched in October 2010 at the Paris Motor Show by a consortium of the Organisation for Economic Co-operation and Development (OECD) International Energy Agency (IEA) and eight countries – namely, China, France, Germany, Japan, South

Table 5.4 One Hundred Years of Development of Battery Electric Vehicles

Year	Vehicle	Capacity	Battery Technology	Battery Capacity (kWh)	Range (km)	Top Speed (km h^{-1})
1911	Detroit electric car	2 people	Nickel–iron	30[a]	128	32
1960	*EZGO* golf cart	2 people	Lead–acid	5[a]	30	20
1997	Toyota electric *RAV4*	5 people	Nickel–metal-hydride	27	130	126
1998	GM *EV1 Generation 2*	2 people	Nickel–metal-hydride	26	160	128
2001	*REVAi/G-Wiz*	2 people	Lead–acid	14[a]	77	45
2007	Smiths *Edison* van	2 metric tons	Sodium–nickel-chloride	64	150	60
2008	Tesla *Roadster*	2 people	Lithium-ion	53	320	160
2010	Nissan *Leaf*	5 people	Lithium-ion	24	160	145

[a]*Estimated.*

Table 5.5 Advantages and Disadvantages of Battery Electric Vehicles

Advantages	Disadvantages
• Simplicity of electric drive-train	• Mass and volume of batteries
• High torque at low speed (good acceleration)	• Short vehicle range
• Silent in operation	• Fear of running out of charge
• No exhaust and no local pollution	• Long recharge times (hours)
• Independent of petroleum, thus diversifies the energy base of transport	• Requires access to a charging point; comparatively few public charging points yet available
• Relatively low running costs	• High cost and uncertain lifetime of batteries
• No electricity consumed when stationary	• Most batteries not suited to high and low ambient temperatures.
• Overnight charging helps to load-level the electricity supply network	• Problems of supplying ancillaries, e.g. heating and air-conditioning

Africa, Spain, Sweden and the USA. The aim is to achieve rapid market development of electric and plug-in hybrid electric vehicles (BEVs, PHEVs) around the world to reach a combined target of about 20 million on the road by 2020. According to the IEA, this target would put global BEV + PHEV stock on a trajectory to exceed 200 million by 2030, and one billion (10^9) by 2050. The initiative is seen as a key element for the entire global economy to achieve the IEA target to reduce carbon dioxide emissions in 2050 to 50% of those in 2005 (the so-called Blue Map scenario; see Section 9.2). Such an ambition requires, for decades to come, the cooperation of all those individuals worldwide who make value judgements about vehicle purchase. It also depends heavily on the electricity being sourced from renewable or nuclear sources.

5.3.3.2 Battery electric buses

At present, the majority of buses deployed in urban transit schemes are powered by diesel engines. They consume considerable quantities of oil-based fuels, generate environmental pollution at street level and, in a wider context, contribute substantial amounts of carbon dioxide to the world inventory of greenhouse gases. On a more positive note, most bus manufacturers and urban transit operators have worked hard to eliminate emissions of nitrogen oxides and particulates through engineering advances.

To reduce the consumption of hydrocarbon fuels and the concomitant environmental impact, some transit authorities have turned to mains electric buses, such as trolleybuses, or to battery electrics. The latter, though, are mostly small in size and limited in range. The preferred mode of improving fuel economy and reducing emissions appears to be through the introduction of hybrid electric buses – as described above in Section 5.2.6.

A unique development is taking place in the city of Adelaide, South Australia, where a solar photovoltaic array has been installed on the roof of the central bus station. Solar electricity is used to power a demonstrator battery electric bus – the *Tindo* – that carries passengers around the city; see Figure 5.12 (note: 'Tindo' is an aboriginal name for the sun). This is the world's first electric bus to operate directly on solar energy. A further notable feature of the bus is that it is equipped with 11 ZEBRA battery modules, which are manufactured in Switzerland; see Chapter 7, Section 7.7. It should also be mentioned that Adelaide has put in place a number of other sustainable transport initiatives as part of its Solar City

FIGURE 5.12

Tindo solar electric bus on duty in Adelaide, South Australia

project (e.g. the council's parking inspectors use electric motorcycles) and it is also the home of the World Solar Challenge; see Section 5.4 below.

5.3.3.3 Electric two-wheel vehicles

Electric two-wheelers (E2Ws), which include electric bicycles ('e-bikes') and electric scooters, constitute the BEV sector that has achieved the greatest number of sales to date. It has been estimated that there were 120 million E2Ws in China as of early 2010; see Figure 5.13. By 2015, it is expected that this number will have risen to 150 million, with sales in other parts of the world also expanding rapidly. Positive attributes of E2Ws are that they are inexpensive personal transport, and their zero tailpipe emissions; high energy efficiency on a plug-to-wheel basis; and easy, inexpensive recharging.

In general, e-bikes incorporate a hub motor of 180–250 W, a controller, and a 36-V battery of 12-, 14- or 20-Ah capacity. By contrast, electric scooters are somewhat scaled up with 350–500 W motors and 48-V batteries. Top speeds are around 30 and 40 km h^{-1} for e-bikes and e-scooters, respectively, and each has a range per charge of up to 25 miles (40 km), which is more than adequate given that the mean daily bicycle commute in China is around 6 miles (10 km). The vast majority of the batteries used are valve-regulated lead–acid (VRLA) types that can be removed from the bike for charging (which takes six to eight hours) from a standard domestic electrical outlet. Lithium-ion batteries would provide longer range or reduced battery weight, but so far have made little penetration of the market in China because of the cost differential. This situation may change in the future as the average income rises in Chinese households. The VRLA batteries withstand the deep-cycling service (down to as low as 10% SoC) for between one and two years and must then be replaced. The acceptable performance of VRLA batteries in E2W applications is both driving down battery cost and stimulating the development of improvements in bicycle/scooter technology that may later be employed in three- or four-wheeled BEVs. The charging of larger batteries in such vehicles will, however, prove to be less convenient than it is for E2Ws.

FIGURE 5.13

Electric bicycles in China (a) Linixia City Daxia River esplande electric bike shop (b) Buddhist Monk on an electric bike; (c) Woman on an electric bike

(Images sourced from Wikipedia and available under the Creative Commons License)

The introduction of E2Ws has been successful in China for two main reasons: (i) there has been legislation to ban the use of petrol-powered motorcycles in the centres of several of the larger cities; (ii) rising standards of living have resulted in a surge in the demand for inexpensive private transport. If the trend to increased personal wealth continues, however, it will result in aspirations for automobiles rather than two-wheeled vehicles. Ironically, the continued increase in the E2W market is also now threatened by one of the factors that helped to launch this form of transport. Specifically, legislation is now being put in place in some cities to outlaw electric bikes in favour of public transport because of the reduced rider safety and the impairment of traffic flow that results from the presence of large fleets of e-bikes on the streets. Thus legislation can be a strong factor in the introduction of a new technology – such as that for the electrification of road vehicles – in either a positive or a negative sense.

After China, the next-largest E2W market is currently Japan. The E2Ws preferred in Japan are different from those in China in that they are typically pedal-assist-type vehicles and they are powered by either nickel–metal-hydride or lithium-ion batteries. Electric bicycles have not sold so well in other parts of Southeast Asia – possibly because the performance of the VRLA, the most affordable battery type, degrades rather quickly in regions where temperatures are very high throughout the year.

5.4 Solar cars

The Australian '*Tindo*' bus (*v.s.*) is the first serious road vehicle to be powered by solar energy, albeit pre-generated and packaged electricity. There are obvious limits to the area of photovoltaic cells that can be mounted on a vehicle and therefore it may be unrealistic to dream of solar-powered transport, even in a sunny climate and using photovoltaic (PV) cells of much improved efficiency. The possibility exists, however, for 'solar-assisted' cars, whereby roof-mounted PV arrays would contribute towards keeping the vehicle's batteries charged during hours of daylight. Already such a system is employed on canal boats in Europe to maintain battery charge when the craft is moored and the engine is not running. Such uses of solar energy would be a modest, but direct, contribution of renewable energy to the transport sector.

It is, in fact, possible to power a vehicle exclusively by solar energy provided that the vehicle is ultra-lightweight, high-efficiency PV cells are used and weather conditions are favourable (e.g. little headwind and/or crosswind, high insolation levels). The realization and pioneering demonstration of this fact by two Australian adventurers gave rise to the World Solar Challenge (WSC) – an endurance competition for solar cars that has been run regularly since 1987 across the outback of Australia from Darwin in the north to Adelaide in the south, a distance of 1870 miles (3010 km) along the Stuart Highway, one of the world's most remote long-distance routes. The object of the event (apart from winning!) is to act as a test-bed and showcase of new technology for PV cells, lightweight electric motors and batteries. When the challenge was announced, it was taken up enthusiastically by mechanical engineering teams from companies both large and small, and by universities from around the world. Even schools and individuals have submitted entries. The regulations of the WSC are laid down carefully, as regards both equipment and conduct, to ensure fairness in judging the technology employed. The principal requirement, however, is both daring and simple – once the event starts, the solar cars must be powered only by direct sunlight.

The use of solar cars is simply a means to an end – nobody has any delusion that, in the future, family cars will be driven directly by solar energy (especially in temperate climates). Even so, many of the world's major automotive companies have taken part in the WSC and the experience is believed to have helped in the development of components for both electric and hybrid electric vehicles. More specifically, the beginning of a renaissance in the electrification of conventional road vehicles has been attributed to the running of the first WSC in 1987 when *Sunraycer*, a car entered by General Motors Corporation (USA), surprised the world with its flawless cruise across a continent; see Figure 5.14(a). Completing the course at an average speed of 66.92 km h^{-1} and with a commanding lead of 600 miles (965 km) in front of the second-placed car, *Sunraycer* made road transport history.

Over the years, the performance has improved steadily, largely as a result of developments in solar cells and in batteries. Solar cars store energy in cells/batteries to provide supplementary power to climb hills, accelerate, and keep moving in adverse conditions. The cells/batteries must be rechargeable and the vehicle must travel along the entire course with the same make and number as were fitted at the start. For practical reasons, the cars commence the event with batteries that have been charged from the mains.

As demonstrated by the experiences and outcomes of the WSC, the cars are playing a valuable role in charting the progress of rechargeable battery performance. *Sunraycer* used silver–zinc cells, which boasted the highest specific energy at that time. Lithium-ion batteries were first used in the 1996 WSC and, as confidence in the fledgling chemistry grew, they became the technology of choice for most competitors. Since 1999, when *Aurora 101* from Australia took line honours with a silver–zinc pack (see Figure 5.14(b)), each winner of the subsequent seven meetings of the WSC has employed a lithium battery pack. In 2005, *Nuna 3* from the Netherlands completed the course in 29 hours and 11 minutes at an average speed of 102.75 km h^{-1}; see Figure 5.14(c). It was the first time that a solar car had bettered the elusive 100 km h^{-1} barrier. By comparison, *Sunraycer* had taken 44 hours 54 minutes to cross the continent – such is the march of progress! It should be remembered, however, that the performance of a solar car is determined not only by its inherent technology but also on a fuel supply that depends on the vagaries of the weather. The latter was brought into stark reality during the WSC in 2011 when the cars had to battle dust storms, high winds, bushfires and thunderstorms such that the winner and defending champion, *Challenger 2* from Tokai University in Japan (see Figure 5.14(d)), recorded an average speed of only 91.54 km h^{-1}. In the 2013 WSC, a solar car from the Eindhoven University of Technology in the Netherlands with a driver and three passengers used only 64 kWh of electricity to cover the course in a time of 40 hours 7 minutes.

Enormous enthusiasm for the challenge of designing and competing solar-powered cars has been shown by young engineers, both professional and amateur, from around the world. The contest not only acts as a test-bed for improved components, but also serves to inspire young people to take up a career in engineering.

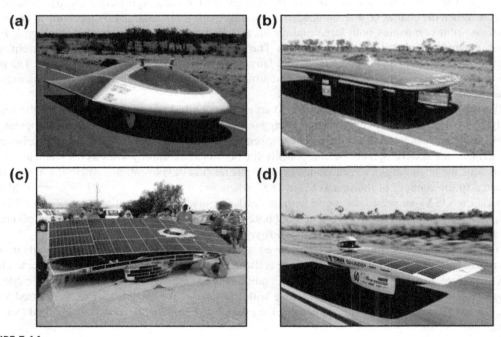

FIGURE 5.14

World Solar Challenge winners: (a) *Sunraycer*, USA, 1987; (b) *Aurora 101*, Australia, 1999; (c) *Nuna 3*, the Netherlands, 2005; (d) *Challenger 2*, Japan, 2011

(Images reproduced with permission from World Solar Challenge)

5.5 Benchmarks of progress towards cleaner and more efficient vehicles

5.5.1 Energy efficiency

A 2008 study by the Massachusetts Institute of Technology (MIT) into the energy efficiency of mid-sized American passenger cars, both present-day and projected to 2035, led to the values shown in Table 5.6. The figure reported for the 2008 US family car – namely, 8.9 L per 100 km – equates to 31.6 miles per UK gallon, which is less economical than most cars driven by Europeans. The projections for the spark-ignition (SI) engine to 2035, based on anticipated engineering advances, show an improvement in fuel consumption of 38% for the naturally aspirated engine and 45% for the turbocharged engine. Again, these projections have already been achieved for some cars. For instance, the Ford *Fiesta* fitted with the three-cylinder 1-L turbocharged Eco-boost® engine is claimed to consume only 4.3 L per 100 km (combined cycle) compared with the 4.9 L per 100 km projected for a turbo spark-ignition car in 2035. Of course, the Ford *Fiesta* is a rather smaller car than the authors of the MIT report had in mind, but a compact car may become the vehicle of choice for many people in an era of high petroleum prices.

The same situation prevails for diesel cars. The projected figure for 2035 of 4.7 L per 100 km (60 miles per UK gallon) has already been achieved in medium-sized European diesel cars. This is also about the same economy reported by the US Environmental Protection Agency for the 2012 Toyota *Prius* petrol hybrid for a combination of urban and highway driving. At present, diesel-engined cars and full hybrids running on petrol return similar fuel economies. The new *Prius* plug-in hybrid will presumably be more economical. Finally, the projection for the BEV, in petrol-equivalent terms, is substantially better at 1.7 L per 100 km. There are, however, many other factors to be considered when opting for a BEV, as already listed in Table 5.5.

In summary, remarkable progress is being made in improving the fuel efficiency of all classes of vehicle and there is undoubtedly scope for yet further advances – not only through the progressive introduction of additional electrical functionality but also by the avoidance of the energy wastage that is embodied in hydrocarbon fuels. At present, a considerable amount of energy is lost as heat in the exhaust gases and in the engine-cooling system, as shown schematically in Figure 5.15. Thus, with the

Table 5.6 Tank-to-Wheels Energy Consumption of a Typical US Passenger Car and Projected Values for Various Future Options in 2035

Configuration	L per 100 km
SI engine (2008)	8.9
SI engine (2035)	5.5
Turbo SI (2035)	4.9
Diesel (2035)	4.7
HEV (2035)	3.1
PHEV (2035)	2.2
BEV (2035)	1.7

SI = spark ignition. For PHEVs and BEVs, the electricity consumption is expressed as petrol-equivalent units.

growth in the electrical demands placed on the vehicle, there is active research under way to explore the possibility of using a thermoelectric system for the direct conversion of waste heat into electric power to supplement the output from the ISG. A project mounted within the 7th Framework Programme of the European Commission targets the recovery of 3 kW of electrical power under full-load conditions, and 1 to 2 kW under partial load.

5.5.2 Carbon dioxide emissions

One of the arguments regularly advanced in favour of BEVs is that they are free of exhaust emissions, i.e. no pollution and no greenhouse gases. This assertion is true at the level of electricity supply to traction effort at the wheels ('battery-to-wheels') but, as mentioned above in Section 5.3.3, it takes no account of the emissions that arise in generating the electricity at the power station ('power-station-to-battery'). The nature and level of these emissions depend on the fuels employed. The worst offender is coal, a variable commodity that releases acid gases (sulfur dioxide and nitrogen oxides) during combustion, as well as numerous other undesirable elements, including arsenic and mercury. Under the EU's Large Combustion Plant Directive, all large coal-fired power stations within the region have a choice of fitting flue-gas desulfurization equipment to their stacks or closing operations before the end of 2015.

Natural gas is much cleaner-burning than other fossil fuels and has a higher hydrogen-to-carbon ratio, so that it releases far less carbon dioxide to the atmosphere per unit of electricity generated.

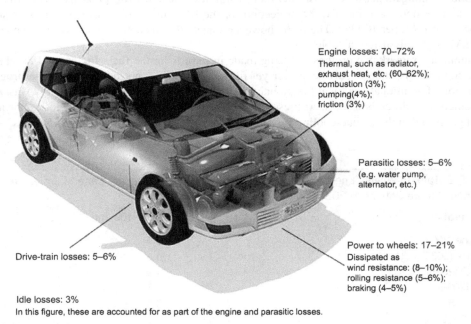

Engine losses: 70–72%
Thermal, such as radiator, exhaust heat, etc. (60–62%); combustion (3%); pumping(4%); friction (3%)

Parasitic losses: 5–6% (e.g. water pump, alternator, etc.)

Power to wheels: 17–21%
Dissipated as wind resistance: (8–10%); rolling resistance (5–6%); braking (4–5%)

Drive-train losses: 5–6%

Idle losses: 3%
In this figure, these are accounted for as part of the engine and parasitic losses.

FIGURE 5.15

Destinations of fuel energy (average values) for a medium-class car in urban areas

(Courtesy of the Department of Energy, USA)

Biofuels combust to carbon dioxide and water but, because they derived their carbon from the atmosphere in recent times and are replaceable (at least, in principle), they are not counted as generators of greenhouse gases (GHGs). The remaining means of generating electricity are nuclear power, a controversial technology, and the renewable sources notably: hydro, wind and solar. These are regarded as essentially free of GHG emissions.

In order to calculate the effective amount of carbon dioxide emissions attributable to PHEVs or BEVs, it is necessary to take account of the mix of fuels employed by the power station that is supplying the electricity. This will vary from country to country, and even from utility to utility. Calculations show that for utilities heavily biased towards coal, BEVs are little better than conventional ICEVs in terms of carbon dioxide emissions. At the other extreme, BEVs based solely on nuclear or renewable electricity are almost free of GHG emissions.

In addition to the carbon dioxide derived from the fuel, a full analysis for road transport should also include the GHGs liberated during the manufacture of the vehicle itself. Road vehicles are made mostly of steel, aluminium, plastics and glass, while BEVs may also contain sizeable amounts of lead, nickel or lithium in their batteries. By taking account of (i) the quantity of each material in a vehicle, (ii) the amounts of GHG liberated per tonne during its manufacture and (iii) the mass of petrol or diesel used in handling and transporting the materials, it is possible to calculate the 'embedded' quantities of GHG in the vehicle. Clearly, this is a vehicle-specific exercise and, when completed, the embedded GHG value needs to be amortized over the total distance travelled by the vehicle during the course of its life. This will enable computation of the quantity of GHG per km that must be added to that released from combustion of the fuel.

Attempts have been made to carry out calculations for future transport scenarios, in which assumptions are made about the evolution of vehicle design and power-trains. At best, however, such analyses serve to provide little more than broad guidance for the future direction of automotive design. Nonetheless, the data obtained in the MIT study cited above do indicate that for PHEVs and BEVs when covering average mileage during their lifespan, the embedded carbon dioxide in the materials of the car are equivalent to emissions of around 20–30 g CO_2 per km. This quantity is not very different from the embedded level for ICEVs. The liberation of GHGs, however, is only one of many factors that the engineer has to take into consideration when designing a vehicle that will find favour with the buying public. Cost is always likely to be a dominant factor.

5.6 Road transport in transition

Targets for reductions in emissions have been set in the member states of the EU, in the USA and in China – and have become progressively more severe with the passage of time; see Section 2.7, Chapter 2. The UK, for example, has signed up to an 80% reduction in overall carbon emissions by 2050 (based upon the 1990 figure). This is a huge challenge, the implications of which many people have yet to grasp.

In an attempt to meet the targets that are to be mandated, automotive companies have begun by reducing engine sizes and fitting turbochargers to compensate for the reduced power output (i.e. 'downsize-and-boost' technology). This first step is unlikely to be sufficient to meet the progressively tightening legislation that is foreseen, except possibly for the smallest cars, i.e. those that are not very profitable for the manufacturers. For larger vehicles, the focus is on hybrid systems.

Present vehicles, however, are costly and may still be unable to meet the emissions targets, especially when account is taken of the embodied carbon (i.e. that evolved during manufacture). Nevertheless, progressive electrification of road vehicles is the way forward. Accordingly, to develop new technology that is both sustainable and affordable, automotive engineering will continue to evolve from a largely mechanical engineering discipline to one based upon electrical engineering and computer control technology.

Mains Electricity Supply for Charging Vehicle Batteries

6.1 Why is electricity supply relevant to road transport?

At first sight, it may appear strange to include a review of the electricity industry in a book on road transport. The justification for this review, however, is clear. If plug-in hybrid electric vehicles (PHEVs) and/or battery electric vehicles (BEVs) are ever to become major components of the road transport scene, then it will be necessary to ensure that there is a sufficient, reliable and readily-available supply of electricity. It is essential, therefore, to maintain a watching brief on developments in both the generation and the distribution of this medium of energy.

The rate and extent of uptake of electric vehicles that are propelled, either partially or wholly, by mains electricity will be influenced significantly by competition from internal-combustion-engined vehicles (ICEVs). A key determinant will therefore be the availability and price of petrol and diesel, irrespective of whether these two motor fuels are obtained from conventional sources or from synthetic sources such as shale oil, gas-to-liquids technology or coal-to-liquids conversion. Other important factors that will have an impact on the market for PHEVs and BEVs are:

- political actions, in particular legislation, in the field of climate change and the price levied on carbon emissions;
- progress in carbon dioxide capture and storage (so-called 'sequestration');
- the price of electricity;
- advances in battery and fuel cell technology.

In the final analysis, the acceptability of electric vehicles to the consumer as a form of transport is a key consideration. The outcome depends not only on vehicle performance, but also on the relative capital and running costs at any particular location and point in the future.

In view of the uncertainties attached to each of the factors listed above, it is virtually impossible to forecast with reasonable accuracy the rate of uptake of PHEVs and BEVs. Moreover, the situation is unlikely to be uniform across the world. To take just one example, countries or regions with an abundance of cheap hydroelectricity (e.g. Norway, Canada, China, Iceland, and Tasmania in Australia) will have a greater incentive to introduce vehicles that depend on mains electricity than those without such resources.

There are many different ways to generate electricity on a large scale. Traditional routes are the combustion of coal to raise steam, the burning of natural gas in gas turbines, the exploitation of nuclear fission and using water pressure (hydroelectricity). In recent times, other forms of renewable energy, notably wind, biomass and solar, have been added to the range of technologies for electricity generation. In many countries, there is agreement that the energy-base of electricity generation needs to be diversified. Such an approach provides security of supply in an uncertain world and the means to respond to targets that have been set for reductions in carbon dioxide emissions. The security issue is heightened by the impending closure of many large coal-fired and nuclear stations that are approaching their end of life, and by the political decision of certain countries to withdraw from, or reject, nuclear generation. The question then arises as to whether there will be sufficient electricity to fulfil the requirements in the decades to come, even before considering the impact of a large fleet of electric vehicles drawing current from the mains. A related question is whether the transmission and distribution networks in a particular country would be adequate to meet the new demands.

Clearly, the electricity supply situation will vary greatly from country to country. The potential ability of a few representative, but quite different, countries to provide adequate electricity for the projected large numbers of PHEVs is examined in this chapter. First, however, a brief review of electricity-generation statistics and methods is appropriate.

6.2 Electricity – a driving factor in the world economy

Electricity is highly valued as an extremely adaptable and clean form of energy. So many tasks of everyday life can be accomplished only by electricity. Electric lighting, electric motors and electronic devices in general (notably televisions, computers and mobile phones) are all integral to modern society. In 2010, 17.7% of the world's final consumption of energy was supplied as electricity. As an energy vector, electricity is versatile in production and versatile in application.

The contributions of different energy sources to world electricity generation in 1973, 2001 and 2010 are listed in Table 6.1. The data show that the combined output has experienced a 3.5-fold increase over this period of 37 years. Approximately 41% of the total electricity produced is consumed in industry, less than 2% in transport and the remainder in all other sectors of the economy.

The individual amounts of electricity generated in 2011 by the top 10 producers are presented in Table 6.2. There is an enormous disparity between one country and another in the consumption of electricity per capita. In kWh, the rate ranges from 52 621 for Iceland, through 11 919 for the USA, 10 238 for Australia, 6017 for Russia, 5467 for the UK and 3493 for China, down to 498 for India, 110 for Sudan and 7.6 for Afghanistan. Clearly, there is huge scope for increased electricity usage in much of the world.

At the present time, growth in electricity production is most pronounced in developing economies. In East and Southeast Asia, the annual growth rate per capita is 6.5%, and in North Africa it is 4.8%. These data contrast with those for the developed world, where the corresponding rates are 0.6% for Australia and 0.4% for Western Europe. In North America, the per capita consumption is even declining as a result of conservation measures (building insulation, high-efficiency lamps, etc.) and the effects of an economic recession.

Most electricity is derived from the combustion of fossil fuels, especially coal and gas, although nuclear energy and hydroelectric energy both make sizeable contributions. These latter two energy sources are clean in the sense of not liberating significant amounts of carbon dioxide or other pollutants into the atmosphere during operation. Notable changes have also taken place in the proportions contributed by each of the fuels used. In percentage terms, as seen in Table 6.1, coal has remained about the same, gas has increased significantly, oil has decreased markedly, and nuclear is in decline over the past decade as older power stations have closed and fewer new ones have been built. Hydroelectricity (including pumped storage plants) is the principal source of electricity from renewable energy, with an 82.9% share (equal to 16.0% of global electricity). Wind energy, the second largest renewable since 2009, accounts for a further 8.3%; it outperforms the biomass sub-sectors of solid biomass, liquid biomass, biogas and renewable household waste (total 6.3%). These generators are followed by geothermal power (1.6%), solar power (photovoltaic and solar–thermal plants, 0.8%) and marine energies (tidal, wave energy and ocean–thermal, 0.01%). As might be expected, the distribution of energy sources differs greatly between countries.

Table 6.1 Percentage of Fuel Shares of World Electricity Generation in 1973, 2001 and 2010

Fuel	1973	2001	2010
Coal/peat	38.3	38.7	40.6
Oil	24.7	7.5	4.6
Natural gas	12.1	18.3	22.2
Hydro	21.0	16.6	16.0
Nuclear	3.3	17.1	12.9
Other[a]	0.6	1.8	3.7
Total (%)	100.0	100.0	100.0
Total (TWh)	6115	15476	21431

[a]Other includes geothermal, solar, wind, biofuels and waste, and marine energy.

Table 6.2 Major Producers of Electricity in 2011

	TWh	% of World Total
PR China	4700	21.3
USA	4308	19.6
Japan	1104	5.0
Russian Federation	1052	4.8
India	1006	4.6
Germany	615	2.8
Canada	608	2.8
France	564	2.6
South Korea	520	2.3
Brazil	501	2.3
Rest of world	7040	32.0
World total	22018	100.1

Source: BP Statistical Review of Energy, June 2012.

The suitability of the electricity production status for the widespread uptake of PHEVs and BEVs with minimal release of carbon dioxide also varies greatly from one country to another and depends not only on the primary energy mix but also on the transmission and distribution systems that are available.

6.3 Generation and distribution of electricity

6.3.1 Coal-fired electricity plant

The world's prime fuel for electricity generation is coal, although this situation is now changing as more gas-fired stations are being built. In traditional coal-fired stations, the coal is first crushed to a fine powder and then blown into a combustion chamber to raise steam that powers a steam turbine. Such stations are large, typically 1–4 GW_e output, but their overall energy efficiency is low, generally in the range of 35–40%. Integrated gasification combined-cycle (IGCC) generation, which is currently under development, offers an alternative and more efficient coal-based means of electricity production. The IGCC process has the potential to increase thermal efficiencies to over 50% with greatly reduced emissions of greenhouse gases. The pulverized coal is first gasified by treatment with steam in a limited supply of air at high temperature to produce a mixture of hydrogen and carbon monoxide (so-called 'water gas'). After the product gases have been cleaned and purified, they are subjected to the 'water-gas shift reaction', which is a catalytic process that converts the carbon monoxide to carbon dioxide. The latter may then be separated in a form suitable for sequestration (e.g. in geological structures), while the hydrogen is fed to a gas turbine to generate electricity. The exit gases from the gas turbine, still hot, pass through a steam turbine to produce further electricity. The IGCC process is not only more efficient than conventional coal combustion, but also holds out the prospect of convenient carbon capture and storage (CCS). The process is generally referred to as 'clean coal technology'. If coal is to have a future in electricity generation, it will have to be via IGCC combined with CCS.

6.3.2 Oil-fired and gas-fired electricity plant

As shown in Table 6.1, the burning of oil produced only 4.6% of the world's electricity in 2010. The appreciable decline in the annual percentage in recent years is largely due to the increasing use of gas as old oil-fired plant becomes obsolete. Most oil-producing countries have gas resources as well and thus have little reason to use oil, a more valuable commodity, to generate electricity. Nevertheless, oil-based generation is unlikely to fade completely; it is ideally suited to small states and communities where the electricity requirement is modest and there is no native gas or coal; for example, in island communities. Such generation is also employed in small-scale combined heat and power (CHP) schemes, and for fast start-up of large coal and nuclear stations (standby operation) when no gas is available.

The growing use of gas to generate electricity is based on:

- the discovery and exploitation of major new gas fields;
- the relative cleanliness of natural gas as a fuel;
- the development of high-efficiency combined-cycle gas turbines;
- the construction of pipelines and liquid natural gas (LNG) carriers to convey the gas to market;
- the advent, particularly in the USA, of shale gas as a new major source of supply.

Combined-cycle gas turbines (CCGTs) are widely employed in the production of electricity from natural gas. The gas is burnt in a high-temperature gas turbine that is coupled to a generator. As with the coal-based IGCC process, the exhaust gas from the turbine is used to raise steam, which is then fed to a conventional steam turbine and a further generator. Not only is gas much cleaner than coal in terms of pollutants (no sulfur dioxide release and greatly reduced emissions of NO_x), but also the quantity of carbon dioxide released is more than halved. A comparison of the advantages and disadvantages of using coal or natural gas to generate electricity is given in Table 6.3.

6.3.3 Nuclear power

The release and harnessing of the immense energy locked in the nucleus of the atom was one of the great scientific achievements of the 20th century. Now, in the 21st century, a debate still rages as to

Table 6.3 Advantages and Disadvantages of Electricity Generation by Coal or by Natural Gas

Advantages	Disadvantages
Coal	
• Large reserves of low-cost coal are available • Well-established industry • Indigenous fuel source for many countries	• Carbon-rich fuel leads to extensive liberation of carbon dioxide • High sulfur content of many coals leads to pollution by sulfur dioxide and to acid rain
Natural gas	
• Natural gas is widely available • Methane has the highest H:C ratio of all the hydrocarbon primary energy sources and therefore the least generation of carbon dioxide per kWh • Natural gas is distributed by buried pipeline or LNG tanker with no environmental impact • By using a gas turbine combined with a steam turbine, high efficiencies are achieved • Natural gas contains almost no polluting impurities	• It may be argued that, where supplies are limited, natural gas is too valuable a fuel to be used for electricity generation and should be reserved for space heating and chemical manufacture • For a country with indigenous coal, but no natural gas, the use of imported gas for electricity generation will place a strain on that nation's balance of payments • Carbon dioxide is still produced, albeit less than that from the burning of coal

whether nuclear energy should be seen as 'clean energy' or 'dirty energy'. Proponents of nuclear power argue that a nuclear reactor gives off no gaseous pollutants, liberates no greenhouse gases and is completely contained. Also, the reserves of uranium and thorium are such that nuclear energy may be regarded as 'sustainable', if not 'renewable'. Opponents point to the radioactive fission products that are inevitably formed in the reactor and that have to be disposed of safely. Also, the 1979 near-disaster at Three Mile Island in the USA, the 1986 tragedy at Chernobyl in the Ukraine and the 2011 destruction of the Fukushima reactors in Japan caused by a tidal tsunami have all demonstrated the very real risk of major reactor accidents. In addition, there is the possibility that nuclear materials will be diverted by countries or terrorist organizations for the purpose of making nuclear weapons. Advocates of nuclear power argue that all three major accidents were caused by special circumstances, that lessons have been learnt, and that new designs of reactor and installation are far safer today.

The construction of nuclear power stations in the USA and in Europe was at its peak in the 1970s, but subsequently, in some countries (USA, Germany, Sweden, for example), there has been growing public opposition to the introduction of new nuclear capacity and the associated radioactive waste depositories. In other countries (France, Japan, Korea), such opposition has been relatively muted and major programmes of nuclear construction have taken place, although those in Japan have stopped in the wake of the disaster at the Fukushima I Nuclear Power Plant. With the increasing realization of the importance of anthropogenic carbon dioxide with respect to climate change and the limited options there are for sustainable energy sources, nuclear energy is once more being considered in certain countries.

Table 6.4 Major Producers of Nuclear Electricity in 2010

Country	Production TWh	% of World Total	Installed Capacity GW
USA	839	30.4	101
France	429	15.6	63
Japan	288	10.4	49
Russian Federation	170	6.2	24
South Korea	149	5.4	18
Germany	141	5.1	20
Canada	91	3.3	13
Ukraine	89	3.2	14
PR China	74	2.7	11
UK	62	2.2	11
Rest of world	424	15.5	51
Total	2756	100.0	375

The principal nuclear nations and their respective outputs of nuclear electricity in 2010 are shown in Table 6.4. The countries that are most nuclear oriented (expressed as a percentage of national electricity output), and therefore most vulnerable to a closure of their nuclear programmes, are France (76%), Ukraine (47%), South Korea (30%), Japan (26%), Germany (23%) and the USA (19%). For these countries, especially France, it is totally unrealistic to assume that, in the short-to-medium term, nuclear electricity could be replaced by any other form. It is significant that three of these countries (France, South Korea and Japan) have very little in the way of indigenous fossil fuels. The introduction of nuclear power was therefore undoubtedly inspired by a desire to minimize imports of oil and gas. Germany relies heavily on both coal and nuclear power for its electricity, since it has no domestic oil or gas, but has nevertheless announced an intention to terminate nuclear operations, as has Sweden. Presumably they will rely on imported gas. By contrast, the UK plans to build new reactors, to replace those that are being retired, and there are strong nuclear programmes continuing in the Far East.

6.3.4 Electricity from renewable energy

The many and diverse forms of renewable energy that are being exploited today, albeit on a comparatively small scale, have the potential to contribute a greater share of the world's primary energy. It is convenient to describe these energy sources in terms of the following three categories:

 (i) those that give rise to heat – whether through combustion (biomass), through solar heating or through extraction of energy from the earth (geothermal) – at a temperature that is sufficiently high for raising steam to generate electricity;
(ii) those that may be harnessed to generate electricity via the medium of mechanical energy without first going through a steam cycle; these sources include hydro, wind, tidal and wave energy;
(iii) those that employ solar radiation directly to produce electricity – namely, photovoltaic and photo-electrochemical systems.

It may be noted that most forms of renewable energy, with the exception of geothermal and tidal energy, are derived indirectly from solar energy.

In 2010, combustible renewables (mostly wood) and waste constituted 10% of the world's primary energy supply and thereby made biomass the fourth major source (after oil, coal and natural gas), i.e. greater than both nuclear power and hydroelectricity. In terms of electricity generation, however, biomass is a minor player; see Table 6.1. There is potential for it to assume a larger role, but only on a local basis through small CHP schemes. The technology to burn wood at all scales from the domestic room heater up to quite large power stations is well established. The factor that limits the size of the power plant is the distance over which it is economic to transport forestry waste. Many such generating units will lie in the range 5 to 30 MW_e, which is small compared with traditional thermal power stations.

The second category of renewable energy – the one that yields mechanical energy directly – is already contributing significantly to global electricity generation via the harnessing of hydro (including pumped hydro), 16%, and wind energy, ~3%. The scope for expanding hydroelectricity is limited by geographical factors, but the exploitation of wind generation is growing rapidly, especially in Europe and parts of the USA. Marine energy sources (tidal, waves) currently make a negligible contribution globally.

The collection and use of solar energy, although still modest in comparison with the other sources of energy, are expanding rapidly in concert with ongoing advancement of the technology. Photovoltaic (PV) cells and panels are becoming more efficient and their cost is falling as a consequence of mass production. Most PV installations are comparatively small scale; their electricity may be consumed locally or fed back into the grid.

In summary, it can be said that renewable sources of electricity (with the exception of hydro), comprise units that are small individually, but which when replicated in large numbers are capable of making a useful contribution to present-day electricity supplies, with the potential to grow substantially in the years ahead.

6.3.5 Combined heat and power (co-generation)

The combined heat and power (CHP) concept, or 'co-generation', is a simple means of increasing the overall efficiency of a power plant by utilizing the waste heat in the exhaust gases rather than discharging them to atmosphere while still warm. Thus, fuel is used at high efficiency and overall emissions of carbon dioxide are minimized. In order to obtain waste heat that is useful, it is normally necessary to raise the temperature of the exhaust gases, which reduces the efficiency of electricity generation but increases the overall efficiency. The latter may be defined as the ratio of the sum of the electrical and heat output energies to the input heat of combustion. The overall efficiency for CHP is typically in the range 70 to 90%, as opposed to 35 to 55% for conventional power plants. At maximum efficiency, the ratio of heat output to electricity output is around three. Co-generation is particularly appropriate for applications that can accept this ratio, e.g. in factories or farms that operate drying processes, or in hotels where the demand for hot water is high.

Co-generation plants are usually small compared with conventional power stations and are often sized to suit an enterprise that requires a fairly constant quantity of electricity and heat for its operation. In the UK, almost half of the CHP schemes have an electrical output of less than 100 kW_e, whereas those larger than 10 MW_e account for over 80% of the total installed CHP capacity. Applications for larger CHP units (i.e. megawatt size) are found in various commercial activities, e.g. the chemical and petrochemical, paper and board, food and drink, and iron and steel industries. Such a unit might, for example, generate around 40 MW_e of electricity and 120 MW_{th} of heat to support the operation of a petrochemical plant. Smaller units (20 kW_e upwards) are used to supply heat and power to hotels, hospitals and apartment blocks. Natural gas is the most common fuel and is burnt in gas engines or in micro gas turbines. In Scandinavia, CHP is widely employed in district heating schemes. Several European countries, e.g. Denmark, Finland and the Netherlands, are said to produce one-third or more of their total electricity by co-generation.

Micro-CHP units are now being manufactured for domestic service to replace conventional gas boilers. These operate by means of the Stirling cycle (i.e. as an external-combustion engine). A typical example is shown in Figure 6.1. In addition to providing hot water for central heating, baths, showers, etc., present models also generate around 1 kW_e, which would be ideal for overnight charging of a PHEV, or even a small BEV. Micro-CHP units can be wall-mounted and require no more maintenance than a standard domestic boiler. Systems with greater electrical output would undoubtedly become available for buildings that can use larger quantities of heat. It remains to be seen when, or in what circumstances, micro-CHP would be attractive for charging electric-vehicle batteries, but the idea of independence from the grid has its appeal, particularly in countries or regions where the mains supply is erratic or unreliable.

FIGURE 6.1

An example of a micro-CHP unit

(By Courtesy of Urban Green Energy)

6.3.6 Distributed generation

The concept of local, small-scale generation of electricity is by no means new. Indeed, early in the 20th century, before the days of the cross-country grid, most electricity supply was of this nature. With the advent of the grid, however, local generation became confined largely to isolated communities and islands. There is currently a resurgence of interest in local generation, referred to as 'distributed generation'. The activity began with battery backup systems to provide uninterruptible power supplies, as well as with small-scale CHP systems, but has grown to encompass a range of other technologies, e.g. engine–generator sets; small hydroelectric plants; micro gas turbines; and renewable electricity based on wind, biomass, landfill gas, photovoltaic cells and fuel cells; together with various storage technologies. Typically, generating plant is dispersed across the network, near to the consumers, rather than concentrated at one location; hence 'distributed generation'. The systems may operate independently of the grid, but there is an increasing trend for them to be grid-connected, so-called 'embedded generation'. The growing interest in the decentralization of electricity generation is driven by considerations of overall energy efficiency, by an increasing desire to introduce renewable energy sources and, more recently, by deregulation of many electricity markets.

Distributed generation is considered to offer the following benefits:

- enhanced system reliability when standby generators are installed;
- the high energy efficiency of CHP plants;
- cost-effective peak shaving;
- lower grid transmission losses, typically a 10 to 15% saving;
- ability to provide voltage support and power-factor correction;
- deferred investment in upgrading the transmission and distribution system;
- facilitation of the introduction of renewable forms of electricity.

Evidently, this is a powerful set of incentives for the uptake of distributed generation.

Local electricity generation, of whatever kind, is particularly important for people living in remote regions, far away from a mains supply. These include, for example, many in rural Africa and India, in the outback of Australia, in areas of northern Canada and Siberia, and in thousands of islands spread throughout the world. Such people, where they can afford it, at present generate their own electricity domestically by using small petrol- or diesel-fuelled generators, or by means of wind turbines or photovoltaic panels. In poorer villages, there may be a single generator to provide sufficient electricity to illuminate a communal meeting centre with associated television or film projection facilities, or to operate a refrigerator to store medicines. In the context of road transport, BEVs are particularly attractive for use on small islands, where distances are short and batteries could be recharged using distributed generation.

6.4 Electricity availability in selected countries: contemporary case studies

In this section, five countries, which differ greatly in size, prosperity and fuel supply, are considered to explore how readily their respective electricity generation and distribution facilities might cope with a major influx of electric vehicles that draw power from the mains. In some countries, the conversion to vehicle fleets that do not depend on the availability of fossil fuels can be easily envisaged. In other cases, no such change in road transport appears likely for a very long time.

6.4.1 Iceland

Iceland is considered first because it is a country that has no fossil fuel resources of its own, but abundant renewable energy. It has a small population (319 000, i.e. no more than that of a medium-size city) and a mountainous country terrain with strong rainfall for most of the year. The streams that cascade down from the mountains provide an ample source of hydroelectric power that is harnessed to produce around 75% of the electricity supply. Iceland is located above the Mid-Atlantic Ridge, a major fault line in the earth's crust, and is thus noted for its volcanic activity and geothermal hot springs. The thermal energy from these two natural phenomena provides the remaining 25% of the electricity, as well as all the hot water required for domestic and office heating. With hydro and geothermal power to spare, Iceland has no real need of wind or wave power. The import of oil is largely confined to its use in road transport, aviation, shipping (fishing), agriculture and industry.

In 2010, Iceland generated 16.6 TWh of electricity, of which over 80% was consumed by industry, predominantly for the manufacture of aluminium. The electrolytic extraction of aluminium is an energy-intensive business and therefore, in economic terms, the process is best conducted where electricity is both plentiful and cheap. In recent years, this argument has been reinforced by the concomitant saving in carbon emissions to the atmosphere through the use of renewable forms of electricity. The size and importance of the aluminium operations are demonstrated by the fact that, in 2010, the electricity generated per capita in Iceland was 52 MWh; by comparison, the corresponding amount in the UK was just 6–7 MWh.

Iceland imports more than twice as much oil per capita as the UK, mostly for use in various forms of transport. It is the consumption of oil by *road* transport that the government wishes to target for conversion to a renewable base. This would greatly reduce the country's carbon emissions (small though they are on a global basis) and, at the same time, improve its balance of payments. Synthetic fuels are a possibility (e.g. hydrogen), but so also is hydroelectricity to power electric vehicles directly.

The population of Iceland is largely centred on the capital, Reykjavik, in the southwest of the island. Consequently, much of the road travel is semi-urban with short journeys and, in practice, should be ideal for BEVs, although (as always) these are not suited to occasional long journeys. The government strategy, therefore, has been to leverage the country's experience of electrolysis in the aluminium industry to produce hydrogen electrochemically for use in fuel cell vehicles (FCVs). The total potential resource of hydro and geothermal power is clearly in excess of current usage. Given that the population of Iceland is not expected to grow greatly, there should be no difficulty in generating adequate electricity to deliver the hydrogen required for a fleet of road vehicles powered by fuel cells. This development would make Iceland largely free of oil imports, apart from those needed to supply fishing vessels and aircraft, and it would then become one of the world's first nations to be almost completely powered by renewable energy. A pilot study was carried out between 2001 and 2005 in which three public-service buses powered by fuel cells were operated in Reykjavik and a refuelling station was set up to dispense hydrogen that was produced electrolytically. The pilot study has yet to be taken further. Given that BEVs are already on the market and that FCVs are still some years away from being economically viable and commercially available (see Chapter 8), it may be that the Icelandic population will opt for the former technology first.

Iceland is, of course, exceptional. As a small and developed country with a low population density and abundant sources of renewable energy, it lies at one end of the energy spectrum of nations. At the other end are under-developed countries with high population density, low gross domestic product

(GDP) and few energy resources, either fossil or renewable; examples are Bangladesh and the Philippines. Between these extremes lie the developed countries of Europe, each with its own energy characteristics, e.g. Germany and Poland with coal and wind resources, Norway with oil and gas, France with nuclear power and hydroelectricity, Spain and Portugal with solar energy.

The UK is fortunate in having offshore oil and gas (the latter is a diminishing resource, but still substantial), shale gas (not yet exploited), coal (no longer mined in quantity, but potentially available), nuclear power, and a good supply of wind energy. Thus, it is of interest to consider next the UK and its ability to provide electricity to a large fleet of electric vehicles.

6.4.2 United Kingdom

The UK is a relatively small country in terms of area, with a high population (63 million). Overall, there are around 260 people per km^2. This number varies greatly on a regional basis, however, with parts of Scotland and Wales being sparsely populated, and with the southeast of England and the Midlands having much higher population densities. Electricity is generated mostly in large, central power stations, coupled together in a national grid. These stations operate on fossil fuels (coal or natural gas) or nuclear power, although there are one or two remaining oil-fired stations of smaller size. The national electricity flow in 2011 is shown schematically in Figure 6.2. The source of the energy for generation (expressed in TWh) is given on the left-hand side of the diagram, and where the electricity was consumed on the right-hand side.

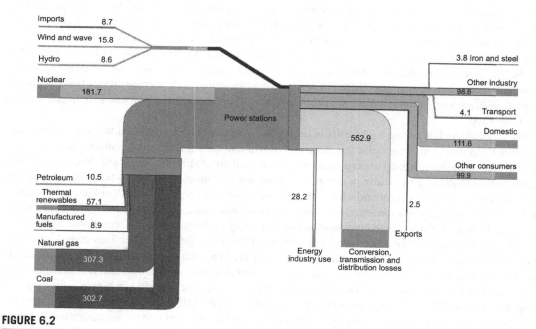

FIGURE 6.2

Chart of UK electricity flow (TWh), in 2011

(Image courtesy of UK Dept. Energy and Climate Change: 'Digest of Energy Statistics 2012')

The total electricity generated was 381 TWh, of which only 1% was used in the transport sector (mostly electric trains and London Transport). Almost half of the electricity (46%; 175 TWh) was derived from the use of gas. By comparison, coal and nuclear sources accounted for 28% and 16%, respectively. Renewable forms of energy (mostly biomass and wind energy) accounted for just 6.8% – a very different situation to that in Iceland. Of the total energy input for electricity generation (929.8 TWh), 61% was lost in conversion, transmission and distribution. The major part of this loss stems, inevitably, from the conversion of heat energy to electricity. Coal-fired power stations generally have efficiencies in the range of 35 to 40%, whereas gas-fired counterparts that employ modern CCGT technology operate at 50 to 55% efficiency. (Power-station efficiency is defined as the ratio of the electrical output, measured in MWh, divided by the calorific value of the fuel supplied, expressed in the same units.) Many of these base-load stations are very large – namely, 2–3 GW and 1–1.5 GW for coal-fired and nuclear stations, respectively. The total installed generating capacity of the UK is almost 90 GW. Over the past few years, the maximum instantaneous load met at any time in the year was 73–78% of the installed capacity. Allowing for outages (breakdowns and maintenance), this demand is approaching the realistically achievable generating capacity, while leaving a margin for safety.

The 'plant load factor' is a measure of how intensively the power station has been used – that is, the amount of electricity actually generated in the course of a year as a percentage of that theoretically possible if it were to be operated continuously at the maximum rated value. The load factor achieved in practice is controlled by the demand (very weather-dependent), by competition between utilities, by the respective costs of operating stations of varying ages and using different fuels, and by the extent to which a station is unavailable due to breakdowns or maintenance. On average, coal-fired stations have load factors of around 40%, whereas those for gas-fired CCGT plant and nuclear stations are around 60%. By contrast, onshore wind generators can only manage load factors of 20–30% because of low average wind speeds. The more recent offshore wind farms perform better, typically 40%, as wind speeds are generally higher and more consistent at sea. There is also a seasonal variation, with average speeds being higher in winter than in summer, although there are times in winter when atmospheric pressure is high and there is very little wind.

Where does the UK obtain fuel for its major power stations? The nation used to have a substantial coal-mining industry, but most of it has been shut down and only a few mines remain open. Many of the coal seams are buried deep underground and working them is difficult, dirty and dangerous. Often the seams are shallow, 1 m or less in depth, which means that much rock has to be mined in order to reach the coal and miners frequently have to work in a doubled-up position. When the coal is finally brought to the surface, it is more expensive than imported open-cast coal. Also, much of it is of poorer quality and has a higher average content of sulfur, typically in the range 1–3%, compared with 1% or less for imported coals. Under the Large Combustion Plant Directive set by the European Union, the levels of sulfur dioxide in discharge gases must be controlled by fitting equipment for flue gas desulfurization. For these reasons, most coal used in British power stations is now imported. It is brought from as far afield as Russia, Indonesia, Australia, and from wherever it is available in high quality and at a competitive price. When the large coal-fired power stations in the UK reach their end of life, it appears unlikely that they will be replaced by coal-based units, unless and until carbon sequestration is developed and becomes commercially attractive.

Given these issues with coal, several of the major UK generators are now busily building new, efficient plant based on combined-cycle gas turbines. Natural gas comes mostly from the North Sea and by pipeline from Norway, Russia and the Caucasus, although some is now imported as liquid natural gas (LNG) in refrigerated tankers from terminals in the Middle East or North Africa (Algeria, Nigeria). The expectation is that more gas will be obtained by pipeline from Russia and from countries around the Caspian Sea. In addition, new terminals for the import of LNG by bulk tankers are being constructed. Further into the future, indigenous gas may be derived from shale sources identified in Lancashire and other parts of the country, although at present this technology is immature and controversial.

The UK government has announced its intention to permit a new series of nuclear power stations to be built and it is unlikely to be deflected from this course of action by the disaster at the Fukushima I Nuclear Power Plant in Japan. At the same time, numerous wind farms are being erected, both onshore and offshore, despite their high cost and relatively poor load factors, and despite objections from some sections of the community based on adverse visual, noise and ecological impacts. The construction of nuclear power stations and wind farms is being driven by the need to meet targets for reduced carbon emissions, and financial incentives are in place to encourage these developments. There are also subsidies for householders to install photovoltaic ('solar') panels; initially these were generous but have been reduced in line with the falling cost of solar cells. Altogether, major changes are now in progress in the UK electricity industry to ensure that a sustainable supply will develop and continue through the coming decades. Looking further ahead, new generations of nuclear reactor are on the drawing board, e.g. high-temperature gas-cooled reactors and fast-breeder reactors, and it is even possible that fusion reactors may finally become practicable towards the end of the century. In summary, there should not be a shortage of electricity in the UK in the medium term, at least to meet currently projected demands.

Given the above background, it is instructive to assess the consequences of a significant proportion of the UK fleet of road vehicles having to rely on the mains electricity supply for its fuel. This is a two-part issue – namely, the adequacy of (i) the generating capacity and (ii) the transmission–distribution capacity. At present, both the installed generating capacity and the grid are able to meet the maximum national demand with a margin for comfort. The electricity generation situation will, however, change over the next few years as existing coal-fired and nuclear stations reach their end of life and new plant is progressively introduced. It is anticipated that about 25% of existing capacity will have to close and be replaced within the next 10 years. This situation represents a major challenge to the industry. The UK government has announced plans to ensure that no further coal-based generation will be permitted without carbon capture and storage (CCS). The emphasis will be on providing a level playing field for companies to invest in gas-fired generation (preferably with CCS), in new nuclear plant and in renewables. Together with a floor price guaranteed for carbon, it is expected that these measures will help the UK to meet its targets of 15% renewables (mostly wind and hydro) by 2020 and 80% reduction in carbon emissions by 2050. In view of the complex technical, economic and political constraints on the energy market, doubts have been expressed that these targets will be met.

At present, the number of light-duty vehicles in the UK is around 31 million. As an exercise, assume that about one-tenth of this fleet (3 million) becomes mains-powered electric vehicles (i.e. a combination of PHEVs and BEVs) that are primarily used for commuting, shopping and other short journeys. If all of these vehicles were simultaneously plugged into domestic sockets at the same time and charged overnight, thereby drawing up to 3 kW of power each (the worst-case scenario), then the maximum added demand would be 9 GW. This load represents an additional 13–14% to the annual peak load of 65–70 GW (*v.i.*) and lies within the capability of the present installed capacity of almost 90 GW. Thus the extra loads should be acceptable, bearing in mind that the peak demand almost never occurs in the late evening or at night-time, i.e. during the predominant period for charging electric vehicles. To be doubly sure, it might prove necessary to have special domestic sockets that are powered only after 9.00 p.m. when the peak evening demand has passed. Most owners of electric vehicles would, in any event, wish to avail themselves of cheap late-night tariffs when electricity usage is low.

Whereas the above analysis relates to the overall national scene, it is equally important to consider the regional situation in densely populated areas such as around London, the Midlands and the north-west. The grid is perfectly capable of transmitting electricity throughout the country to meet peak demand, although in certain localities the distribution system may have to be strengthened to handle the peak load. There should be more than sufficient time to make any necessary infrastructure improvements because the purchase of mains-dependent electric vehicles is only likely to build up slowly over a period of years. In a developed economy, such as the UK, it is also possible that the demand for electricity for conventional applications may decline as a consequence of increases in the unit price and the introduction of 'smart' electricity meters.

A problem arises for all those aspiring purchasers of electric vehicles who live in apartments or in terraced houses with no garages or driveways. Forced to leave their vehicles in the street overnight, these drivers would not normally have personal recharging facilities and it would therefore be necessary to establish public outlets wherever street parking is allowed. Alternatively, drivers would have to replenish their batteries during the day at public or work-based stations and, no doubt, at high tariffs. At the purely local level, it is clear that adequate deployment of recharging stations does not yet exist and a major challenge will be to install the necessary infrastructure across the country, or at least in cities, where electric vehicles appear to be most appropriate. Until that happens, it seems improbable that car owners without private recharge facilities will elect to purchase electric vehicles. The problem of ensuring ready access to electricity is likely to be even more acute in continental Europe where far more people live in apartments. By contrast, in North America and Australia, where space is less of a premium, people often live in detached houses on individual plots and with personal garages.

Infrastructure issues will clearly differ from country to country, as dictated by the nature of the embedded electricity supply network. France, for example, is heavily dependent on nuclear and hydro-electric power. One of the characteristics of nuclear reactors is that they are best suited to base-load operation rather than to load-following because they are not easy to ramp up and down. Consequently, the French electricity authorities should welcome an increase in off-peak (overnight) consumption as it would allow a growth in base load. By contrast, countries that have (or will have) no nuclear power and are heavily dependent on fossil fuels and erratic wind energy (Germany, for example) may find it more difficult to accommodate sizeable numbers of electric vehicles without increasing carbon emissions.

Obviously, other countries that plan to intensify the supply of solar electricity will not be able to employ this energy for overnight battery charging. The choice will lie between using non-solar electricity or recharging batteries during hours of daylight, for instance when drivers have parked their vehicles at workplaces or at shopping complexes. There is scope for adjusting the hours of peak demand by means of differential tariffs in all parts of the world. It is mostly in those regions where the network is already overloaded and unreliable that serious problems will be encountered.

6.4.3 India

Having discussed electricity supply in a small, developed country (Iceland) and a typical medium-size European nation (UK) consideration is now given to India, which differs in almost all respects. India is a large, developing sub-continent with a huge population (1.2 billion (1.2×10^9); 17% of the world total). Although there are many major cities, a large part of the population is widely distributed in villages and engaged in agriculture, rural crafts and small-scale manufacturing.

Because of its huge population, India has the fifth largest electricity generating system in the world (185.5 GW of installed capacity in 2011, i.e. more than twice that of the UK). With a population of around 20 times that of the UK, clearly India still has a long way to go to reach European levels of development. One-third of the rural dwellers (some 300 million people) still have no access to mains electricity. Nationally, almost half of their households are without electricity, although this deficiency is unevenly spread across the 28 states. Even in regions serviced by a grid, the domestic electricity consumption per capita is only around 100 kWh per year in rural areas and approaching 300 kWh in towns and cities. By comparison, the average domestic consumption per head for people living in the UK is 1.5–2 MWh per year.

The Indian electricity grid struggles to keep pace with demand; voltage shedding and blackouts are frequent when demand exceeds supply. Such outages have a significant impact on the manufacturing industry, as well as on domestic activities. The International Energy Agency (IEA) estimates that as India industrializes further and prospers, it will need to add 600–1200 GW to its generating capacity by 2050 – three to six times the present installed capacity and similar to the entire installed capacity of the 27 states of the EU. Much of this extra electricity will be required to power domestic cookers (to replace traditional units which burn biomass and therefore are highly polluting), as well as to run refrigerators and freezers to keep food fresh in a hot climate. Fuelling electric vehicles is well down the priority list.

At present, India has a mixed portfolio of generating plant. Coal-fired stations account for around 55% of installed capacity, hydro power 21%, renewable energy (mostly wind) 12%, natural gas 10% and nuclear 2–3%. The composition of the new plant to meet the above-mentioned projection of the IEA is open to conjecture. There is probably little further scope for hydro power, and new coal-fired plant will not be popular until CCS becomes an economic proposition. There is a limited amount of biomass in India and that available will be required for heating and cooking. For these reasons, India is actively expanding its nuclear facilities, with several power stations in the course of construction. From its 2011 base of 4.8 GW of installed capacity, the Nuclear Power Corporation of India plans to reach 63 GW by 2032. This is a massive expansion in just 20 years, but is still only 5–10% of the added capacity that the IEA foresees as necessary by 2050. To meet the target would require the installation of about 2 GW of new plant (of all kinds) every month for the next 35 to 40 years – a formidable challenge! The programme set by Nuclear Power Corporation includes the building of six reactors, of total capacity 9.2 GW, at Koodankulam on the southern tip of India. The first two of the reactors are nearing completion, but commissioning is held up by objections from the local populace. Similar opposition to nuclear power is encountered in many other countries and constitutes a major uncertainty in planning for future electricity supplies.

In addition to new generating capacity, India will have to enlarge and extend its national grid, both to carry much greater electrical loads and also to provide electricity to all its citizens. At present, the distribution system is said to be inefficient, with network losses that exceed 32%, i.e. well above the world average of less than 15%. Renewal of the grid will represent as great a challenge, and possibly the same capital investment, as that to be encountered with expansion of the generating capacity.

Given this situation, what are the implications for sales of electric vehicles in India? The starting point is that today the vast majority of Indians do not own personal mechanized transport and have to be content with bicycles or, in the countryside, with oxen to draw carts. To own a small car is the ambition of most Indians, but is still far from being realized by the bulk of the population. The Tata Motor

Corporation is introducing a small, basic ICEV – the *Nano* (see Section 4.3.1.2, Chapter 4) – at a target cost price of about £1000 (US$1600), but this will still be beyond the means of the majority of people. Moreover, if most Indians were indeed able to afford such a car, the inadequate road network would soon become totally saturated. A major expansion and improvement of the road system in India is now taking place, but the prime aim is to smooth the flow of present traffic and to speed up inter-city journeys, rather than to accommodate vastly more vehicles.

Mains electric vehicles (PHEVs, BEVs) are appreciably more expensive than basic ICEVs. Accordingly, it is improbable that this technology will command a significant share of the Indian vehicle market in the foreseeable future. Electric vehicles may prove attractive to some well-to-do individuals for use in cities or for commuting, but doubtless will be few in number and will not require increased generating and distribution capacity beyond that planned for the next few decades. In short, there should be an adequate supply of electricity in India to fuel the anticipated small fleet of electric vehicles. More likely, a major growth in the use of electric bicycles (*v.i.*) will occur, but this again would not make an impossible demand on the supply network.

Although foreign-made HEVs and BEVs are available in India today, sales are expected to have minimal impact on the conventional automotive market. Most Indians have to be concerned with the costs of motoring. Consequently, as in the rest of the world, the drive is towards small, more efficient ICEVs that emit fewer pollutants. The exception might be for hybrid versions of buses and trucks. In order to reduce urban pollution, the city of New Delhi passed legislation in 2004 that made the use of engines fuelled by compressed natural gas (CNG) mandatory for all public-service vehicles, which included taxis. It is conceivable that this initiative might foreshadow the introduction of hybrid buses, although at present such vehicles are too expensive without government grants being made available. If the pollution legislation were extended to private cars then it is possible that sales of electric vehicles might take off, but capital cost considerations would still favour CNG-fuelled conventional vehicles.

6.4.4 China

China is another developing country; it has the world's largest population (1.35 billion) and is industrializing rapidly. Much of the rest of the world depends heavily on China for manufactured goods, on account of its massive workforce and low labour rates. As China prospers, however, its people will look for greater remuneration for their work and the fulfilment of such aspirations will permit countries that currently have higher labour rates to compete in manufacturing.

The population of China is spread unevenly across its vast area, with many of the largest cities and much of industry concentrated in the eastern and southeastern coastal regions. Such a distribution is in sharp contrast to that in India, where the population density is more uniform. China has extensive reserves of coal and a substantial hydroelectric capability, but unfortunately these energy sources are located at a considerable distance from the centres of population. Specifically, much of the coal is located in the north of the country and the hydro reserves are in the southwest and west where rivers descend from the Himalaya Mountains. Instead of a fully-integrated grid, China has a number of regional networks and such diversity limits the scope for electricity to be transmitted across the nation. Large tonnages of coal are transported by train from the northern coalfields to power stations in the southeast. To maximize the use of clean hydro power, China is developing ultra-high-voltage (UHV) transmission lines (800 and 1000 kV) to convey hydroelectricity from the west to the east of the country. By 2020, the capacity of the UHV network is expected to be about 300 GW.

Although not all of the Chinese who live in rural areas have access to mains electricity, the country is now the largest generator of electricity in the world; production overtook that of the USA in 2011. The installed capacity is in excess of 1000 GW and, in recent years, has been increasing by around 10% per annum. The rate of growth is expected to slow somewhat during the present decade to reach about 1600 GW by 2020. Chinese electricity was made up of 79% from coal and peat, 16% from hydro, and 5% nuclear. Because of its predominant dependence on coal and peat, China will be limited in attempts to reduce greenhouse gas emissions while growing its economy. Indeed, many new coal-fired power stations are under construction. Even so, the country is also vigorously expanding its hydro and nuclear generating capacity to meet at least some of the anticipated electricity demand in a carbon-free manner. A major hydroelectric dam has been built on the Yangtze River (Three Gorges, 18.2-GW capacity) and another on the Yellow River (Xiluodu Dam, 15.8 GW). There is considerable scope for further hydroelectric projects; estimates show that only 25–30% of the economically feasible resource is currently being exploited. These massive river-damming projects do, however, impose major disruption on the respective local community and ecology, as well as on agricultural activity.

China also has a large programme to expand its nuclear electricity generation from around 10–11 GW installed capacity in 2011 to at least 60 GW by 2020, and then 200 GW by 2030. The focus is on building these stations in the coastal industrial belt, remote from the coalfields and where the electricity is most needed. Reactor technology developed by American and French companies is being employed.

As indicated in Table 2.2, Chapter 2, China produced far more road vehicles in 2012 than any other country, with a particular emphasis on cars. Most international automotive companies now have factories situated in China to produce their own brands. There are also numerous indigenous Chinese manufacturers, notably SAIC Motor Corporation Ltd (formerly Shanghai Automotive Industry Corporation) and Chery Automotive Co, Ltd. In 2009, China had 62 million registered road vehicles of which 26 million were privately owned cars (cf. the USA with around 250 million road vehicles).

Battery electric cars are also growing in significance in China. At least a dozen different Chinese companies have developed such vehicles, which are mostly small four-seat saloons or hatchbacks equipped with lithium traction batteries (based on the use of lithium iron phosphate; see Section 7.6.1, Chapter 7). Some models are already available internationally in showrooms; recent examples are shown in Figure 6.3. Clearly, there will be great competition between these companies and their different vehicle designs. Furthermore, there are concerted efforts to target world markets for urban BEVs at affordable prices.

Given the country's rapid industrialization and associated increase in prosperity, the understandable desire for personal transport, the rise of the automotive industry to meet that need, and the rapid expansion of the highway system, it is clear that China is a mega-growth market for road vehicles. Add to this scenario the government's aim to reduce petroleum imports, the phenomenal expansion in electricity generation and a nascent BEV industry, and then it is possible that BEVs will prove to be real competition for ICEVs in China. Indeed, with mounting attention on clean energy, the proliferation of BEVs in Chinese cities may well be faster than elsewhere in the world, although the intrinsic limitations of the technology may still restrict their acceptability outside the urban environment. In the next couple of decades, it seems probable that the burgeoning electricity network will be able to accommodate the requirements of a BEV market in most parts of the country. If electrically-powered vehicles really do prove popular internationally, then China will be in the vanguard.

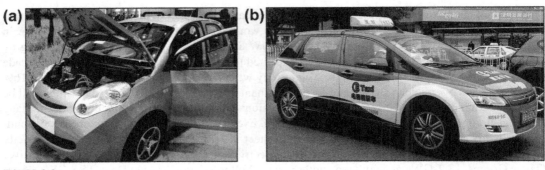

FIGURE 6.3

Hatchback versions of battery electric cars made in China. (a) An EV version of the Chery Automotive Company Supermini. (b) BYD Company E6 Electric Taxi in Shanghai

(Images sourced from Wikipedia and available under the Creative Commons License)

6.4.5 United States of America

The USA is an example of a large nation with a high standard of living overall, a correspondingly high consumption of energy, a well-developed automotive industry and an extensive road network. The country consists of 50 separate states, each of which has a high degree of autonomy in making its own laws, subject only to the overarching authority of the federal government in certain areas of national importance. State governments are responsible for most matters pertaining to highway construction, driving regulations, and the licensing of both vehicles and drivers. Electricity generation and interstate transmission is carried out by public utilities who are responsible to the federal government, but the distribution to consumers is mostly the role of local companies that each operate under a state licence. At both federal and state levels, administrations also seek to stimulate some advanced automotive technologies through subsidy schemes.

The US government imposes low taxes on motor fuels compared with those set by European countries. This policy results in relatively low petroleum prices so that there is little incentive to drive small, economical cars. Until the 1990s, the average car purchased in the USA was a six- or eight-cylinder 'gas guzzler'. As the international price of oil has risen, however, this practice has gradually changed and many Americans have turned away from sports utility vehicles (SUVs), a modern form of 'gas guzzler', and elected instead to purchase smaller and more efficient, but still luxurious and high performance, Asian and European cars. This proved to be a serious setback for the 'Big Three' US car manufacturers (General Motors, Ford and Chrysler), who had placed their faith in SUV that had formerly constituted a highly profitable line of business. At first, the smaller vehicles were imported, but now they are often made in the USA by foreign companies. The domestic manufacturers had to re-tool to introduce competitive models of their own, both for the home market and to compete in world markets. These difficult circumstances have been compounded by the international financial problems that have arisen since 2008 and a concomitant decrease in consumer spending.

The advent of petrol–electric hybrid cars, in particular the Toyota *Prius*, has proved popular in the USA and significant numbers have been sold. Until now, these models have not consumed mains electricity. There has been very little enthusiasm for BEVs, except for a few small commercial vehicles and buses. Nevertheless, a modern BEV sedan in the form of the Nissan *Leaf* was introduced into the USA in December 2010. Plug-in hybrids have only recently become available in the form of the General Motors *Chevrolet Volt* (2010), the latest model Toyota *Prius Plug-in Hybrid* (2012*)*, the Ford *C-Max Energi* (2012) and the Honda *Accord Plug-in Hybrid* (2013). The jury is still out on the prospects for these mains-charged electric vehicles.

The overall energy scene in the USA is well-balanced. For example, in 2009 it consisted of 37% petroleum, 25% natural gas, 21% coal, 9% nuclear power and 8% renewable energy (hydro power, biomass, wind). Electricity generation is based primarily on a mix of coal, natural gas and nuclear, but with a significant contribution from renewables. The consumption of electricity in 2010 was 4354 TWh and is predicted to rise to over 5200 TWh by 2035. In 2008, the Massachusetts Institute of Technology produced a report in which, after a comprehensive analysis, was advanced a forecast for the road transport scene in the USA in 2035, in terms of various possible scenarios. In the 'hybrid strong scenario', the market share of PHEVs grows to 15% of new light duty vehicles. The total electricity demanded by these vehicles would then be 59 TWh, or 1.1% of the anticipated electricity generation in 2035. The authors concluded, therefore, that the impact of PHEVs on the electricity grid will be small and give no cause for concern.

6.5 Recharging electric vehicles

A charger is, in essence, a transformer that converts alternating current (AC) line voltage to the voltage required by the traction battery, followed by a rectifier to convert AC to direct current (DC). Some BEVs have had the charger installed on board the vehicle. The advantage of this approach is that the only connection required is to a standard domestic outlet. On the other hand, the benefit is offset by the problem of accommodating a heavy and bulky charger within the vehicle. For this reason, the majority of BEVs have traditionally relied on external chargers, although the above-mentioned Nissan *Leaf* is now available with an optional a 3.3-kW unit.

There is also the issue of workplace recharging facilities, whether in private ownership of the employer or in public parking places. A slow recharge during the working day would permit commuting from distances up to the full range of a single battery charge. Typically, this would be at least 100 km and more than enough for most daily commuters. Similarly, installations at railway stations or airports would cater for drivers who expect to be away for appreciable periods. Finally, there is a need for public sites for those who wish, in a single journey, to exceed the range provided by one complete charge of the battery. These outlets must include fast-recharge facilities so that the driver is not inconvenienced by having to wait for an unreasonable time. Supermarket car parks are ideal locations; almost everyone uses them and a weekly shopping trip to buy food for a family takes around an hour, which is sufficient for a fast-charge boost. Similarly, restaurants, cinemas, theatres and motorway service stations are all places where the visitor could take advantage of a quick battery 'top-up'. The essential requirements here are for a high-rate charger and for a battery that is able to accept charge at that rate without overheating. It should be noted, however, that such powerful chargers will probably be expensive so their introduction may not be rapid. Some provision would have to be made for terminating a recharge when complete in the absence of the driver, so that the installation would be available for another user.

Another factor to be borne in mind by the driver is that the vehicle range on a full charge quoted by the automotive manufacturer relates to a new battery and to a vehicle driven at a reasonable speed. If the car is driven at high speed, where frictional losses are greater, or if electrical ancillaries are used extensively (e.g. heater, air-conditioning, headlamps, windscreen wipers), then the range attainable will be reduced. Similarly, as the battery ages, its performance may be expected to deteriorate before it eventually fails. All these are factors to be recalled when deciding whether a particular journey between recharge points is feasible. Range limitation will be a continuing concern for BEV owners as they contemplate making a journey longer than their usual daily routine. This issue has been referred to in the literature as 'range anxiety'.

Home charging can be accomplished simply by plugging a dedicated charger into a standard domestic socket, which, in principle, will provide a current of up to 13 A in many countries. In practice, the output may be no more than 10 A when account is taken of other loads in use on the same circuit. A better arrangement will be to have a dedicated spur from the fuse box that is independent of the rest of the domestic circuitry. Proprietary installations are available that can be mounted either internally in a garage or outside for recharging in the driveway. These are weatherproof and electrically safe. With such a facility, a current of 15 A can be drawn from a 240-V supply, and thereby reduce considerably the time of recharge. The device monitors the battery voltage and cuts out when top-of-charge is reached, so protecting the battery from overcharge as well as conserving electricity. A security code keypad is fitted to protect access.

Public recharge points pose a more complex problem as it is necessary to monitor who uses them, to record the consumption of electricity, and to make arrangements for billing the customers. Also, a larger current is required for fast charging. Networks of charging points are being set up across some regions of Europe; those in Britain each provide a current of 32 A, equal to 7 kW on a single-phase 240-V circuit and 21 kW when on a three-phase circuit. A driver will need to pay a monthly access fee and a small charge each time a unit is used. Installations are taking place in at least two supermarket chains, a network of highway restaurants and a series of national car parks. Also the UK government has introduced its 'Plugged-in Places' scheme whereby selected regions (initially, London, Milton Keynes, east and northeast England, and Scotland) are to be supported to establish collectively up to 2500 recharge points in a first phase, with more to follow.

Similar public/private co-operative efforts are reported from other countries. For instance, the city of Amsterdam has entered into an agreement with Nissan and Renault to encourage sales of BEVs in the Netherlands. Initially, 250 public charging bays are to be constructed in the city, with the number rising later to around 2000. The goal is to have BEVs constituting 5% of the car fleet (10 000 vehicles) in Amsterdam by 2015. The initiative is being encouraged by giving a significant subsidy to businesses that choose to buy an electric vehicle rather than an ICEV.

Meanwhile, over 5000 charging stations have been installed in the USA. California, a pioneering state for the introduction of electric vehicles, has about 1500 of these facilities, which are spread fairly uniformly across the different counties. Given the size of the state, this is still a relatively small number, but it is one that is growing. Elsewhere, several hundred public recharge stations have been established in each of the states of Florida, Michigan, Texas, Oregon and Washington. The two last-mentioned states are situated in the northwest of the country where a large amount of hydroelectricity is generated.

There are two possible schemes for charging an electric-vehicle battery – namely, conductive charging and inductive charging. By far the more common at present is conductive charging, whereby a stout copper cable connects the battery charger with the vehicle. Of course the design of plug on the charger cable must be compatible with the socket provided on the vehicle that connects with the battery. It is also desirable that these shall be of a universal design, e.g. like universal serial bus (USB) plugs on a computer, so that all makes of electric vehicle are compatible with all makes of charger. In the USA, the Society of Automotive Engineers (SAE) has developed a technical standard (J1772-rev. 2009) for a conductive charge coupler to be used with electric vehicles; it is expected that this specification will be adopted throughout North America. The device, shown in Figure 6.4(a), is suitable for charging at either 120 V or 240 V and is designed to carry a maximum current of 80 A, i.e. up to 19 kW of power. The connector also incorporates a pin to permit electronic communication and control between the battery and the charger. Major automotive companies in Japan and Europe, as well as the USA, have now accepted the SAE standard. Examples of conductive charging stations in different parts of the world are shown in Figure 6.5.

Inductive charging is a second method of charging an electric-vehicle battery and is similar to that employed in some electric toothbrushes. The technique involves no physical wire connection between the charger and the battery; rather, it relies on electromagnetic induction. The vehicle has an inductive coil fitted to the battery that can be brought into close proximity with a similar coil attached to the charger. A high-frequency (HF) current passing through the latter induces a corresponding HF current in the former coil that is then rectified and charges the battery. Induction charging was employed in the General Motors *EV1* in the 1990s and in the Toyota *RAV4-EV*, but then fell out of favour. For the *EV1*, the hand-held primary coil was inserted into a slot at the front of the car. A modern version of this approach is to have the primary coil located on the ground and the secondary coil attached to the underside of the vehicle. To charge the battery, the EV is simply parked over the coil for the requisite period.

An advantage of inductive charging is that it is said to be safer as there are no exposed terminals to create a potential hazard. The main disadvantage is that it is less efficient than conductive charging (higher losses), although some enthusiasts are attempting further development of the technology. A possibility for the future is to have an HF cable buried in the surface of the road and for vehicles to pick up power inductively as they move along above the cable. In principle, this form of wireless system would allow smaller, lighter batteries and longer vehicle range as charging would be continuous – 'charge as you drive'. This concept has been investigated by the Korean Advanced Institute of Science and Technology, but is still far from being a practical and economic proposition.

(a) 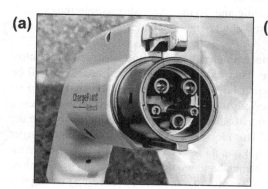 **(b)**

FIGURE 6.4

(a) Typical EV charging plug. (b) Nissan Leaf undergoing charging

(Images sourced from Wikipedia and available under the Creative Commons License)

FIGURE 6.5

Charging stations for battery electric vehicles: (a) Israel; (b) San Francisco; Parts a and b sourced from Wikipedia and available under the Creative Commons License. (c) Paris, photo author's own

Many utilities provide night-time power, when demand is low, at a preferential rate so as to encourage its use. Domestic tasks such as clothes-washing and drying and dish-washing may conveniently be carried out overnight. Re-charging electric-vehicle traction batteries falls in the same category.

Because of the unpredictability or unreliability of much renewable energy, the operation and control of the grid will become increasingly difficult as more wind or solar energy is utilized. This will be particularly so at peak periods during the day. It should be possible to arrange for chargers to be switched off remotely by the grid controller for short periods to help smooth the peak demand. An even more radical idea is to allow power, when needed, to flow backwards from plugged-in vehicles to the grid and thereby assist the utility in meeting high loads. This requires the use of a bi-directional charger and grid-integrated vehicle controls. Given a high enough population of PHEVs and BEVs, this methodology could be a significant means of meeting peak demand and would aid the introduction of more renewable electricity. At present, this scheme – known as 'vehicle to grid (V2G)' – is being investigated by the University of Delaware and the US company AC Propulsion, but there is no guarantee that it will come to fruition.

6.6 De-regulation of electricity markets

Traditionally, many countries have had a state monopoly for the generation and distribution of electricity. In recent years, however, there has been a move towards privatization of this industry and the introduction of competition so as to promote efficiency and lower the price of electricity – so-called 'de-regulation' or 'liberalization' of the electricity market. The UK has been foremost in this field in Europe and the actions taken are therefore discussed here as a case study of de-regulation. First, the generating function was split into three major companies (two fossil and one nuclear) that competed with one another to deliver the supply to a transmission company, and numerous regional electricity companies purchased electricity wholesale from the generators and marketed it retail to consumers. This arrangement persisted for some while before the two fossil-based generators were further subdivided when they sold off some of their power stations. New companies tendered for and were awarded licences to build gas-fired power stations. At the same time, the retail market was de-regulated and the retail companies were permitted to supply customers nationwide and thus were no longer confined to their regional origins. Finally, the major generators bought up some of the retail companies, thus integrating vertically. The UK now has a fully privatized and competitive electricity market in all aspects except for cross-country transmission, where one company (National Grid) has a monopoly. Other countries are proceeding along somewhat similar paths, albeit at different rates.

Is de-regulation a good thing? From the viewpoint of the original concept of promoting efficiency and lowering the price of electricity it has certainly been a success. Moreover, the UK government has rid itself of the chore of running a monolithic state enterprise and the need to raise taxes to invest in new power stations and transmission lines. Private capital is employed instead. In the longer term, however, some problems loom large and arise from the fact that private companies tend to focus on short-term return on investment and have little interest in ventures that will not return a profit for many years. Nor does the national interest enter much into their calculations. This situation militates against the use of nuclear energy since such power stations are capital-intensive to build and have a long lead-time (typically seven to ten years) for construction. It is therefore unlikely that, for purely financial reasons, private capital would invest in building new nuclear stations without major government incentives, or at least long-term guarantees. Rather, companies would opt for combined-cycle gas turbine technology, which is comparatively cheap and quick to build, efficient to operate, and is expected to yield profits after four to five years. Such a strategy is acceptable so long as there is an unlimited supply of natural gas at low cost. While this may be true for some years yet (see Section 3.2.4, Chapter 3), there is always political uncertainty over security of supply by pipeline from far distant countries, particularly in an uncertain future. The UK government is anxious to have a balanced portfolio of power stations and not to rely too heavily on one technology or one fuel and is therefore giving encouragement and incentives to build a new generation of nuclear power stations.

In the USA, the state of California has experienced some severe power shortages that have led to load-shedding and blackouts. In part, this is attributable to the rapid growth in electricity demand that has arisen from the mushrooming of computers and industrial equipment. Nevertheless, blame can also be attached to the commercial situation that arose following de-regulation of the electricity market. Whenever there is a power shortage, the spot market price for electricity rockets. Utilities are obliged to buy at high prices and then sell to customers at the lower prices set down in the existing supply contracts. Clearly, this is an unstable business practice. The long-term solution may be to construct more power stations and transmission lines, but a faster and cheaper solution may be to build distributed generation plant (micro gas turbines and/or CHP systems) or storage facilities (pumped hydro, batteries and/or flywheels) to meet the peaks in demand. Large consumers who buy directly from the spot market have an incentive to install their own storage facility, or even to shut down their operations at times of peak price. Customers on long-term fixed-price contracts could even sell the unused electricity back into the spot market at a profit! In Los Angeles, a power company has offered financial incentives to its largest customers to lower their energy consumption at peak times by installing their own distributed energy-storage systems.

The private ownership of the electricity industry also militates against renewable forms of energy as these too are more expensive than gas-fired power stations. So long as supplies of cheap gas are available, private industry is only likely to invest in renewables in response to government subsidies or the imposition of a sizeable carbon tax. For this reason, the UK government introduced the "non-Fossil Fuel Obligation" (later the 'Renewables Obligation'), whereby generators using fossil fuels are charged a tax that is used to provide financial support to those who produce electricity from renewable energy. This support is given both to large-scale operations (e.g. offshore wind farms) and to those households that install photovoltaic solar panels on their roofs.

Distributed generation and energy storage also have to fit within national electricity frameworks, and will only be adopted to the extent that the incumbent participants in the industry, together with new entrants to the market, see these two activities as profitable. The challenge facing a government is how to legislate and regulate the electricity supply so as to achieve long-term sustainability without sacrificing the benefits that stem from a privatized industry.

Batteries and Supercapacitors for Use in Road Vehicles

7.1 Fundamentals of energy storage in batteries

Batteries in road vehicles, as in other applications, store chemical energy in the form of reactive species produced during 'charging' and are designed to release this energy as electrical power during 'discharging'. In order to appreciate the extent to which the various candidate battery systems can fulfil the requirements of road vehicles, it is first necessary to understand how batteries work. In this section, a brief review is presented of the basic concepts of battery operation for the benefit of those readers who are not familiar with this application of electrochemical science.

Many chemical reactions involve an exchange of electrons between one reactant species and another. For example, in the following simple reaction between a metal (M) and a non-metal (X):

$$M + X \rightarrow M^+ + X^- , \tag{7.1}$$

the reactants are initially uncharged. In the product (MX), however, the metal has lost an electron to the non-metal and so carries a unit positive charge to form a 'cation', whereas the non-metal has gained the electron to produce an 'anion'. The general term 'ion' applies to any atom that has become electrically charged and more than one electron may be involved in a reaction. The product is typically a salt or an oxide.

A simple example of such a reaction is that of zinc with oxygen, where two electrons are transferred to form zinc oxide:

$$2Zn + O_2 \rightarrow 2Zn^{2+} + 2O^{2-} . \tag{7.2}$$

This is the electrochemical reaction that takes place in a hearing-aid battery. Essentially, the same reaction also occurs in the common disposable battery that is used in torches (flashlights), radios, etc., although in this case the oxygen is present as an oxygen-rich compound, manganese dioxide (MnO_2), and not free as in air.

An 'electrochemical' reaction involves the transfer of both mass and electronic charge and the unit within which a single such reaction takes place is termed an 'electrochemical cell'. The mass transfer (of ions) takes place through a medium that conducts ions, but not electrons, and is known generically as the 'electrolyte'. In the above-mentioned cell, Zn and MnO_2 are the reactants and potassium hydroxide (KOH) solution serves as the electrolyte. The electronic charge is transferred through an external circuit to perform useful work. The open-circuit voltage (i.e. no current flow) exhibited by a single electrochemical cell is generally rather small. Therefore, when it is necessary to build up high-voltage units, this is achieved by connecting several cells in series to form a 'battery'. Although, strictly speaking, a battery is a multi-cell array (connected either in series or in parallel), in common usage many single cells are called 'batteries'; this practice is followed here, so that the term 'battery' is used to encompass single cells as well as multi-cell devices housed in a single container.

There are two broad categories of battery. 'Primary' batteries are designed for a single discharge, after which they are discarded. Examples of this type are the manganese-dioxide batteries mentioned above and the lithium batteries employed in watches and other portable electronic devices. 'Secondary' batteries, on the other hand, can be recharged by passing a current in the opposite direction to that flowing on discharge. Batteries for road-vehicle applications are required to be rechargeable, and hence are of the secondary type.

The essential components of an electrochemical cell are a positive electrode, a negative electrode and an electrolyte. The convention used to distinguish the two types of electrode polarity in electrochemical cells is outlined in Box 7.1. During cell operation, ions are created and consumed at the two electrode | electrolyte interfaces by oxidation–reduction reactions. The positive electrode can be an oxide, a sulfide or some other compound that is capable of being reduced during discharge of the cell (e.g. MnO_2), while the negative electrode is generally a metal or an alloy that is capable of being oxidized (e.g. Zn). These substances are referred to as the 'active-materials'. Both electrodes must be electronically conducting in order to convey electrons to and from the reaction sites. In cases where the electronic conductivity of the positive electrode is inadequate, a conductor (such as carbon) is intimately mixed with the active-material. The electrolyte provides the internal circuit between the electrodes, and is an ionizing liquid or solid that has a high and selective conductivity for the ions that take part in the electrode reactions. The electrolyte must be a non-conductor for electrons in order to avoid 'self-discharge' of the cell. Other components of a cell include metallic 'current-collectors' which conduct current to and from the electrodes, and 'separators' which ensure that the electrodes do not touch each other and create a short-circuit. Where a solid electrolyte is employed, this serves also as the separator. In some cell designs, each electrode polarity may consist of a number of 'electrode plates' joined in parallel.

The voltage of cells used in energy storage, V, is of the order of 1 to 3 V and, since the energy-storage systems in road vehicles are required to provide considerably higher voltages – from 12 V in conventional internal-combustion-engined vehicles (ICEVs) up to 600 V in some battery electric

---

BOX 7.1 TERMINOLOGY USED IN THE OPERATION OF ELECTROCHEMICAL CELLS/ BATTERIES

In an electrochemical cell, the electrode that exhibits the more positive potential is the positive; the other electrode is the negative. The terms 'cathode' and 'anode' are also used, and denote electrodes at which cathodic and anodic processes occur, respectively. A cathodic process involves the flow of electrons from the electrode to reduce solution species, i.e. a so-called 'reduction reaction'. An anodic process is the reverse situation in which solution species are oxidized and electrons are transferred to the electrode, i.e. an 'oxidation reaction'. A cell that spontaneously produces a voltage is a 'galvanic cell' and, in this case, the positive electrode is the cathode; see Figure 7.1(a). A cell that is driven by the application of an external voltage is an 'electrolysis cell' and the positive electrode is now the anode; see Figure 7.1(b). Naturally, this leads to some confusion with secondary batteries because the electrodes that serve as cathodes on discharge become anodes on charge, and vice versa; see Figure 7.1(c).

Where the terms anode and cathode are used for batteries, it is normal to employ the terminology of the galvanic cell, i.e. the positive electrode in each cell is the cathode and the negative is the anode. It is best, however, to confine the terms cathodic and anodic to currents and use 'positive' and 'negative' to characterize electrodes. Not only does the latter terminology remain the same on charge and discharge, but it is also particularly appropriate because the terminals of commercial cells and batteries are clearly marked + and −.

vehicles (BEVs) – several (or many) cells must be connected in series in an assembly to form a battery. The final voltage of the battery is NV, where N is the number of cells connected in series. The capacity (Ah) remains that of the single cell. For example, the familiar 12-V automotive starter battery consists of six 2-V lead–acid cells that are series-connected in a single enclosure. In large high-voltage batteries, such multi-cell units are sometimes termed 'modules' or 'monoblocs'. A 'battery pack' is then a collection of these modules that can be joined together in series and/or in parallel configurations to yield the desired voltage and current capability. The parallel wiring of units (cells or modules) increases the capacity of the battery but retains the unit voltage.

Examples of reactions involved in the discharge of electrochemical cells that are used in hybrid electric vehicles (HEVs) and/or BEVs – viz., nickel–metal-hydride, lithium-ion and sodium–metalchloride – are given in Table 7.1. The central column shows the electrolyte medium that separates the positive and negative electrodes, together with an indication of the ion that moves between the two electrodes. In each case, for the discharge reaction to proceed, electrons must pass round a circuit external to the cell and, in so doing, they perform useful work. The reaction involves two separate processes: (i) oxidation at the negative electrode with the release of electrons, and (ii) reduction at the positive electrode, which involves ions that have passed through the electrolyte and electrons that have travelled via the external circuit. For the electrochemical reaction to take place, it is necessary that the reactants are at a higher energy state than the products so that there will be a spontaneous release of energy once the ions and electrons are brought together. This energy is then available to perform useful work which

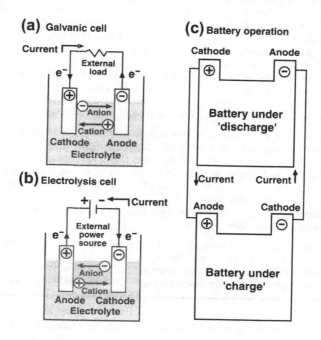

FIGURE 7.1

Illustration of cell/battery terminology

Table 7.1 Examples of Electrochemical Cells Used in the Batteries of HEVs and/or BEVs (Discharge Reactions Shown)

Positive Electrode Reaction	Mobile Ion/Electrolyte	Negative Electrode Reaction
$e^- + NiOOH + H_2O \rightarrow OH^- + Ni(OH)_2$	OH^-/aqueous KOH	$MH + OH^- \rightarrow M + H_2O + e^-$, where M is a complex alloy
$xe^- + xLi^+ + CoO_2 \rightarrow Li_xCoO_2$ or: $xe^- + xLi^+ + FePO_4 \rightarrow Li_xFePO_4$	$Li^+/LiPF_6$ in alkene carbonates	$Li_xC_6 \rightarrow 6C + xLi^+ + xe^-$
$2e^- + 2Na^+ + NiCl_2 \rightarrow 2NaCl + Ni$	Na^+/beta-alumina[a]	$2Na \rightarrow 2Na^+ + 2e^-$

Lead–acid is not included because its reaction scheme is fundamentally different from those shown in that the electrolyte takes part in both the discharge and the charge reactions. Lead–acid cells are described in Section 7.4.
[a]Beta-alumina is a class of ceramic oxides that sustains sufficiently high conductivity of sodium ions at temperatures of ~ 300 °C for each material to function as both an electrolyte and a separator in the cells of secondary batteries.

BOX 7.2 ELECTRODE POTENTIAL AND CELL VOLTAGE

The electrodes in a cell that is providing no current (i.e. is on open-circuit) each have a 'reversible potential', E_r, that is determined by the electrochemistry. The 'standard electrode potential' is the reversible potential of an electrode with all the active-materials in their standard states. Usual standard states specify a pressure of 101.325 kPa for gases and unit activity for elements, solids and 1 mol dm^{-3} solutions – all at a temperature of 298.15 K. The difference in reversible potential between the two electrodes is the reversible cell voltage, V_r, and the difference in the standard potentials is the standard cell voltage, $V°$.

When current flows through a battery, there is departure from equilibrium conditions and the useful work performed by the battery is less than the thermodynamic maximum value. The shift in potential of an electrode away from the reversible (equilibrium) value is termed the electrode 'overpotential'. There are three causes of the restriction in the current flow that is associated with electrode reactions, namely:

(i) 'activation overpotential', which results from limitations imposed by the kinetics of charge transfer at the electrode; this is an intrinsic property of the electrode I electrolyte interface;
(ii) 'concentration overpotential', which results from depletion of reactants in the vicinity of the electrode due to slow diffusion from the bulk electrolyte or across the product layer; this is an extensive property that is dependent on the thickness and porosity of the electrode and the ease of diffusion through it, as well as on mass-transport processes in the electrolyte;
(iii) 'resistance overpotential', which results from the ionic resistance of the electrolyte and the electronic resistance of the other cell components.

The terms 'potential' and 'overpotential' should be used only for single electrodes, with the corresponding terms 'voltage' and 'overvoltage' reserved for cells and batteries.

In summary, overpotential refers to an individual electrode reaction and is the difference in the values of the actual electrode potential and the equilibrium (reversible) potential for the reaction under consideration.

is manifest as the current through the external circuit. The driving force for the external current derived from a cell is the difference in the potentials of the two electrodes. This is the cell voltage; see Box 7.2 for a more detailed explanation. During recharge, the reactions shown in Table 7.1 are driven in reverse and the reactant species are restored.

7.2 Key criteria for candidate batteries

7.2.1 Capacity

The capacity of a battery refers to the amount of electricity that it can supply, in terms of coulombs (ampere-seconds, As) or ampere-hours (Ah), before it is fully discharged. The theoretical capacity of a cell is given by the expression:

$$Q = x\,nF, \tag{7.3}$$

where x is the number of moles of reactant consumed in the complete discharge of the cell, n is the number of electrons transferred per molecule during the reaction and F is the Faraday constant (96 485 coulombs).

7.2.2 Energy

The amount of energy stored in the battery is the product of the capacity and the voltage at which the capacity is provided, i.e. Ah × V, and is quoted in terms of watt-hours, Wh. For road-vehicle duty, the energy-storage capability should be provided with a minimum weight penalty. In essence, the battery should have a high 'specific energy', i.e. a high energy output per unit mass, Wh kg^{-1}.

In vehicles, the volume occupied by the battery is equally important so that the battery should also have a high 'energy density', i.e. a high energy output per unit volume, Wh L^{-1}.

Finally, a high 'coulombic efficiency', i.e. {charge out}:{charge in}, is an asset as this preserves primary energy. This is also known as the 'ampere-hour efficiency' and is usually expressed as a percentage.

7.2.3 Power

In many vehicle battery operations, it is not the total amount of energy that matters but the rate at which it can be provided – that is, the power achievable. This parameter is the product of the current and the voltage, A × V, and the units are watts. As with storage energy, the weight and volume of the battery are important factors in vehicle design, so that the respective key parameters are specific power (W kg^{-1}) and power density (W L^{-1}). It should also be noted that the power that a battery can accept on charge is just as important as that available during discharge.

7.2.4 Charge-acceptance

It is essential that the battery can be charged efficiently, especially at the high rates involved in regenerative braking; see Section 5.2.1, Chapter 5. Currents that exceed the rate that can be accommodated by the normal charge reaction of the cell chemistry in question will be diverted into parasitic secondary reactions and will leave discharged material in place on the plates. Efficient high-rate charging is particularly important for HEVs; see Section 7.3.1.

7.2.5 Battery life and modes of failure

Obviously, good operational life is a prerequisite for all batteries if they are to be economically viable. This parameter may be defined chronologically (i.e. years of use, so-called 'calendar life'), or in terms of the number of duty operations (e.g. charge–discharge cycles) that the battery will sustain before its function (e.g. capacity) falls to a set percentage of its initial value (so-called 'cycle-life'). When the battery is used in partial-state-of-charge (PSoC) cycling, as in HEVs, the number of deep cycles is no longer relevant and the lifetime yardstick that is adopted is the total amount of charge passed in and out of the battery in terms of the number of multiples of the battery capacity ('the equivalent number of capacity throughputs').

To identify the prime determinants of battery life, it is first necessary to distinguish between 'catastrophic' failure, as characterized by a sudden inability of the battery to function, and 'progressive' failure, as demonstrated by a gradual decline in capacity during service.

Catastrophic failure is usually easy to diagnose and characterize. The following problems are commonly encountered:

- incorrect cell design and component selection, e.g. short-circuits due to plate and separator faults;
- poor quality control during battery manufacture, e.g. open-circuits arising from loose plates/ terminals, and/or attack from contaminants;
- incorrect operation of batteries ('abuse'), e.g. undercharging, overcharging, vibration, high ambient temperature, entry of harmful foreign species, prolonged storage with insufficient recharging;
- external and/or internal damage, e.g. broken containers and covers, damaged terminals, electrolyte leakage.

Clearly, there is a certain degree of overlap between some of these categories, and battery failure may have indeed resulted from some combination of effects.

'Progressive' failure is more difficult to predict and explain. Often, it represents a subtle deviation from optimum performance that originates from micro-scale changes in plate properties brought about by the influence of manufacturing variables and service conditions. Parameters that depend upon the materials, processing and design of the battery are known as 'inner parameters', while those determined by the conditions of battery use are referred to as 'outer parameters'. The classification is as follows.

Inner parameters:

- the chemical composition and physical properties of the active-materials;
- plate thickness;
- composition and processing conditions of current-collectors;
- composition (including additives) of electrolyte;
- choice of separator;
- cell design.

Outer parameters:

- storage time before use;
- frequency and current density of discharge;
- depth of discharge;
- stand time in a state of partial or complete discharge;
- current and voltage of recharge;

- degree of overcharge;
- temperature;
- uniformity of concentration and maintenance of liquid electrolytes.

Both sets of parameters will act (to varying degrees) to cause the eventual failure of the battery.

Much research and development effort has been devoted to the study of the mechanisms that limit the useful life of batteries (the 'failure modes') and thereby to find means to extend useful life. These failure modes include: (i) corrosion, (ii) the development of short-circuits as a result of the growth of metal 'dendrites' that penetrate the separator and (iii) the progressive loss of contact between the active-material and the current-collector.

It is important that cells in the long strings of high-voltage batteries remain closely matched at all states-of-charge. If a balance is not maintained, then some cells in the string may be damaged through overcharging or undercharging. This requirement is particularly critical for long strings of lithium-ion cells, and is achieved through the careful monitoring of each cell and, if found necessary, charge equalization by a battery-management system (BMS).

7.2.6 Thermal issues

Batteries generate heat during charge–discharge cycling and this must be dissipated to the environment to prevent the battery temperature from rising continuously. The heat effects originate from the following sources:

- a change in the entropy, ΔS, of the cell reaction – a fundamental thermodynamic entity that represents the unavoidable heat emission (energy loss) or heat absorption (energy gain) that is associated with electrochemical reactions; energy loss in one direction means energy gain when the reaction is reversed, hence the phenomenon is commonly referred to as the 'reversible heat effect';
- energy losses caused by overpotential (see Box 7.2);
- ohmic resistance ('joule heating').

The reversible heat effect is strictly connected with the amount of material that reacts and thus does not depend on discharge or charge rates. The quantity, which is the product of the absolute temperature, T, and the entropy change of the reaction, ΔS, has a positive value for some cell reactions and a negative value for others. In the lead–acid system, for example, the value of $T\Delta S$ is negative and the reversible heat effect exercises a cooling during discharge and a heating during charge. For the nickel–metal-hydride battery, however, the sign of $T\Delta S$ is positive and the effect contributes heating during discharge and cooling during charge. By contrast, joule (resistive) heating, which is the dominant effect at all but the very slowest of reaction rates, occurs during both charging and discharging.

The net amount of heat generated within a battery must be balanced by the sum of the heat-dissipation processes – conduction, convection and radiation – if the temperature of the battery is not to rise. Proper heat management will ensure that the battery temperature does not exceed a safe level and will maintain all the cells within as small a range of temperatures as possible. The penalties for allowing the development of temperature gradients are non-uniform states-of-charge and states-of-health between cells, either of which can result in premature cell failure.

7.2.7 Cost

Finally, a *sine qua non* is acceptable cost. For any given battery, the key factors to be considered are initial price, maintenance costs and operational life. The cost of production is highly variable from battery to battery; it is dependent both on the materials involved (for example, nickel is much more expensive than lead) and on the procedures involved in manufacture. Maintenance costs are uniformly low for most of the batteries that are used in road-vehicle applications. The determinants of battery life have been discussed above in Section 7.2.5.

7.3 Battery duty in different road vehicles

Batteries in different categories of road vehicle are required to perform widely disparate duty cycles. In conventional ICEVs, the battery must power the starter motor and the lights, provide the spark for the ignition of fuel in the cylinders (petrol engines), and supply energy for ancillary electrical functions. This duty is referred to as SLI – starting, lighting and ignition – and the battery is maintained at close to top-of-charge so that the operational service may be met reliably. In recent years, the number of electrical devices to be powered in road vehicles has increased rapidly and consequently the demands made on the battery have burgeoned. If this trend continues, it may be necessary to replace 12-V electrical networks and batteries with 48-V systems. In all forms of HEV, batteries must perform a duty that is far more demanding in terms of power than the traditional provision of energy for SLI operations. In BEVs, the requirement is for the battery to supply large amounts of energy for propulsion throughout the journey and this can only be sustained for long distances by drawing deep discharges from high-voltage batteries.

The major challenges imposed on the battery by the shift in technology towards hybrid electric and all-electric vehicles are summarized in Table 7.2. The data show that engine cranking in a conventional ICEV requires the SLI battery to deliver high discharge currents of short duration; the currents are typically 15 times the one-hour rate ($15\,C_1$), i.e. the four-minute rate. By contrast, replacement of the charge used in cranking occurs at a modest rate, at around the one-hour rate ($1\,C_1$). The ultimate failure modes of lead–acid SLI batteries are sufficiently understood and well managed so that most designs are now able to give four-to-six years of service in temperate climates. Discharge

Table 7.2 Typical Discharge and Charge Rates of Batteries in Different Vehicle Duties

Duty	SLI	Deep-Cycle	HRPSoC
Vehicle	Conventional motor car	BEV	HEV
Range of SoC	85–90%	20–100%	50–70%
Maximum normal discharge rate[a]	$15\,C_1$	$1\,C_1$	$15\,C_1$
Maximum normal charge rate	$1\,C_1$	$0.5\,C_1$	$> 30\,C_1$

[a]*Battery charge and discharge rates are expressed in terms of the rate to charge or discharge the battery fully in one hour. This is given the symbol C_1. A half-hour rate is $2\,C_1$ while a five-hour rate is $C_1/5$. The faster a battery is discharged, the lower is its working voltage and the less is the energy obtained. Hence, capacity has to be specified in terms of the discharge rate employed.*

rates are much lower in BEVs and seldom exceed 1 C_1 and, as in SLI operation, the batteries are normally recharged at a modest rate, i.e. $< 0.5\ C_1$. In HEVs, including all stop–start vehicles that make use of regenerative braking, the batteries are subject to high-rate partial-state-of-charge (HRP-SoC) operation. The charge rate can be very high, e.g. up to 30 C_1 (the two-minute rate), because power returns of up to 30 kW are commonly experienced.

The various batteries that have been employed in BEVs and HEVs possess quite different properties. Historically, BEVs have almost always been powered by lead–acid traction batteries. When compared with more modern types of battery, lead–acid has several disadvantages but one overriding advantage – relatively low cost. It is this factor that has motivated ongoing research to adapt the mature technology and thereby render it suitable for use in modern electrified drive-trains where, as just mentioned, the demands on the battery are quite different from those encountered in the traditional SLI application.

7.3.1 Dynamic charge-acceptance

Dynamic charge-acceptance (DCA) is the rate (power) that the battery will accept on charging. As described in Section 5.2, Chapter 5, the first stage in the progressive improvement of vehicle fuel economy involves the introduction of 'stop–start' operation, which causes a marked increase in the demands made on the battery. Prior to this development, the battery was charged by a discrete alternator and was required to support only one engine-start event for each journey (up to around 6 kW for a few seconds for a medium-sized car), plus lighting, ignition and other electrical loads (to a maximum of 2 kW continuously). Once the stop–start system is in operation, the average journey will involve several engine restarts and on each occasion around 7 Wh will be withdrawn for a vehicle equipped with a conventional starter and alternator, although only about 1 Wh for a vehicle with an integrated starter–generator (ISG). More importantly, the total of all the continuous loads must be borne by the battery alone during engine-off periods. As a result, batteries in stop–start vehicles will experience deeper discharge than their counterparts in earlier vehicles. Further, as the replacement of the extra charge that is withdrawn must depend on the output of the internal combustion engine (ICE) alone, the battery may spend a significant fraction of the time in an undesirably low state-of-charge. The duration of this period can be reduced if vehicles employ regenerative braking, i.e. the capture of energy, in the form of electricity, which would otherwise be lost as friction in braking, but which is only available for a few seconds and at very high rates. A key parameter then is 'dynamic charge-acceptance' (DCA) – the term used to describe the way in which a battery can accept charge that arrives as a result of kinetic energy recovery as opposed to other forms of charging.

A vehicle that is designed to make use of regenerative braking energy normally incorporates an ISG rather than a separate starter and alternator. With present technology, the energy can arrive in currents of 100 to 200 A, but the uptake is limited by the charge-acceptance of the battery. This depends to a large extent on the battery voltage. Hybrid electric vehicles with high-voltage battery packs accept tens of kW, whereas 12-V batteries admit only about 1 kW. Although the total energy stored in full hybrids and 12-V stop–start hybrid vehicles (SSVs) is about the same (~ 1 kWh; see Table 5.1, Chapter 5), the configurations of the two types of battery are quite different. The cells in the HEV battery (~ 6 Ah) are exposed to charge rates of 30 C_1 or more during DCA, whereas those in the SSV (up to 100 Ah) experience rates of only 1 to 2 C_1.

The higher the DCA of the battery, the greater will be the fraction of the available braking energy that can be recovered. Thus, the amount of current that can be accommodated by the charge reaction, as opposed to being wasted, is a measure of the effectiveness of the process. In the case of lead–acid batteries, the energy recovered depends to some considerable extent on the immediate history of the battery (i.e. charged, discharged or rested), but there remains the central problem of limited charge acceptance at the high currents created by regenerative braking; see Section 5.2.1, Chapter 5. A system that conditions the battery by means of a periodic full charge appears to be an effective way of sustaining DCA, but this is not an expedient solution.

7.4 Lead–acid batteries

The fundamental elements of the lead–acid battery were set in place some 150 years ago. In 1859, Gaston Planté was the first to report that a useful discharge current could be drawn from a pair of lead plates that had been immersed in sulfuric acid and subjected to a charging current; see Figure 7.2(a). Later (1881), Camille Fauré proposed the concept of the pasted plate. Since then, the cell reactions of the battery have not undergone any further radical change. The most commonly employed design has 'flat plates'; see Figure 7.2(b). These are prepared by coating pastes of lead oxides and sulfuric acid on to conductive lead or lead-alloy 'grids', which act as current-collectors. The pastes are then 'formed' electrolytically into the active-materials. One alternative cell design uses positive plates in which the active-material is contained in tubes, each fitted with a coaxial current-collector; see Figure 7.2(c). Such 'tubular plates' serve to prevent shedding of the material during battery service.

7.4.1 Principles of operation

The lead–acid battery has undergone many developments since its invention, but these have involved modifications to the materials or design, rather than to the underlying chemistry. In all cases, lead dioxide (PbO_2) serves as the positive active-material, lead (Pb) as the negative active-material and sulfuric acid (H_2SO_4) as the electrolyte. The electrode reactions of the cell are unusual in that the electrolyte (sulfuric acid) is also one of the reactants, as seen in the following equations for discharge and charge.

At the positive electrode (plate):

$$PbO_2 + 3H^+ + HSO_4^- + 2e^- \underset{\text{Charge}}{\overset{\text{Discharge}}{\rightleftharpoons}} PbSO_4 + 2H_2O, \quad E^\circ = +1.690 \text{ V}; \quad (7.4)$$

at the negative electrode (plate):

$$Pb + HSO_4^- \underset{\text{Charge}}{\overset{\text{Discharge}}{\rightleftharpoons}} PbSO_4 + H^+ + 2e^-, \quad E^\circ = -0.358 \text{ V}, \quad (7.5)$$

where E° is the standard electrode potential for each reaction, i.e. the electrode is in a standard state; see Box 7.2.

The overall cell reaction is:

$$PbO_2 + Pb + 2H_2SO_4 \underset{\text{Charge}}{\overset{\text{Discharge}}{\rightleftharpoons}} 2PbSO_4 + 2H_2O, \quad V^\circ = +2.048 \text{ V}, \quad (7.6)$$

where V° is the standard cell voltage. It is noteworthy that this voltage is the highest for any type of commercial battery that has an aqueous electrolyte.

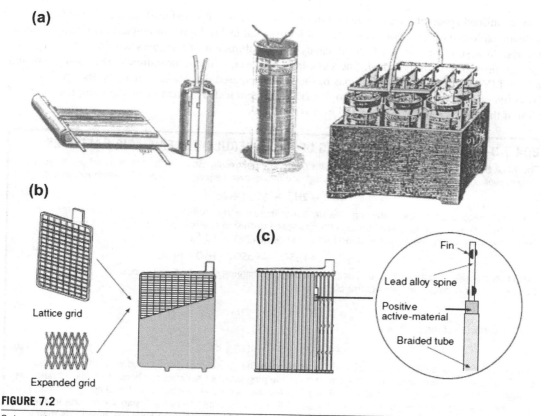

FIGURE 7.2

Schematic representation of: (a) Gaston Planté's cell and battery; (b) flat plate; (c) tubular positive plate

During discharge, HSO_4^- ions migrate to the negative electrode and react with the lead to produce H^+ ions and $PbSO_4$, which (significantly) is non-conductive. This reaction releases two electrons and, thereby, gives rise to an excess of negative charge on the electrode that is relieved by a flow of electrons through the external circuit to the positive electrode. At the latter, the lead of the PbO_2 is also converted to $PbSO_4$ and, at the same time, water is formed. Thus, $PbSO_4$ progressively develops in equal quantities at both the positive and the negative electrode, with a concomitant weakening of the electrolyte solution. The decrease in electrolyte concentration (or relative density) provides a convenient means for determining the degree of discharge that has taken place or, conversely, checking the state-of-charge of the cell. The lead–acid battery is unique in this regard.

The above processes are reversed on charging. As the cell approaches full charge, the majority of the $PbSO_4$ will have been converted back to lead or PbO_2 with replenishment of the sulfuric acid solution. Further passage of current will give rise to electrolysis and the evolution of hydrogen at the negative electrode and oxygen at the positive. The gases are released in stoichiometric proportions (i.e. twice as much hydrogen as oxygen) and, with traditional cell designs (so-called 'flooded-electrolyte' batteries), this results in a loss of water from the cell electrolyte solution.

In a confined space, the gases released during charging of a flooded lead–acid cell could also constitute an explosive hazard. For many years, scientists and technologists attempted to develop 'sealed' batteries. At first, efforts focused on the catalytic recombination of the gases within the battery; this approach proved to be impractical. Success came, however, with the invention of the valve-regulated lead–acid (VRLA) battery. Two versions of the battery entered the market – namely, the 'gel' and the 'absorptive-glass-mat (AGM)' technologies (*v.i.*). A technical explanation of the complex mode of action of the VRLA battery is presented in Box 7.3.

BOX 7.3 FUNDAMENTAL PRINCIPLES OF VALVE-REGULATED LEAD–ACID BATTERIES

The VRLA battery is designed to operate by means of an 'internal oxygen cycle' (or 'oxygen-recombination cycle'). Oxygen evolved during the latter stages of charging, and during overcharging, of the positive electrode, that is:

$$H_2O \rightarrow 2H^+ + \tfrac{1}{2}O_2 \uparrow + 2e^-, \qquad (7.7)$$

transfers through a gas space to the negative electrode where it is reduced ('recombined') to form water, (Equation 7.8). Diffusion of oxygen through the gas space (diffusion coefficient 0.18 cm^2 s^{-1}) is very much faster than through the electrolyte solution (diffusion coefficient 9×10^{-6} cm^2 s^{-1}):

$$Pb + \tfrac{1}{2}O_2 + H_2SO_4 \rightarrow PbSO_4 + H_2O + heat. \qquad (7.8)$$

Two other reactions must be taken into account during the charging of a VRLA cell – namely, the evolution of hydrogen at the negative plate:

$$2H^+ + 2e^- \rightarrow H_2, \qquad (7.9)$$

and the corrosion of the positive grid:

$$Pb + 2H_2O \rightarrow PbO_2 + 4H^+ + 4e^-. \qquad (7.10)$$

Thus, the charging of a VRLA cell is potentially more complex than the charging of its flooded counterpart. For a VRLA cell, thermodynamic and kinetic conditions allow the progress of six separate reactions at significant rates: two charge reactions (the reverse of Equations 7.4 and 7.5) and four secondary reactions, i.e. Equations 7.7–7.10.

The oxygen cycle, defined by Equations 7.7 and 7.8, shifts the potential of the negative electrode to a less negative value and thus decreases the rate of hydrogen evolution to a much lower level (i.e. much less than in the older, flooded design of battery). A one-way pressure-relief valve is provided to ensure that even the small amounts of hydrogen produced do not generate a high pressure within the battery – hence the term 'valve-regulated'. Since the plate is simultaneously on charge, the lead sulfate produced by Equation 7.8 is immediately reduced to lead via the reverse of Equation 7.5. This restores the chemical balance of the cell, i.e. in stoichiometric terms, the net sum of Equations 7.7 and 7.8 and the reverse of Equation 7.5 is zero. Thus, part of the electrical energy delivered to the cell is consumed by the oxygen-recombination cycle and is converted into heat rather than into chemical energy.

There are two alternative designs that provide the gas space in VRLA cells. One design has the electrolyte solution immobilized as a gel, whereas the other has the electrolyte solution held in an absorptive-glass-mat (AGM) separator. Gas passes through fissures in the gel, or through channels in the AGM.

The traditional use of antimony as an alloy with lead to provide grids with the necessary strength and castability is no longer practised for VRLA cells because this element lowers the hydrogen overpotential and therefore encourages gassing at the negative electrode. Care needs to be taken against the introduction of other elements that might act similarly. Excessive gassing at either the negative or the positive electrode can result in selective discharge of the respective electrode. Lead–calcium–tin alloys are preferred by manufacturers of VRLA batteries for standby-power ('float') duties and lead–tin alloys for cycling applications.

If the oxygen cycle is worked too hard then substantial heat is generated and charging of the negative plate becomes difficult so that a progressive build-up of lead sulfate (so-called 'sulfation') commences from the bottom of the plate where the acid concentration tends to be highest. The function of the oxygen cycle is subtly linked to the microstructure of the separator material (for AGM designs) and to the nature of the charge algorithm applied, especially near top-of-charge.

The basic discharge–charge reactions of the lead–acid cell involve dissolution–precipitation mechanisms that, collectively, are known as the 'double sulfate theory', as first suggested by Gladstone and Tribe in 1882. For example, discharge and charge of the electrodes may be considered as dissolution into, and electroplating out of, dilute solutions of lead ions. It should be noted that the cell is able to operate effectively as an energy-storage device by virtue of the following three critical factors:

- contrary to thermodynamic expectations, the liberation of hydrogen from sulfuric acid solution by lead takes place at only a negligible rate, i.e. there is a high hydrogen overpotential;
- the high oxygen overpotential at the positive electrode allows $PbSO_4$ to be converted to PbO_2 before appreciable evolution of gas commences;
- although the solubility of $PbSO_4$ in the electrolyte solution is sufficient to promote the electrode dissolution–precipitation reactions, the value is so low that there is little migration of the material during charge–discharge cycling and, hence, a high degree of electrochemical reversibility is maintained.

It is interesting to reflect that were it not for these three attributes, there would be no rechargeable lead–acid batteries and, throughout the 20th century, engines would either have had to be started by hand-cranking or by using nickel–cadmium batteries, which are considerably more expensive and incorporate large quantities of toxic cadmium.

The capacity (Ah) exhibited by a lead–acid battery when discharged at a constant rate depends on a number of factors. These include the design and construction of the cell, the cycling regime (history) to which it has been subjected, its age and maintenance, and the prevailing temperature. As with batteries of all other chemistries, the realizable capacity is dependent on the rate of discharge; for example, the capacity obtained from a 30-min discharge is only a fraction of that from a 10-h discharge. Moreover, the cell voltage is much reduced and, thereby, results in an even greater reduction in available energy (Wh). As cycling (or life) proceeds, a number of processes ('failure mechanisms') can degrade further this limited performance. These are reviewed in Section 7.4.3.

7.4.2 Manufacturing processes

7.4.2.1 Grids

Traditionally, lattice-type grids for flat plates are cast from molten lead alloys, either singly or in pairs joined by their lugs (also known as 'tabs'), and are quench-cooled; see Figure 7.2(b). Automated casting machinery is quite sophisticated and much development work has been devoted to optimizing the process for the provision of defect-free grids of different alloy compositions. With modern casting machinery, 15 to 20 pairs can be produced each minute. An alternative procedure is to start with a reel of alloy foil that is then slit and expanded to form a diamond mesh structure, which is amenable to pasting; see also Figure 7.2(b). A growing fraction of grid production for VRLA batteries is now undertaken by means of a third method – through punching the lattice pattern out of a rolled sheet of lead alloy.

Over the years, there has been a substantial reduction in the thickness of grids – from more than 2 mm in the 1960s to about 0.8 mm today. Several factors have combined to make this practice both possible and acceptable – notably, superior casting technology and improved charge-control systems for batteries. Overcharge can cause severe corrosion of the positive grid (v.i.) that results in structural weakness and ultimate failure. Therefore, it was customary to employ thick grids to ensure adequate battery life. Better voltage control restricts the extent of corrosion and, together with the ready availability of defect-free castings and rolled lead sheet of high quality, thinner grids are now practicable.

7.4.2.2 Pasted plates

The active-material for pasted plates is prepared by first reacting lead ingots with air in a ball mill, or molten lead with air in a furnace. The resulting powder – known as 'leady oxide' – is composed of lead monoxide (PbO) and unreacted lead particles ('free-lead'). Next, the leady oxide is combined with sulfuric acid solution to form a paste. During this procedure, a significant proportion of the leady oxide converts to various basic lead sulfates These compounds serve to consolidate and strengthen the paste, a process that may be likened to the setting of cement. Certain minor additives are included in the mix for the negative lead plate, i.e. barium sulfate lignosulfonates and carbon black. Collectively known as 'expanders', the additives improve the low-temperature performance of the battery and extend the cycle-life. The paste is fed from an overhead hopper on to cast grids, or on to continuous expanded-metal grids, that pass horizontally underneath.

In the next stage of manufacture, the plates are flash-dried and then 'cured'. The latter process consists of putting the stacked plates in an enclosure of controlled temperature and humidity for a given period. Such conditioning allows the further development of basic lead sulfates and the oxidation of free lead in the active mass to proceed to near completion. Curing increases the mechanical integrity of the active mass, and thereby provides plates with the desired 'handleability'. Finally, the plates are mounted in the battery together with the separators, welded into groups of like polarity, and 'formed' by charging electrochemically to produce lead dioxide at the positives and 'spongy' lead metal at the negatives.

7.4.2.3 Tubular plates

Tubular plates are normally used in traction batteries on account of their longer life under deep-discharge conditions. In a tubular cell, the positive plate is constructed from a series of vertical lead-alloy spines (or 'fingers') that essentially resemble a comb; see Figure 7.2(c). These act as current-collectors and are inserted into tubes made from woven, braided or felted fibres of glass or polyester. The tubes may be either mounted individually or joined together in a row (the 'gauntlet' design) with spacing equal to that between the spines. The tubes are sealed at the base with plastic caps which are mounted on a common bar. The active-material is packed into the annulus between the spine and the tube wall. A mixture of leady oxide and red lead (Pb_3O_4) is commonly used for the positive plates, and conventional pastes for the negatives. After filling with oxide, the positive plates are soaked in sulfuric acid solution (the so-called 'pickling' process) to convert the majority of the lead oxides to lead sulfate With the tubular design, it is not possible to shed active-material, except in cases of severe battery misuse where splitting of the tubes may occur.

7.4.2.4 Spiral-wound plates

A more recent cell configuration, aimed at high-power applications, has a single pair of positive and negative plates that are interleaved with glass-mat separators and wound together in a cylindrical can (the 'spirally wound', or 'jellyroll' design). Ironically, this arrangement mimics that originally used by Planté; cf. Figure 7.2(a). Instead of placing the positive and negative plates alternately side-by-side, a pair of long plates may be wound into a spiral to create a cell that is cylindrical in shape; see Figure 7.3(a). The two electrodes each have several current take-off tabs, as shown in Figure 7.3(b), in an arrangement that spreads out the contours in potential and invests such cells with good high-rate (high-power) capability, as is required in HEVs. It should be noted, however, that the manufacture of spiral lead–acid cells is more complex than the production of flat-plate designs.

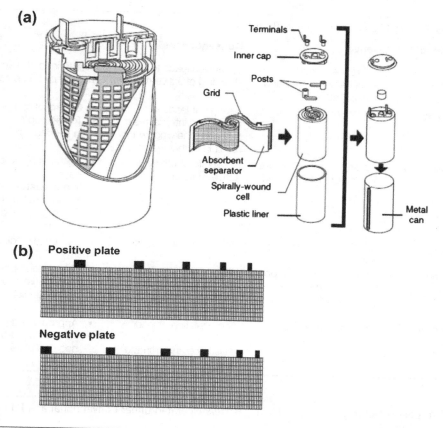

FIGURE 7.3

(a) Spiral-wound design of a lead–acid battery; (b) grids with multiple tabs

7.4.3 Failure mechanisms and remedies

The factors that limit the life of a lead–acid battery and result in ultimate failure can be quite complex. The dominance of one over another is bound up with the design of the battery, its materials of construction, the quality of the build and the conditions of use. The most common failure mechanisms are summarized in Table 7.3, together with remedies that can be adopted.

7.4.4 Battery management

The rapidly expanding use of electrical functions in place of less-efficient mechanical equivalents within road vehicles requires batteries with high power, good ability to withstand extensive periods of operation at a partial state-of-charge (PSoC) and long service life. Compared with conventional float and cycling duties, PSoC duty brings both an advantage and a disadvantage. On the credit side, VRLA batteries enjoy a significant increase in lifetime storage capability (i.e. an increase in the total amount of energy that can be stored and delivered during the life of the battery) when subjected to PSoC

Table 7.3 Failure Modes of Lead–Acid Batteries

Failure Mode	Cause	Consequences	Remedies
Positive plate expansion	Difference in molar volumes of charged and discharged reactants	Expansion leads to shedding of active-material	Constraint of the active-material by compression
Water loss by electrolysis	Overcharge	Drying out leads to increase in internal resistance and heating, which result in further water loss	• Control of charge regime • Topping up of electrolyte in flooded cells
Acid stratification	Dense acid sinks to lower regions of plate (flooded cells)	Non-uniform utilization of active-material	• Agitation by deliberate gassing • Immobilization of electrolyte in VRLA technology • Use of passive mixing device
Incomplete charging	Electrode not reaching adequate potential	Decline in cell capacity	Apply periodic equalization charge
Positive grid corrosion	Various factors, e.g. poor castability, impurities, elevated temperature	Rise in resistance together with grid expansion; followed ultimately by grid disintegration	• Use of appropriate alloys • Reduce time spent at high potential (top-of-charge)
Negative plate sulfation	Charge is too rapid	Accumulation of lead sulfate leading to rise in resistance and ultimate failure	• Appropriate grid design • Addition of carbon to negative plate

Table 7.4 Energy Delivery of a Typical VRLA Battery Operated under Conventional and Partial-State-of-Charge Conditions

Operating Regime	Number of Cycles	Total Energy (Ah)[a]
Conventional, 10% DoD	6000	86 000
Conventional, 40% DoD	1250	71 500
Conventional, 80% DoD	700	80 000
PSoC (40 to 70% SoC)	5500	235 950

[a]Battery capacity of 143 Ah.

regimes at modest rates of charge and discharge, as opposed to regular cycling; an example of this beneficial effect is given in Table 7.4. Therefore, operation over a reduced range of depth of discharge (DoD), as required by the new vehicle protocols, might be expected to lead to a longer cycle-life than in the case of the deep-discharge duty experienced with BEVs; see Figure 7.4. This advantage is more than offset, however, by a challenge to the charge efficiency of the negative plate under this cycling regime. Normally, the efficiency is high until charging is almost complete, so that operation in a PSoC mode ought to be favourable, as anticipated by the data in Table 7.4. It appears, however, that the charge efficiency is diminished at the very high rates of current return experienced in HEVs (see Table 7.2),

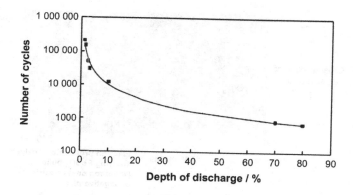

FIGURE 7.4

Variation of cycle-life with depth of discharge of a typical spiral-wound VRLA battery

(Graph from A. Cooper L. T. Lam P. T. Moseley D. A. J. Rand in Valve-Regulated Lead–Acid Batteries, *Elsevier, 2004)*

even when the plate is not at a full state-of-charge. This is a manifestation of the problem of poor dynamic charge-acceptance, which has been discussed in Section 7.3.1.

Lacking a construction that is purpose-designed for high-rate PSoC duty (HRPSoC), VRLA batteries typically lose capacity rapidly during such operation – even at very modest rates of charge. For example, a decline in performance is observed at a rate of only 2 C_1. Detailed studies have shown the detrimental behaviour to be associated with a progressive build-up of lead sulfate on the surface of the negative plate, together with the evolution of hydrogen; see Figure 7.5. Since the battery is not brought to a full state-of-charge during PSoC duty, no routine method is available for the removal of this lead sulfate. Nevertheless, recent research has resulted in two developments in cell design that have provided valuable means of tackling the 'negative plate sulfation'; these are described in the following two sections.

7.4.4.1 *Grid design for high-rate partial-state-of-charge duty*

At the extraordinary levels of power demanded in HRPSoC duty (see Table 7.2), consideration must be given to optimizing the current-collection function of the grid. It is well established that the performance of lead–acid batteries can be adversely affected by a non-uniform distribution of current over the plates that is a consequence of ohmic losses. This effect becomes more pronounced the higher is the rate of charge and discharge and the larger is the plate. At the highest rates, the drop in voltage down the grid causes an inhomogeneous utilization of the active-material (i.e. that near the current take-off tab is worked harder than that further away) and thereby increases the heat generated in the cell. Non-uniform distribution of current can be ameliorated by increasing the electrical conductivity of the grid.

Research conducted at the Commonwealth Scientific and Industrial Research Organisation (CSIRO) in Australia has demonstrated that the addition of a second tab, symmetrically placed opposite the first, will bring considerable advantage in terms of improved performance from VRLA cells subjected to HRPSoC duty; see Figure 7.6(a). The current passing through each tab is approximately half that which would pass through a single take-off and the extra connection also acts as an additional heat-sink. Both these features safeguard against the development of high operating temperatures within cells and result in more uniform utilization of the plate active-materials. These benefits, in turn, translate into improved power capability and longer cycle-life.

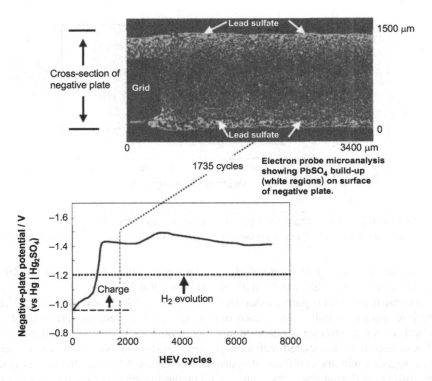

FIGURE 7.5

Build-up of lead sulfate and concomitant hydrogen evolution on negative plates after 1735 cycles of simulated modest HRPSoC duty. (Battery initially discharged at C_1 rate to 50% SoC. Duty cycle: charge/discharge at 2 C_1 pulses of 3% capacity

(Images from A. Cooper L. T. Lam P. T. Moseley D. A. J. Rand in Valve-Regulated Lead–Acid Batteries, *Elsevier, 2004)*

A 36-V battery was built from 18 of the dual-tab cells (see Figure 7.6(b)), and its performance in a demanding cycling regime was compared with that of a 36-V counterpart formed from the single-tab variety. The results showed the dual-tab design to have a pronounced superior durability. Subsequently, in early 2004, a Honda *Insight* HEV was retrofitted with a pack of these batteries and a purpose-built management system underwent successful road trials in the UK; see Figure 7.6(c).

The above improvements result from the fact that the plate has been designed to avoid heterogeneous current distribution. Such research has provided strong evidence that a redesign of VRLA batteries is necessary for the technology to make a significant contribution to the growing HEV share of road transport. Provided that the remarkable benefits associated with the dual-tab design in HRPSoC duty can be attributed to a reduction in resistive ('ohmic') losses, then it can be anticipated that a similar performance improvement could be achieved by a number of other approaches to grid design. These include: (i) altering the aspect ratio of the grid so as to minimize the distance of the tab from the remote corners of the plate (i.e. avoid tall, narrow plates), (ii) increasing the thickness of the appropriate grid wires to reduce the maximum resistance down (across) the grid and (iii) using copper-cored wires of greater electrical conductivity (feasible in the negative plate).

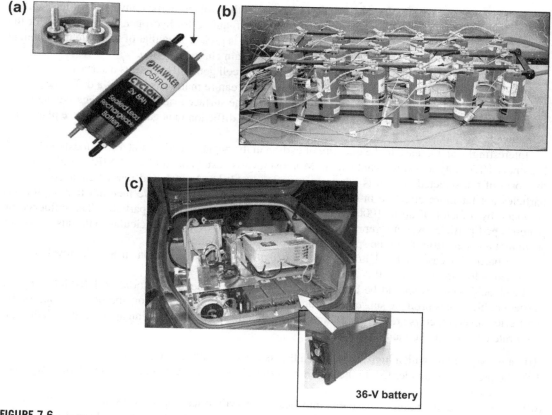

FIGURE 7.6

(a) Spiral-wound cell with dual tabs; (b) 36-V battery of dual-tab cells under test; (c) dual-tab battery system installed in Honda *Insight* HEV

(Images from P. T. Moseley D. A. J. Rand in Encyclopedia of Electrochemical Power Sources, *Volume 4, page 570, Elsevier BV 2009)*

7.4.4.2 Carbon inventory

The special characteristics of HRPSoC duty in HEVs give rise to rapid and destructive accumulations of lead sulfate on the negative plates of lead–acid batteries. This failure mode is manifest as inadequate dynamic charge-acceptance. To determine the origin of the phenomenon and thereby devise means for its suppression, it is necessary to examine and assess the extent to which each of the stages of the charge mechanism imposes a limitation.

The large numbers of charge and discharge events use only a small fraction of the cell capacity and therefore limit the participation of acid to that which is locally available. The fact that these events occur at very high rates and that the cell operates around a PSoC baseline markedly distinguishes this type of duty from other applications of lead–acid batteries. Such exceptional operating conditions focus attention on the surface-area available for reaction and the diffusion of ions in the electrolyte. The latter property influences both the local solubility of lead sulfate and the differential supply of reactants for the competing reactions. Bisulfate ions and protons are both involved in the conversion of lead sulfate back to lead (see Equation 7.5), whereas only protons are required for hydrogen evolution and for the storage of capacitive charge. Since the mobility of protons is greater than that of bisulfate ions then, at high rates, the latter two reactions are favoured over the main charge reaction.

The active-material on the positive plate of a lead–acid cell generally presents a surface-area which is appreciably greater than that of the negative plate. This feature manifests itself as a difference in the flux of electrons to the reaction sites on the two plates. If the surface available at the negative plate can be extended, then the impact of the unequal dependence on diffusion in negative and positive plates can be reduced.

Interestingly, an increase in the quantity of carbon in the negative active-material can extend battery life under HEV duty to a significant extent. Moreover, early results (again from CSIRO) indicated that the form of the selected carbon is of critical importance. It has been found that acicular conducting particles are far more effective in enhancing the conductivity of a resistive medium than equi-axed powders, by a factor of up to 1000, as dictated by the aspect ratio of the particles. The influence of plate-shaped particles was not reported but, by analogy with the case of acicular materials, it would seem that any departure from the spherical shape would be beneficial.

The particular case of the UltraBattery™, which makes use of carbon in a very novel way, is described below in Section 7.10.

Lead–acid batteries should be designed to cope with the special conditions of HRPSoC duty. In particular, high rates produce substantial gradients of local acid concentration and potential across the plate grid. Several factors are worthy of consideration in the development of an appropriate cell design. High-rate charge may be facilitated by:

(i) a negative plate with a high surface-area that is electronically conducting;
(ii) the presence of an element in the active-material of the negative plate that can absorb charge as capacitance;
(iii) the presence of a second phase that impedes the growth of lead sulfate crystals;
(iv) a grid design that allows potential to be as uniform as possible across the whole grid.

The first three of these perceived beneficial changes in cell design may be realized by the addition of three different types of additive to the negative plate but the programme of empirical development that has taken place in recent years indicates that different forms of one element, carbon, may perform the required functions adequately.

7.4.5 Summary

In lead–acid batteries for either 12-V or 48-V vehicle systems, the cells in the strings will have to be matched better than in the past, and also assembled and managed in such a way that they will not tend to become unbalanced.

All of the future vehicle concepts in prospect call upon increased electrical capacity and greater power. The battery requirements will not be met by conventional lead–acid products. Adjustments to the grid design and to the surface-area and conductivity of the negative active mass, as well as the deployment of elements capable of reducing the proclivity of the cell to evolve gas (especially hydrogen) during charging, all show promise in advancing the ability of both the VRLA battery and enhanced designs of flooded battery to cope with the rigorous duties that are envisaged.

It remains to be seen whether the benefits that can be gleaned from these design modifications, when employed together, will prove to be sufficient to provide a satisfactory life for lead–acid batteries in new-generation road vehicles. If this goal is indeed achieved, then the rewards, particularly in terms of reduced battery costs for such vehicles, will be considerable.

7.5 Nickel–metal-hydride batteries

In contrast to lead–acid, the nickel–metal-hydride (Ni–MH) cell employs an alkaline electrolyte – namely, an aqueous solution of concentrated potassium hydroxide. The battery represents the culmination of a development of alkaline secondary batteries that passed from traditional nickel–cadmium and nickel–iron to nickel–hydrogen for providing power in spacecraft (in pursuit of an increase in specific energy and life at the expense of cost), and then to the present chemistry for use in BEVs and HEVs, as well as many electronics applications.

Over the years, there has been a growing objection to the involvement of cadmium due to its toxicity, so much so that sales of nickel–cadmium batteries have become greatly restricted. The nickel–hydrogen cell uses essentially the same nickel oxide positive electrode as in nickel–cadmium, but the negative active-material is pressurized hydrogen gas which reacts electrochemically at a Teflon-bonded platinum-black catalyst that is supported on a nickel grid. The use of a platinum electrode and a pressurized container makes the nickel–hydrogen cell too expensive for any application other than aerospace.

During the latter part of the 20th century, the realization that the hydrogen could be stored within the confines of the cell as a metal hydride provided an eminently more practical, affordable and benign power source. The key factor that allowed the final stage in the development of the commercial Ni–MH cell was the identification of suitable metal alloys that could absorb hydrogen reversibly to constitute the negative electrode. Alloys of general formula AB_2 and AB_5, which are based on the $ZrNi_2$ and $LaNi_5$ structures respectively, have the most advantageous combination of high hydrogen-storage capacities and desirable operational parameters.

AB_2 alloys contain titanium, zirconium or hafnium at the A-site and transition metals (manganese, nickel, chromium, vanadium and others) at the B-site. By contrast, AB_5 alloys combine a hydride-forming metal A, usually a rare-earth metal (lanthanum, cerium, neodymium, praseodymium, or their mixture known as 'misch metal'), with a non-hydride-forming element B, such as nickel. The latter is doped with other metals, such as cobalt, tin, manganese or aluminium, to improve materials stability or to adjust the equilibrium hydrogen pressure and temperature required for hydrogen charge/discharge. AB_5-type alloys offer some advantage over AB_2 counterparts in terms of reduced self-discharge.

A common problem with hydrogen-storage materials is the volume change during the charge and discharge processes. This expansion leads to cracking and pulverization of the alloy and thereby renders it more susceptible to oxidation. Much effort has been devoted to adjusting the composition and microstructures of the alloys used in Ni–MH cells in order to prevent such degradation.

Nickel–metal-hydride cells have proved to be very successful in operation, are now widely available commercially in a range of sizes, and have almost completely taken over from nickel–cadmium for many applications. They were once the preferred technology for portable electronic devices (mobile phones, laptop computers, etc.) but, in these uses, they have now largely been replaced by lithium-ion batteries.

7.5.1 Principles of operation

The discharge reactions at the electrodes are shown in Table 7.1. During charge, reduction of water at the negative electrode produces atomic hydrogen that diffuses into the lattice of the alloy to form a metal hydride. This process is reversed during discharge. The reactions at the positive electrode are the same as those in the nickel–cadmium cell, namely the inter-conversion of $Ni(OH)_2$ and $NiOOH$.

The overall cell reaction can be written as:

$$NiOOH + MH \underset{\text{Charge}}{\overset{\text{Discharge}}{\rightleftharpoons}} Ni(OH)_2 + M, \quad V° = 1.3 \text{ V}, \tag{7.11}$$

where $V°$ is the standard cell voltage.

The electrolyte medium is a strong caustic solution mainly based on potassium hydroxide with a concentration of around 30 wt.% (7 mol L^{-1}). Additions of small amounts of lithium and/or sodium hydroxide are made in pursuit of improved charge-acceptance. The cells employ separators that are fabricated from polypropylene/polyethylene sheet materials with surfaces that are treated to ensure wetting by the electrolyte.

7.5.2 Manufacturing processes

Two methods are available for the production of the positive electrode. In the first, a nickel structure with a high surface-area is formed on a nickel-coated steel strip by means of a sintering process and is then impregnated with the active-material (nickel hydroxide) via a precipitation procedure. Alternatively, the electrode substrate can be constructed as a foam material of high porosity by depositing nickel on to a plastic precursor structure, which is then removed by burning. The nickel-hydroxide active-material is then distributed within the pores of the foam by a coating process. The resulting electrode provides a somewhat higher energy-storage capacity and the manufacturing process is less complicated than for the sintered counterpart.

The negative electrode is formed by coating a perforated nickel-coated steel strip with a slurry of an alloy powder, which is capable of reversibly storing hydrogen, mixed with binder materials and carbon powder. The electrode is then dried and sintered.

The preparation of the two electrodes enables cells to be assembled in the discharge state.

For HEV applications, high-power performance has been achieved by optimizing the current-collector through the development of a multi-contact design rather than a single tab, i.e. through adoption of the technique introduced earlier by CSIRO for the spiral-wound lead–acid cell (*v.s.*). High surface-area and low thickness also assist the electrodes in providing high power.

Individual cells may be cylindrical or prismatic in shape. The former geometry is generally chosen for units with low capacity, i.e. up to 10 Ah, and the latter for the larger capacities required by BEVs (up to 100 Ah). The prismatic design enables an efficient packing of modules to form a large battery and, provided that the cells are built with low thickness (high surface-area), good thermal management can be accomplished with either an air or a water cooling system.

7.5.3 Failure mechanisms and remedies

Susceptibility to corrosion in the strong alkaline environment renders the hydrogen-storage alloy the most vulnerable component of the Ni–MH cell. A certain amount of cobalt in the alloy is indispensable for the cell to achieve long life through suppression of the corrosion processes.

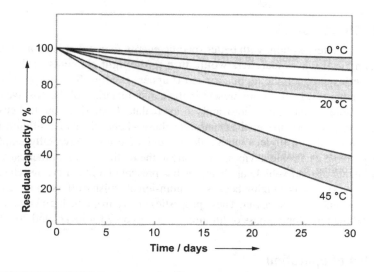

FIGURE 7.7

Self-discharge of a nickel–metal-hydride cell at different temperatures

(With permission from P.H.L.Notten Rechargeable Nickel–Metal Hydride Batteries: A Successful New Concept, ch, 7 vol 281, p 151. NATO ASI Series E. London, ISBN 07923-3299-7)

To protect the metal hydride from damage during overcharging and over-discharging, the cell is configured such that the capacity is determined by the positive electrode. With such a design, the sealed Ni–MH cell is capable of providing a large number of deep-discharge–recharge cycles at high rates. The system does, however, suffer from considerable self-discharge under open-circuit conditions; see Figure 7.7. The extent of this loss of energy is markedly dependent on temperature.

7.5.4 Battery management

As the practical voltage of a single Ni–MH cell is only 1.2 V, a large number of cells (i.e. more than for the 2-V lead–acid system) must be connected in series to reach the voltages (and therefore power) that are required for road-vehicle applications. It is necessary to keep the SoC of the battery at a balanced level so that the system can operate reliably over a long working life. This involves some charge balancing from time to time by means of a controlled amount of overcharge and a subsequent adjustment of the SoC. A BMS initiates the rebalancing whenever it detects an unacceptable discrepancy in SoC between modules.

7.5.5 Summary

Now that cadmium has been banned from most batteries because of its toxicity, the nickel–metal-hydride battery is the only alkaline electrolyte battery in widespread use. Research and development have led to many of its shortcomings being resolved and it has been employed in BEVs and HEVs, and in high volumes for portable power applications. The specific energy of the cell design is 65–80 Wh kg^{-1} for the former applications and 90–110 Wh kg^{-1} for the latter. Thus it is a compromise choice between relatively inexpensive lead–acid, with low specific energy, and lithium-ion with much higher specific energy but with problems that include relatively high cost and questionable safety.

7.6 Lithium-ion batteries

When the search for a battery system with better specific energy than either lead–acid or Ni–MH began in earnest, it was natural that researchers would turn to the light elements in the early part of the periodic table. Batteries in which the mobile ion is Li^+ appear to offer the most promise.

In the lithium-ion battery, Li^+ ions are present in the active-materials of the positive and the negative electrodes, and are able to intercalate into and de-intercalate from (i.e. shuttle between) both host structures, as well as to transport through the electrolyte phase – hence the name 'lithium-ion'; the mechanism is shown schematically in Figure 7.8. Single-cell units are now universally employed in mobile phones and other portable electronic devices. By contrast, the realization of a practical system that satisfies all the requirements of road-vehicle applications has presented a much greater challenge. Nonetheless, the promise of the system is enticing because lithium-ion chemistry offers an attractive combination of high specific energy and specific power. These properties derive from the high voltage of lithium cells, generally 3–4 V, and the low atomic mass of lithium, i.e. 6.9 vs 58.7 for nickel and 207.2 for lead.

7.6.1 Principles of operation

Various materials have been considered for the positive electrode of the cell and have been found to exercise a major influence not only on the power and energy performance of the cell, but also on life, abuse tolerance and cost. In all cases, the materials contain at least one transition-metal element that can change valence state to allow the accommodation of a Li^+ ion without influencing the overall

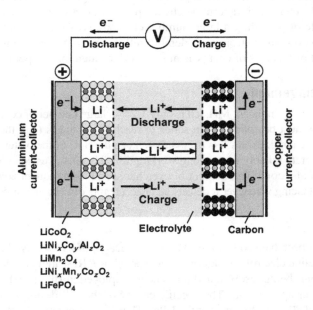

FIGURE 7.8

Shuttle mechanism of lithium-ion cells

(Reproduced by permission of The Royal Society of Chemistry from R.M, Dell and D.A.J. Rand, Clean Energy, 2004)

electro-neutrality. Small cells designed for electronics applications have achieved considerable success through the use of lithium cobalt oxide (or 'lithium cobaltate'; $LiCoO_2$) but for the large batteries required in vehicles the cost of the high cobalt inventory becomes prohibitive. Other compositions that have been explored include oxides with reduced inventories of cobalt, or materials that avoid the element altogether. The four most commonly used are:

- lithium nickel cobalt aluminium oxide (or 'NCA'), $LiNi_xCo_yAl_zO_2$;
- lithium manganese oxide (or 'lithium manganate'), $LiMn_2O_4$;
- lithium nickel manganese cobalt (or 'NMC'), $LiNi_xMn_yCo_zO_2$;
- lithium iron phosphate (or 'LFP'), $LiFePO_4$.

Lithium iron phosphate is an attractively inexpensive option and its high thermal-decomposition temperature renders it safer in use than some of the alternatives. On the debit side, it is an electronic insulator and therefore a second phase material must be incorporated to provide an electronically conducting pathway throughout the electrode volume. This capability is achieved by coating nano-sized particles of $LiFePO_4$ with a very thin layer of carbon. The discharge reactions at electrodes in cells using lithium cobaltate or LFP positives are listed in Table 7.1. There is also a variant of standard lithium-ion technology in which the liquid organic electrolyte is immobilized in a polymer matrix – a so-called 'gelionic' electrolyte. These cells are marketed as 'lithium–polymer' types. This terminology is, however, quite misleading as the gel electrolyte is not a genuine polymer. A more accurate description is 'plastic lithium-ion'.

Each of the candidate materials for the positive electrode of the lithium-ion cell has a distinct set of merits in terms of performance, safety, cost and intellectual property ownership, as summarized in Table 7.5. Consequently, to date, no single composition has been universally accepted. In particular, the commercial development of large-scale lithium-ion batteries is complicated by the fact that each chemistry gives rise to a different electrode potential. This difficulty holds true for both the positive and the negative electrode, as illustrated in Figure 7.9, so that a wide range of cell voltages (the sum of the individual electrode potentials) is possible. A vehicle manufacturer is therefore faced with a decision about which variant of lithium-ion to choose and hence what voltage the cell design must accommodate. The negative materials span a range of potential from zero (lithium metal) to around 1.5 V

Table 7.5 Characteristics of Lithium-Ion Batteries with Different Positive Active-Materials

Positive Active-Material	$LiCoO_2$	$LiNi_xCo_yAl_zO_2$	$LiMn_2O_4$	$LiNi_xMn_yCo_zO_2$	$LiFePO_4$
Mean voltage vs Li/Li+ (V)	3.8	3.9	4.0	3.8–4.0	3.3
Capacity (mAh g^{-1})	150	170	120	130–160	170
Safety	–	0	+	+	+
Cycle-life	> 1000		< 1000		> 1000
Temperature stability (°C)	220	230	350	300	400
Price	High	Medium	Low	Medium	Low

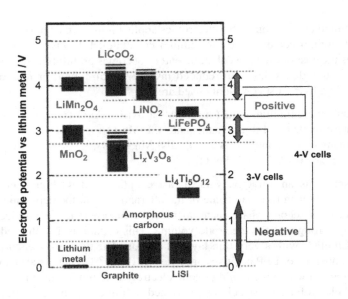

FIGURE 7.9

Potentials of candidate electrode materials vs lithium metal

(Image from G. Gutmann in Encyclopedia of Electrochemical Power Sources, *Volume 1 page 228, Elsevier BV 2010, and re-drawn)*

($Li_4Ti_5O_{12}$) with respect to lithium metal and the positive materials from just above 2 V ($Li_xV_3O_8$) to around 4 V ($LiMn_2O_4$, etc.). Thus, depending on the choice of chemistry for each electrode, the voltage of the resultant lithium-ion could be less than 1 V or as high as around 4 V. The higher-voltage combinations are generally preferred.

Initially lithium metal was used for the negative electrode but it was soon found that this led not only to passivation reactions, which reduced cycle-life and lowered the practical specific energy of the cell, but also to safety problems through the formation of lithium metal dendrites and subsequent short-circuiting of the cell. It has therefore become common practice for the negative to be based on graphitic carbon, into which Li^+ ions can intercalate reversibly. Since negative electrodes operate at a potential outside the electrochemical stability window of the electrolyte components, reductive electrolyte decomposition accompanied by irreversible consumption of lithium ions takes place at the electrode | electrolyte interface when the electrode is in the charged state. The decomposition products build up 'protective layers' that cover the surface of the electrode; see Figure 7.10. This process occurs mainly (but not exclusively) at the beginning of cycling, especially during the first cycle. The protective layers act as a so-called 'SEI' (solid | electrolyte interphase). They are permeable to lithium ions but rather impermeable to other electrolyte components and electrons. This feature is crucial to the continued operation of lithium-ion cells that use graphite negatives as it not only serves to suppress continued decomposition of the electrolyte and exfoliation of graphene layers from the electrode, but also protects the negative electrode from corrosion. It should be noted that while SEI formation takes place mainly in the first few charge–discharge cycles, SEI conversion (i.e. changes in composition and morphology), stabilization and growth will also proceed during further cycling and storage. Therefore, the SEI formation influences both the cycle-life and the calendar life of a cell.

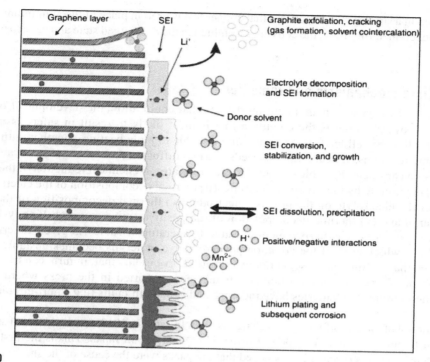

FIGURE 7.10

Reactions of the negative electrode with the electrolyte in a lithium-ion cell. SEI = solid electrolyte interphase

(Image from J Vetter, M Winter and M Wohlfahrt-Meherens in Encyclopedia of Electrochemical Power Sources, *Volume 5 page 394. Elsevier BV 2009)*

Other possible systems for the negative plate are lithium alloys with silicon or tin, which can offer large capacities but, as yet, unproven cycle-life. Lithium titanate ($Li_4Ti_5O_{12}$) is also a candidate – it has very good charge efficiency, calendar life and cycle-life but yields a lower cell voltage. For example, when partnered with a $LiFePO_4$ positive, the output is reduced to ~ 1.5 V; see Figure 7.9. The specific energies of batteries using lithium titanate are also lower than those with the other materials.

The selection of lithium as the key element precludes the involvement of an aqueous electrolyte. Instead, an organic solvent is used, usually an alkene carbonate or a mixture of alkene carbonates, in which is dissolved, at a concentration of ~ 1 mol L^{-1}, a lithium salt such as lithium hexafluorophosphate ($LiPF_6$) or lithium tetrafluoroborate ($LiBF_4$). Whereas such an electrolyte does provide adequate conductivity, it also introduces a flammable component.

7.6.2 Manufacturing processes

The production of lithium-ion cells requires fabrication under controlled atmospheric conditions in an ultra-dry room so as to exclude all traces of water vapour. For the positive electrode, the active-material powder is mixed with carbon and a binder in an organic liquid to form a slurry that is applied to an aluminium foil which acts as both a substrate and a current-collector. Negative electrodes are manufactured in a similar manner, except that copper foil is used in place of the aluminium. As noted above for nickel–metal-hydride, the cells are assembled in the discharged state and have to be charged before use.

7.6.3 Failure mechanisms and remedies

Apart from the existence of manufacturing defects that could allow the development of internal short-circuits, overcharging is the condition that is most likely to result in safety issues. Overcharging of a lithium cell leads to a temperature rise due to joule heating, together with the onset of exothermic reactions. If heat output exceeds thermal diffusion then 'thermal runaway', fire and even explosion can occur. Possible exothermic reactions are: (i) chemical reduction of the (organic, flammable) electrolyte by the negative electrode, (ii) thermal decomposition of the electrolyte, (iii) oxidation of the electrolyte on the positive electrode, (iv) thermal decomposition of the negative active-material and (v) thermal decomposition of the positive active-material. Finally, high-voltage positive electrodes release oxygen at elevated temperatures; Table 7.5 gives the critical temperature above which each of the common positive active-materials becomes unstable. It should also be noted that melting of the separator can create short-circuits that, in turn, can trigger a large output of heat. There have been a number of instances reported in the press where fires have resulted when lithium-ion batteries have incorporated cells with faults (e.g. short-circuits) or have been abused.

Strict control of the cut-off voltage on charge is vital. In January 2013, two incidents of smoke were reported on Boeing *787* aircraft – one at Boston, USA, and one in Japan. This is the first aircraft to employ lithium-ion batteries and it is believed that the packs were the cause of the smoke as the component cells were found to be seriously damaged. The entire fleet of *787* aircraft was grounded pending investigation.

7.6.4 Battery management

When used to power HEVs or BEVs, lithium-ion batteries demand a more comprehensive BMS than other battery types because each cell has to be monitored separately to guard against hazards due to overcharging or over-discharging. A reliable determination of the cell SoC is required for the BMS to function efficiently, and this is possible because of the significant slope in the voltage characteristics of most lithium-ion systems; an example of the rate-dependence of the discharge capacity is shown in Figure 7.11. Supervisory circuits monitor the voltages of the cells and the temperature at representative positions inside modules. The collected information is processed in the BMS to determine SoC, state-of-health (SoH, an arbitrary measure of battery condition) and state-of-function (SoF, the capability, or readiness, of the battery to perform a specific duty). In the event of an imbalance in SoC, charging to equilibrate the relevant cells is managed by the BMS.

Although lithium-ion batteries have somewhat higher energy efficiency than other battery systems, they do require an effective thermal control mechanism to dissipate the waste heat that is generated during operation. Compared with Ni–MH and lead–acid, lithium-ion batteries are more sensitive to heat generation by virtue of their lower heat capacity and, as discussed above, the consequences of uncontrolled temperature rise are more serious because they employ a flammable organic electrolyte.

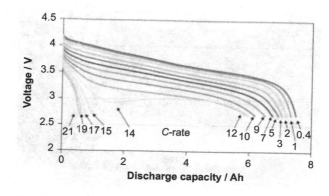

FIGURE 7.11

Discharge characteristics of a 7.5-Ah lithium-ion cell at different rates; the cell has a nickel/cobalt/aluminium oxide (NCA) positive electrode and graphite negative electrode

(Image from P Kurzweil and K Brandt in Encyclopedia of Electrochemical Power Sources, *Volume 5 page 16. Elsevier BV 2009)*

7.6.5 Summary

The technology of lithium-ion batteries has undergone rapid development in recent years as a result of intensive research in universities and the development work in companies around the globe. Large batteries suited to road-transport applications are now commercially available and are being fitted to some BEVs and HEVs. For example, the Nissan *Leaf* (see Figure 5.11, Chapter 5) is powered by a lithium-ion battery with manganese dioxide positive electrodes. The pack has a capacity of 24 kWh and is composed of 192 cells, each of 125-Wh capacity, that are arranged in 48 modules and located low down in the car to enhance stability; see Figure 7.12. The latest model of the *Leaf* has a range of up to 200 km on the New European Drive Cycle, although the range in practice will depend on factors such as the driving pattern and the ambient temperature. As for all types of rechargeable battery, these factors also influence the lifespan of the battery, along with other circumstances such as the depth of discharge employed and the charging rate. Initially, Nissan offered a guarantee of eight years or 160 000 km (100 000 miles) under 'normal conditions of use', but field experience has shown that high ambient temperatures and repeatedly charging at a high rate cause a dramatic shortening of battery life.

Apart from remaining questions over safety, the main disadvantage of the batteries for use in vehicles is their high cost, which may decline as mass-production progresses. The present capital outlay would preclude the replacement of any battery pack that failed during the expected service-life of the vehicle. Nevertheless, large lithium-ion batteries may well feature in several types of road vehicle and there has been considerable effort to evaluate the consequences of their introduction, in terms of a life-cycle analysis. For instance, a study in 2010 by a team at the Argonne National Laboratory considered the energy used and the consequent emissions of carbon dioxide during the manufacture of lead–acid, nickel–cadmium, nickel–metal-hydride, sodium–sulfur and lithium-ion batteries. From an assessment of the resulting cradle-to-gate life-cycle data, it was concluded that, whether on a per kilogram or per watt-hour capacity basis, lead–acid batteries have the lowest production energy, lowest carbon dioxide emissions and lowest combustion-related pollutant emissions. Although there is not yet universal agreement on the conclusions of such studies, it is clear that future limitations on fossil-fuel resources and downward pressure on allowable environmental impact

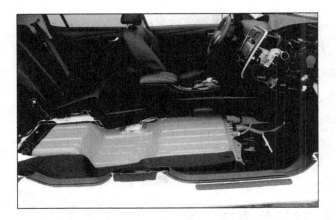

FIGURE 7.12

Lithium-ion battery pack in the Nissan *Leaf*

(Image sourced from Wikipedia and available under the Creative Commons License)

dictate that manufacturing processes for lithium batteries should be improved. For example, some of the active-materials are prepared by solid-state reactions at 700 °C. If the synthesis could be undertaken by processes that require temperatures no higher than 200 °C, then a considerable saving in thermal energy (and emissions) could be achieved.

7.7 Sodium–metal-halide batteries

The sodium–metal-halide battery was developed as a follow-up to the sodium–sulfur battery that was devised in the 1960s by N. Weber and J.T. Kummer at the Ford Motor Company. These two pioneers first recognized that the ceramic popularly labelled 'beta-alumina' possessed a conductivity for sodium ions that would allow its use as an electrolyte in an energy-storage cell provided that the temperature of the cell could be maintained at around 300 °C. At this temperature, the active-materials of the two electrodes (sodium and sulfur are liquids and thus the cell operates with an 'inverse' structure in comparison with most other electrochemical cells that have solid electrode materials and liquid electrolytes. The design of cells in which both of the electrode materials are liquids from which air must be excluded raises serious sealing challenges. The most convenient arrangement is to use the solid electrolyte in the form of a cylindrical tube with one end closed. One of the liquid electrodes resides within the electrolyte tube with electronic contact made through an air-tight seal at the open end, and the other liquid electrode is deployed within an annular space between the outer surface of the electrolyte tube and a co-axial outer container. The sodium–sulfur battery is manufactured in Japan for energy storage in the electricity supply network, but has not proved suitable for use in electric vehicles.

The sodium–metal-halide battery was derived in the early 1980s from the sodium–sulfur design by replacing the sulfur electrode with a mixture of molten sodium chloroaluminate, $NaAlCl_4$, and a solid transition metal chloride, either $NiCl_2$ or $FeCl_2$, or a mixture of the two; see Figure 7.13(a). The technology, which was conceived and developed through collaboration between scientists working in the

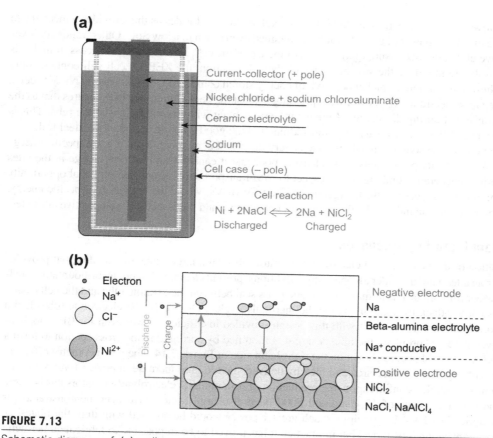

(a)

Current-collector (+ pole)

Nickel chloride + sodium chloroaluminate

Ceramic electrolyte

Sodium

Cell case (− pole)

Cell reaction

$$Ni + 2NaCl \Longleftrightarrow 2Na + NiCl_2$$
Discharged   Charged

(b)

- Electron
- Na^+
- Cl^-
- Ni^{2+}

Discharge / Charge

Negative electrode
Na

Beta-alumina electrolyte
Na^+ conductive

Positive electrode
$NiCl_2$
$NaCl, NaAlCl_4$

FIGURE 7.13

Schematic diagrams of: (a) sodium–nickel-chloride (ZEBRA) cell; (b) charge and discharge reactions

(Images from Encyclopedia of Electrochemical Power Sources, *Volume 4. Elsevier BV 2009)*

UK and in South Africa, became known as the ZEBRA battery. The power source has subsequently proved more practical in terms of operating life, safety and robustness for vehicle applications than the original sodium–sulfur cell, despite the fact that the materials of construction are somewhat more costly.

7.7.1 Principles of operation

During cell discharge, sodium atoms lose electrons to the external circuit and Na^+ ions then pass through the wall of the beta-alumina tube to the positive electrode region where they react with the transition metal halide to form sodium chloride and free metal; see Table 7.1 and Figure 7.13(b). Within the positive electrode volume, the molten $NaAlCl_4$ ensures good transport of Na^+ ions, while electronic contact to the external circuit is achieved with the aid of a nickel current-collector. The Na–$NiCl_2$ cell has a standard voltage of 2.58 V. This is higher than lead–acid (2.05 V), Ni–MH (1.3 V) and sodium–sulfur (2.08 V), but still less than lithium-ion (3-4 V).

The positive electrode generally includes both nickel and iron chlorides as this provides a mechanism by which the discharge power of the cell can be maintained even down to a low SoC. Other minor additives to the positive electrode mix ensure optimum performance of the cell. Even though an excess of nickel is used beyond that required by the stoichiometry of the cell reaction, the ZEBRA cell still contains only about one-third of the weight of nickel per kWh of energy stored compared with that used in Ni–MH cells.

In the negative electrode, a steel-foil wick pressing against the beta-alumina surface ensures that as the sodium is consumed during discharge it remains in good contact with the full height of the tube. This is possible because of the low viscosity of liquid sodium and the good wicking action of the steel foil.

The geometry of the sodium–metal-chloride battery dictates that although it has a good specific energy (up to 120 Wh kg^{-1}), its power density is relatively poor and it cannot readily accept charge at the rates required during regenerative braking. With respect to road-transport applications, the principal opportunity for the technology is to serve as the energy-storage unit for BEVs, where the appreciable specific energy, long life, ability to operate independently of ambient temperature and modest cost are all positive attributes.

7.7.2 Manufacturing processes

The production of the beta-alumina electrolyte/separator tubes on a large scale, to a quality that provides integrity for an adequate life under all operating conditions, presents a major, but not insurmountable, challenge. A further difficult task is to make a hermetic glass seal between the ceramic tube and the cell casing that is resistant to attack by sodium and can be thermally cycled. Originally, the electrolyte tubes had a circular cross-section, but cells built with this design provided insufficient power even for BEV applications. To overcome this problem, the tube design was modified by corrugating the cross-section to form a so-called 'clover-leaf' shape, as shown on the right-hand side of Figure 7.14. The new geometry offers an electrolyte with a greater surface-area and thereby delivers the required increase in power. Even with this apparently more complex configuration, the production of beta-alumina electrolyte/separators has become reliable and operating lives of 10 years and 5000 deep-discharge cycles have been demonstrated for ZEBRA batteries in BEVs. In principle, much greater power would be attained with thin, flat plates, as used in most conventional batteries, but in practice it has proved to be impossible to fabricate a hermetic glass seal around the entire perimeter of the plate. This is an engineering challenge yet to be solved.

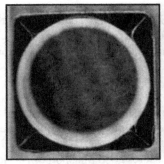 

Original design   **Clover-leaf design**

FIGURE 7.14

'Clover-leaf' design of the beta-alumina electrolyte tube for high power

(Images from Encyclopedia of Electrochemical Power Sources, *Elsevier BV 2009)*

Cells are conveniently assembled in the discharged state (like Ni–MH and lithium-ion) so that the positive electrode volume can be filled with nickel particles and sodium chloride. The air-sensitive materials, sodium and anhydrous nickel chloride, are then generated by charging the cell after it has been assembled and sealed. Na^+ ions pass through the beta-alumina electrolyte to combine in the negative electrode compartment with electrons that pass around the external circuit, thereby forming liquid sodium metal; in the positive compartment Cl- ions combine with nickel metal to form anhydrous $NiCl_2$. This circumvents all the problems associated with the handling and dispensing of liquid sodium. To meet the energy demand of BEVs (several tens of kWh), a parallel arrangement of cells is used in addition to the series connection that is necessary to reach high voltage. Cells are therefore grouped together in modules wired internally in a series–parallel array.

The cells in a sodium–metal-chloride battery that is intended for motive-power duties are housed in a double-skin unit that comprises an inner and outer metal box. The cavity between them is evacuated and filled with thermal insulation; see Figure 7.15. Excellent structural integrity of the container provides a high level of confidence in the safety of the battery against mechanical, thermal or electrical abuse. Indeed, this has been confirmed by an exhaustive series of road and crash tests of BEVs equipped with ZEBRA batteries.

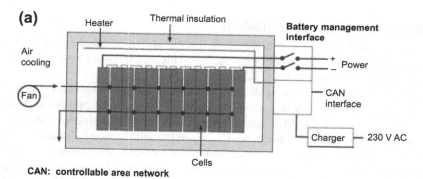

FIGURE 7.15

Design of ZEBRA battery pack

(Images from C-H Dustmann A. Bito in Encyclopedia of Electrochemical Power Sources, *Volume 4 page 326. Elsevier BV 2009)*

7.7.3 Failure mechanisms and remedies

Since the battery must be maintained at elevated temperature, the design must provide the best possible thermal insulation. Typical heat loss from a 20-kWh battery (i.e. suitable for a small BEV) runs at about 120 W and, whenever the battery is not plugged into a mains supply, this energy must come from the battery itself by passing current through an internal resistance heater. In effect, thermal loss is a form of self-discharge, which does not occur electrochemically in this battery. As a result of these heat losses and the consequent drain on the battery, sodium–metal-chloride technology is not considered to be a strong candidate for private passenger vehicles. Rather, it is more likely to be employed in buses and/or delivery vans that are operated in fleets with a high degree of daily utilization and with drivers trained to ensure that the batteries are connected to the mains when not in use. If, inadvertently, the battery is allowed to cool to ambient temperature, it must be re-heated very slowly to avoid mechanical failure.

The battery is supplied as a sealed unit with no maintenance requirements beyond periodic charge equalization. It has been cycled successfully over a period of 15 years with few cell failures. Because the battery operates over a rather wide temperature range (275–350 °C) and is thermally insulated, it may be used at virtually any ambient temperature from sub-zero to tropical. No other traction battery provides such remarkable flexibility.

7.7.4 Battery production and utilization

As long ago as1984, an electric version of the Mercedes *A*-class car powered by a ZEBRA battery was successfully demonstrated on the streets of Derby, UK. This was followed by a series of road vehicle demonstrations both in Germany and in California.

After years of development work in the UK and Germany, sodium–metal-chloride batteries for BEVs are now being manufactured in limited numbers by FZ Sonick, a company in Switzerland that has become part of the Italian FIAMM battery group. Trials are continuing with electric versions of light-duty vans and buses. Several of the buses have been put into service in Italian and Spanish cities (Bologna, Florence, Madrid) and the solar bus in Adelaide, Australia, also operates with ZEBRA modules (described in Section 5.3.3.2, Chapter 5). Elsewhere, there are a number of diesel–battery hybrid buses on the roads that can operate either in full battery mode or in series-hybrid mode with the diesel running efficiently at a constant power level.

More recently, the technology has been taken up in the USA by the General Electric company (GE), which has built a battery production plant in New York State. The initial target market is not traction applications but energy-storage systems for use with telecommunications towers, particularly in remote regions where mains electricity is not available and it is therefore necessary to rely on diesel generators, wind generators or solar photovoltaic panels. Note that, without thermal management, conventional storage batteries with aqueous electrolytes are not suited for use in tropical or arctic conditions. Sodium–metal-halide technology is now being marketed by GE under the trade name 'Durathon' and other applications foreseen are uninterruptible power supplies and battery electric buses.

7.7.5 Summary

Sodium–metal-chloride batteries are remarkable for their tubular ceramic electrolytes and their high operating temperature (~ 300 °C). Although the combination of properties and cost exhibited by such batteries matches the requirements for BEVs fairly well, their prime application is likely to be in stationary energy-storage where they can be supervised by technically competent staff. There may be a concomitant acceptance for BEV use but this will probably be confined to fleets of vehicles that can also be managed by qualified service teams. It is difficult to anticipate the use of these batteries in a large fraction of the future road-vehicle fleet.

7.8 Characteristics of batteries used in hybrid electric and battery electric vehicles

Ideally, electric-vehicle batteries should have minimum mass, i.e. high specific energy, and occupy minimum volume, i.e. have high energy density. The scope of these two parameters for a variety of battery types is shown in Figure 7.16 – the advantages of lithium batteries are immediately apparent. There are, however, many other factors involved in the choice of a battery for a particular vehicle and duty cycle.

Most designs of battery systems can be optimized either for high power, which would render them suitable for use in HEVs, or for high energy content, which would be of value in plug-in hybrid vehicle (PHEV) and BEV applications. The desired battery duty cycles are discussed in Sections 5.2 and 5.3, Chapter 5.

The main properties of different types of battery for use in power applications such as in HEVs are shown in Table 7.6. Lead–acid batteries tend to perform well in PSoC duty because the depth of

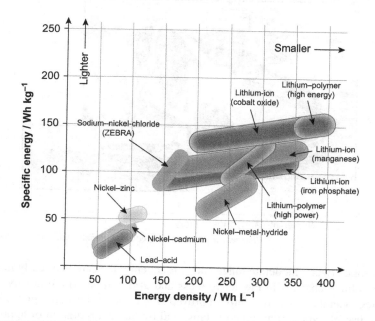

FIGURE 7.16

Comparison of the performance of different battery chemistries in terms of specific energy (per mass of device) and energy density (per volume of device)

(Image re-drawn from P Encyclopedia of Electrochemical Power Sources, *Elsevier 2009)*

Table 7.6 Properties of Batteries Designed for Power Applications (e.g. HEVs)

System	Lead–Acid with Added Carbon	Nickel–Metal-Hydride	Lithium-Ion
Standard voltage (V)	2.05	1.3	4.1
Operating voltage (V)	2.0	1.2	3.6
Specific power (W kg^{-1})[a]	1000	1600	2200
Power density (W L^{-1})[a]	2000	3300	4600
Capacity turnovers in PSoC cycling	10 000	20 000	25 000
Self-discharge rate (% per month)	3	10	3
Safety precautions	Module or battery monitoring	Module monitoring	Single-cell monitoring
Battery price ($ kWh^{-1} present/mass market)	200/150	500/450	1000/500

[a]*As noted in the text, the values for specific power and power density are broad-brush only and depend almost as much on the design of the battery system as on its fundamental electrochemistry. The data are included to provide a guide to the differences between the various types of battery.*

FIGURE 7.17

Toyota Prius cutaway model. Nickel–metal-hydride battery

(Images sourced from Wikipedia and available under the Creative Commons License)

discharge is controlled over a moderate range and consequently the causes of early failure of the positive plate (large volume changes in the active mass and grid corrosion) are largely avoided. Performance in HRPSoC operation is also greatly improved when grid designs are optimized for high rates (e.g. through the use of twin current take-off tabs) and by the incorporation of appropriate forms of carbon in various ways in the negative plate.

The higher power of Ni–MH and lithium-ion batteries gives them an advantage for hybrids, where the highest power is required, but this is offset by the additional cost, which is a factor that will persist even when the economies of mass production take effect. When the first full hybrids were introduced

Table 7.7 Properties of Batteries Designed for Energy Applications (Traction)

System	Lead–Acid	Nickel–Metal-Hydride	Lithium-Ion	Sodium–Metal-Halide (ZEBRA)
Standard voltage (V)	2.05	1.3	4.1	2.58
Operating voltage (V)	2.0	1.2	3.6	2.58
Specific energy (Wh kg^{-1})[a]	25–35	60	140	120
Energy density (Wh L^{-1})[a]	100	200	320	190
Cycle-life (full cycles)	100–1000	1000–2500	800–2500	1300
Self-discharge rate (% per month)	3	10	3	8% per day (thermal loss)
Operating temperature	Ambient	Ambient	Ambient	~ 300 °C
Safety precautions	Module or battery monitoring	Module monitoring	Single-cell monitoring	Battery monitoring
Battery price ($ kWh^{-1} present/mass market)	150/150	500/450	1000/500	600/300

[a]*As noted in the text, the values for specific energy and energy density are broad-brush only and depend almost as much on the design of the battery system as on its fundamental electrochemistry. The data are included to provide a guide to the differences between the various types of battery.*

in the mid-1990s, concerns were expressed over the likely lifespan of the Ni–MH batteries that were installed, both because of the novelty of using batteries for hybrid operation and on account of the high cost of replacement. Toyota claimed that the battery was 'intended to last the life of the car' and offered long warranties. Fortunately for manufacturer and customer alike, such confidence has proved justified, with very few premature failures having been reported. The Ni–MH battery installed in the boot of the Toyota *Prius* (see Section 5.2.2, Chapter 5) is shown in Figure 7.17.

A comparison of the properties of different battery chemistries for the deep-cycle application required in BEVs is given in Table 7.7. The advantages of the lower weights and volumes and longer cycle-lives of the alternatives to lead–acid for this application are clear. For use in extreme ambient temperatures (below −20 °C or above +50 °C), only the ZEBRA battery will perform satisfactorily as it is insensitive to ambient conditions.

7.9 Supercapacitors

Supercapacitors, often also known as 'ultracapacitors', differ from conventional electrostatic and electrolytic capacitors in that they store electrostatic charge in the form of ions, rather than electrons, on the surfaces of materials that have high specific areas (m^2 g^{-1}). In the 'symmetric' design of super-capacitor (see Figure 7.18(a)), the electrodes are usually prepared as compacts of finely-divided, porous carbon that provide a much greater charge density than is possible with non-porous, planar electrodes. Supercapacitors can store vastly more energy than conventional capacitors and most of the storage capacity is due to the charging and discharging of the electrical double layers that are formed at the electrode | electrolyte interfaces. The voltage is lower than for a conventional capacitor, while the time for charge, as well as that for discharge, is longer because ions move and re-orient

(a) **Symmetric supercapacitor**

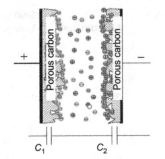

Both electrodes charged/discharged by
reversible adsorption/desorption of ions

$$1/C_T = 1/C_1 + 1/C_2$$

(b) **Asymmetric supercapacitor**

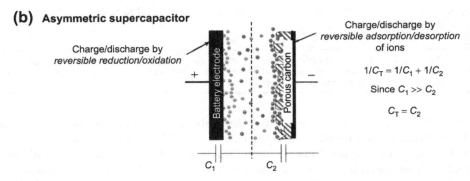

Charge/discharge by
reversible reduction/oxidation

Charge/discharge by
reversible adsorption/desorption
of ions

$$1/C_T = 1/C_1 + 1/C_2$$

Since $C_1 \gg C_2$

$$C_T = C_2$$

FIGURE 7.18

Schematic representation of (a) symmetric supercapacitor; (b) asymmetric supercapacitor

more slowly than electrons. In these respects, the supercapacitor begins to take on some of the characteristics of a battery, although no electrochemical reactions are involved in the charge and the discharge processes.

The 'asymmetric' design of the supercapacitor moves one step closer to a battery; see Figure 7.18(b). Here, an electrode material with a large specific surface-area (i.e. carbon) is combined with a 'battery-like' material that can be reversibly oxidized and reduced over a wide potential range. The energy is stored both by ionic capacitance and by surface (and near-surface) redox processes that occur during charge and discharge. The latter are electrochemical reactions (i.e. faradaic processes) in which surface ions are reduced and oxidized. This enhances the amount of stored energy and the capacitance is twice that of the symmetric counterpart, as described in Figure 7.18, because the total capacitance, C_T, of two capacitors in series, C_1 and C_2, is given by the reciprocal formula:

$$1/C_T = 1/C_1 + 1/C_2 \tag{7.12}$$

Because the ions are confined to surface layers, the redox reactions are rapid (i.e. of the order of a second) and are fully reversible many thousands of times and this makes for a long cycle-life. Also, these devices operate over a wide temperature range (−40 to + 65 °C).

Although supercapacitors can store more energy per unit weight than conventional capacitors, their specific energy is still inadequate for most traction purposes. Thus, vehicle projects that countenance the use of such technology generally also include a rechargeable battery to provide the necessary energy. This strategy presents, however, challenges in terms of additional cost and the extra volume occupied by a second energy/power-storage facility.

7.10 The UltraBattery™

A novel approach to overcoming the problems of HRPSoC duty has been taken by CSIRO through the invention and further development of a radical new design of VRLA battery, in which the negative plate is protected from the deleterious effects of high-rate charge and discharge by sharing the current with an integrated supercapacitor.

The innovative configuration of the CSIRO UltraBattery™ combines a VRLA cell with an asymmetric supercapacitor in a single unit without the need for extra electronic control. The hybrid structure is shown schematically in Figure 7.19(a). The VRLA component, which has one lead dioxide positive plate and one sponge-lead negative plate, is combined with an asymmetric supercapacitor, which is composed of one lead dioxide positive plate and one carbon-based negative plate (i.e. a capacitor electrode). Since the positive plate in the VRLA cell and its counterpart in the asymmetric supercapacitor have the same composition, the two energy-storage devices can be integrated into one unit by connecting internally the VRLA negative plate and the capacitor electrode in parallel. Both these electrodes now share the same positive plate. With this arrangement, the total discharge or charge current of the combined negative plate is composed of two components, namely, the capacitor current and the VRLA negative-plate current. Accordingly, the capacitor electrode can now act as a buffer to share current with the negative plate and thus prevent it from being discharged and charged at high rates. This technology is expected to be less costly and to occupy less volume than would the combination of a conventional battery in parallel with a conventional supercapacitor.

Prototype 12-V units of the UltraBattery™, shown in Figure 7.19(b), were constructed by the Furukawa Battery Company in Japan and were fitted to a Honda *Insight* HEV that successfully completed a 160000-km (100000-mile) test at the Millbrook Proving Ground in the UK; see Figure 7.19(c). Remarkably, the cells of the Ultrabattery™ did not require any form of conditioning during the entire test. This admirable behaviour is in sharp contrast with that of conventional lead–acid cells which, when operating under HEV duty, inevitably demand a periodic 'refreshing' charge to maintain acceptable performance. The battery pack demonstrated very good acceptance of the charge from regenerative braking and remained in an excellent condition throughout. In a further demonstration of the value of the UltraBattery™, a Honda *Civic* hybrid in which the Ni–MH battery had been replaced by an UltraBattery™ recently completed 100000 miles of real-world driving on the roads of Arizona. Production of the UltraBattery™ is in progress in both Japan and the USA.

Several major car companies are currently (2014) understood to be actively working towards introducing lead–carbon batteries into stop–start and/or mild hybrids. An announcement from Kia Motors Corporation, for example, reports that it has preferred a 48-V module of this new chemistry over an lithium-ion alternative because lead–carbon cells: 'require no active cooling, are more easily recyclable at the end of the vehicle's life, and can function much more efficiently in sub-zero temperatures'. Meanwhile, in Japan, the UltraBattery™ is being used as original equipment in the Honda *Odyssey* start–stop van.

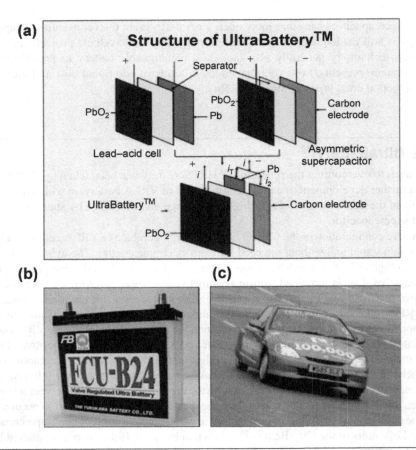

FIGURE 7.19

(a) Schematic representation of UltraBattery™ configuration and operation; (b) UltraBattery™ module designed for HEV applications; (c) Honda *Insight* HEV equipped with Ultrabattery™ pack undergoing field trials at Millbrook Proving Ground, UK

(Images for parts a and b courtesy of CSIRO. Part c, courtesy of the Advanced Lead-Acid Battery Consortium (ALABC))

7.11 Better batteries: future prospects

State-of-the-art batteries provide adequate electrochemical storage for HEVs but fall well short of the specific energy and recharge rates that would be demanded by purchasers of BEVs as mainstream vehicles. The energy available (and therefore the vehicle range) depends critically on the rate at which power is extracted (i.e. the speed of the vehicle and the ancillary electrical loads in use). The specific energy available from different battery chemistries is plotted against the specific power delivered in Figure 7.20 – this is known as a Ragone plot. By way of comparison, the corresponding data for other power sources used in electric vehicles – namely, capacitors and fuel cells – are included together with those for internal-combustion engines.

Drivers who are used to the performance, convenience and cost of modern cars will be reluctant to accept less, although BEVs do hold considerable promise in the mid-term as short-range urban delivery vans and as commuter vehicles. Unless there is a major breakthrough, batteries will continue to limit

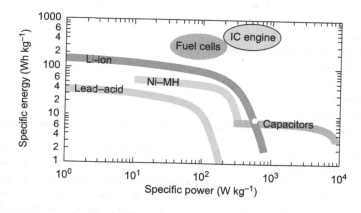

FIGURE 7.20

Ragone plot for different power sources

(Image from P Kurzweil and K Brandt in Encyclopedia of Electrochemical Power Sources, *Volume 5 page 4. Elsevier BV 2009)*

severely the driving range of BEVs between charges to around 100–150 km (62–93 miles) for traditional batteries and about 200 km (125 miles) for lithium-ion batteries, as in the Nissan *Leaf* (*v.s.*). Vehicles powered by Ni–MH technology would be expected to have ranges that lie between these two limits.

The individual lithium-ion cells that are used in BEVs today have specific energies of around 140 Wh kg^{-1}, but this performance is significantly degraded for a whole energy-storage system when packaging, battery management and cooling are taken into consideration. These values do not start to compare with the 13 000 Wh of energy that is contained in one kilogram of petrol (even taking into account the fact that the ICE is far less efficient than the electric motor). The battery is so large that even if the specific energy achievable were to double within the next few years, there would be little improvement in maximum driving range, perhaps to ~ 250 km (155 miles). Moreover, this limit would be markedly reduced if the vehicle were to be driven aggressively and/or there was air-conditioning in use. To meet the requirements of BEVs without compromise, there must be a huge improvement in performance and a reduction in cost.

So what are the prospects that a new, low-cost and safe battery chemistry will emerge to provide a universally acceptable increase in the range of a BEV (to match that of a vehicle on a full tank of liquid hydrocarbon fuel)? The choice of elements that might prove useful tends to be limited to those at the extremes of the electronegativity ranking (i.e. a highly electropositive element paired with a highly electronegative element) to enable the cell to provide a useful terminal voltage. The requirements for high specific energy and specific power focus the search still further on elements with low atomic number (low atomic weight).

In fact, the list of criteria that a battery chemistry must satisfy in order to prove useful as an energy-storage unit in a road vehicle includes all of the following attributes:

- reversibility of the cell reactions;
- ability to operate over a range of ambient temperatures;
- micro-structural stability through the discharge–recharge cycle;
- adequate electronic conductivity in the electrodes and adequate ionic conductivity in the electrolyte (low internal impedance);
- high specific energy, specific power, energy density and power density;
- long cycle-life;

- full materials recycleability;
- safety in use and abuse;
- affordability (both materials and manufacture);
- long shelf life, i.e. low self-discharge;
- good high-rate capability;
- low heat generation;
- high coulombic efficiency;
- suppression of side-reactions, no electrode poisoning, no metal deposition or dendrite formation.

Thus the task facing battery scientists is rendered extraordinarily difficult by the requirement to meet this long list of criteria simultaneously, many of which are inter-related. Adjustments intended to improve one of the performance parameters very often result in the degradation of another parameter.

To find a battery with marked improvement over what is possible with lithium-ion is a major challenge. It is clearly necessary to investigate totally new compositions and new materials. With respect to the positive electrode, the most attractive element by far is oxygen as it is available gratis from the air and does not have to be carried around. As noted above, hearing-aid batteries (non-rechargeable) often have a zinc negative electrode and an air positive. Recharge would pose a major obstacle. The second most attractive option for a positive electrode is sulfur which is abundant and cheap, and not too heavy. The lithium–sulfur battery is already widely viewed as one of the 'next generation' of devices that will offer a significant increase in specific energy (by a factor of two to five) over contemporary lithium-ion. By combining a lithium-metal negative electrode with a positive electrode composed of a conductive sulfur composite (see Figure 7.21(a)), the theoretical specific energy of lithium–sulfur (based on active-material only) is 2600 Wh kg^{-1}. To date, however, few prototype devices have been able to sustain more than half of the available discharge capacity, due to the tendency for the reduction products of sulfur (a range of lithium polysulfides) to migrate out of the positive electrode. On reaching the lithium electrode, these compounds can be reduced further, before travelling back to the positive electrode, and thereby a redox shuttle process is established. Recently, the most promising strategies for dealing with this key problem are focusing on new electrolyte materials in which the solubility of the polysulfides is greatly lowered. The most effective materials also show an improvement in the mobility of lithium ions. These and other significant developments are promoting optimism among battery scientists that lithium–sulfur could become a commercial reality, well before the arrival of the even-more-complex lithium–air battery.

One possible design of the lithium–air cell is shown diagrammatically in Figure 7.21(b). An organic electrolyte is used only on the negative lithium electrode side, and an aqueous electrolyte on the carbon positive side. If the two half-cells are separated by a solid electrolyte which allows only lithium ions to pass through, then the two electrolyte solutions do not mix with each other while the charge–discharge reactions proceed smoothly. The discharge reaction product is lithium hydroxide (LiOH), which dissolves in the aqueous electrolyte so that clogging of carbon pores is avoided. Furthermore, as water and nitrogen do not pass through the solid electrolyte (the partition wall), there are no unwanted reactions with the metallic lithium electrode. During charging, corrosion and degradation of the air electrode are prevented by using another positive electrode, EC, exclusively for charging. Such concepts are at a very early stage of development. The lithium–air cell suffers from the very poor kinetics of the oxygen electrode and from instability of the electrolytes in the electrochemical environment. All variants of lithium–air cell tested to date have lost capacity rapidly on cycling.

Sodium is a possible alternative to lithium. It is conceivable that a system operating with sodium ions at room temperature, analogous to lithium-ion and, unlike the sodium–nickel-chloride battery, with no need for a ceramic separator, may become available and eventually prove less expensive. There

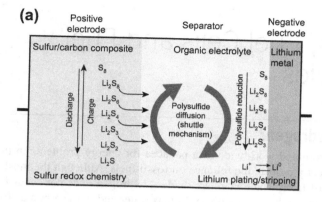

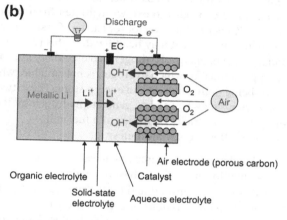

FIGURE 7.21

(a) Shuttle mechanism of a lithium–sulfur cell. Courtesy of DLR; (b) proposed configuration for a lithium–air cell
(Courtesy of the National Institute of Advanced Industrial Science and Technology (AIST))

are still performance challenges, however, in that the theoretical specific energy is lower than for lithium-ion and long cycle-lives have yet to be demonstrated. Nevertheless, considerable advances in materials science have been reported in recent years; these include new electrode materials into which sodium ions can be intercalated reversibly. The research draws heavily on electrolyte and electrode materials developed for lithium–ion batteries over the past few decades. It is likely that success in this venture will lead first to the commercialization of small batteries for electronic devices, well before large electric-vehicle batteries might become viable. The attraction of sodium as a battery material, compared with lithium, is its lower cost and greater availability.

As the hunt for new battery chemistries continues, there are constant reminders that the higher the energy density (both gravimetric and volumetric), the greater is the challenge of providing stable cycling and acceptable safety. Bitter experience has shown that the road from a demonstration of principle in a university laboratory to a fully validated commercial battery suitable for vehicle use is a long and hard one, with many practical materials science problems likely to be encountered along the way. Any totally new system will take many years to perfect and high expenditure will be incurred en route.

Hydrogen, Fuel Cells and Fuel Cell Vehicles

8.1 Why use hydrogen?

Hydrogen is being promoted worldwide as a panacea for energy problems in that it may eventually replace, or at least greatly reduce, the reliance on fossil fuels. Although the most abundant element in the universe – the stuff from which stars are made – hydrogen does not occur freely on earth, but is predominantly found in combination with oxygen as water and with carbon as fossil fuels. Chemical, thermal or electrical energy has to be expended to extract hydrogen from these sources. Hydrogen is therefore not a new form of primary energy, but a vector (or carrier) for storing and transporting energy from any one of a myriad of sources to where it may be utilized. In this respect, it is analogous to electricity, which is also a secondary form of energy. Hydrogen and electricity are complementary: electricity is used for a multitude of applications for which hydrogen is not suitable, whereas hydrogen, unlike electricity, has the attributes of being both a fuel and an energy store. These two energy vectors are, in principle, interconvertible; electricity may be used to generate hydrogen by the electrolysis of water, while hydrogen may be converted to electricity by means of a fuel cell.

Specifically, hydrogen has the following key attributes:

- it can be derived from fossil and non-fossil sources (renewable or nuclear energy);
- it can serve as an alternative fuel for internal-combustion engines;
- it is ideal for use in fuel cells for road transport and for distributed energy supply;
- it is oxidized cleanly to water with no emissions of greenhouse gases;
- when obtained from water using renewables, the fuel cycle is closed and no pollutants are released in the overall process.

The proposal to use hydrogen as a sustainable medium of energy has become known as the 'hydrogen economy'; the overall scheme is illustrated conceptually in Figure 8.1. The upper part of the diagram is generally referred to as the transitional phase, during which hydrogen is produced from fossil fuels; the lower part relates to the long-term post-fossil-fuel age, when hydrogen will be manufactured from renewable energy sources and used as a storage medium and as a super-clean fuel. Not unexpectedly, the building of a hydrogen economy presents great scientific and technological challenges in production, delivery, storage, conversion and end-use. In addition, there are many policy, regulatory, economic, financial, investment, environmental and safety questions to be addressed.

8.2 Hydrogen as a fuel

Hydrogen was first identified as a distinct entity in 1766 by the British scientist Henry Cavendish (1731–1810) after he produced the gas by reacting zinc metal with hydrochloric acid. This gas he called 'inflammable air'. In a paper to the Royal Society in London, Cavendish provided exact

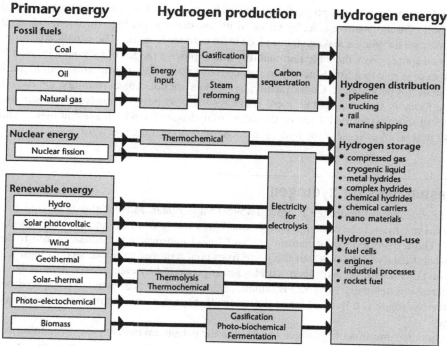

FIGURE 8.1

A sustainable hydrogen economy

(Courtesy of the Royal Society of Chemistry from D.A.J. Rand and R.M. Dell, Hydrogen Energy: Challenges and Prospects, 2008)

measurements of the weight and density of the gas. He also found that water was formed when 'inflammable air' was ignited with a spark in ordinary air, but continued to hold the belief that, in its production, the gas was released from the metal itself rather than from the acid. It was the French chemist Antoine-Laurent Lavoisier (1743–1794) who first appreciated that water is a compound substance formed from a combination of inflammable air and oxygen. Indeed, Lavoisier gave hydrogen its name by coupling together the Greek words 'hydro' meaning 'water' and 'genes' meaning 'genesis'; hence, the 'maker of water'.

If hydrogen is to be employed as an energy vector and a non-polluting fuel, then it is necessary to take account of its basic physical properties. Hydrogen is a colourless, odourless, tasteless and non-toxic gas. It is the lightest of all molecules, with a density of only 0.0899 kg m^{-3} at normal temperature and pressure, i.e. 7% of the density of air. Consequently, hydrogen has the best energy-to-weight ratio ('heating value') of any fuel, but its energy-to-volume ratio is poor.

Hydrogen has a wide range of flammability; the lower and upper flammability limits in air are 4 and 75 vol.%, respectively. Since the lower explosive limit of hydrogen in air (13 vol.%) is higher than the lower flammability limit, hydrogen generally burns rather than explodes. When mixed with pure oxygen in a 2:1 molecular ratio of hydrogen to oxygen and then ignited, hydrogen detonates violently. The energy required to ignite a hydrogen–air mixture is exceptionally low – only one-fifteenth of that needed for natural gas.

The combination of physical properties exhibited by hydrogen has safety implications for the use of the gas in bulk. On the positive side, by virtue of its low density and high diffusivity in air, hydrogen soon disperses safely when a leakage takes place in an outdoor location. The situation is quite different in an enclosed space; given the wide flammability and explosive ranges of hydrogen in air, a leakage is very likely to give rise to a fire. Since hydrogen contains no carbon, it combusts with a non-luminous flame that does not radiate heat. Consequently, bystanders are not subject to injury. On the other hand, because the flame is almost invisible, there is always the possibility of inadvertently straying into its path and being seriously burned. Clearly, the safety of hydrogen from both a technological and societal perspective will be a key issue if the hydrogen economy is to be taken forward.

8.3 Present uses for hydrogen

The world production of hydrogen is around 45–50 Mt per year. Most of this is derived from natural gas by steam reforming; the remainder is obtained principally from oil and coal by partial oxidation processes. The transformation of natural gas and liquid hydrocarbon feedstocks into hydrogen is a straightforward catalytic process, but the route from coal requires an initial step of high-temperature gasification. Only 4% of the hydrogen world-wide is generated by electrolysis, invariably when there are special reasons that make this route economic, e.g. where there is a surfeit of cheap hydroelectricity, or when the hydrogen is a by-product of the chlor-alkali process for the manufacture of chlorine and caustic soda.

The major consumption of hydrogen has been in the petroleum industry for the refining and upgrading of crude petroleum, and in the chemical industry for the manufacture of ammonia (e.g. for fertilizers), methanol and a variety of organic chemicals. Other important uses are found in the food industry for the hydrogenation of edible plant oils to fats (margarine) and in the plastics industry for making various polymers. Lesser applications occur in the metals, electronics, glass, electric power and space industries. To date, very little hydrogen has been used as a fuel, except in rockets for launching satellites and spacecraft.

8.4 Hydrogen from fossil fuels and biomass

8.4.1 Reforming of natural gas

Natural gas consists predominantly of methane (CH_4) with small quantities of other gaseous hydrocarbons. It is widely available, easy to handle, and relatively cheap (at present). Compared with other fossil fuels, natural gas is the most desirable feed for making hydrogen; it has the highest hydrogen-to-carbon ratio and this serves to minimize the quantity of carbon dioxide produced when it is burnt or reformed to hydrogen.

The reforming of natural gas to manufacture hydrogen is a mature industrial process, in which methane is reacted with steam over a nickel-based catalyst at around 900 °C and at elevated pressure:

$$CH_4 + H_2O \ (gas) \ \rightarrow \ CO + 3H_2 . \tag{8.1}$$

The resulting mixture is known as 'synthesis gas' (or 'syngas') because it may be used for the preparation of a range of commodities that include various organic chemicals. Syngas may also be converted to liquid fuels by the Fischer-Tropsch reaction in what has become known as the 'gas-to-liquids' (GTL) process. When the object of reforming natural gas is to produce hydrogen, the syngas is further processed using the water-gas shift (WGS) reaction. This converts the carbon monoxide to carbon dioxide by an additional reaction with steam over a catalyst at a much lower temperature:

$$CO + H_2O \ (gas) \ \rightarrow \ CO_2 + H_2 \ . \tag{8.2}$$

Summing Equations 8.1 and 8.2 yields:

$$CH_4 + 2H_2O \ (gas) \ \rightarrow \ CO_2 + 4H_2 \ . \tag{8.3}$$

Thus, each molecule of methane gives rise to four molecules of hydrogen, of which two molecules originate from the steam. In reality, the practical yield of hydrogen is less than this because some of the methane is usually burnt externally to provide the heat needed for reforming (Equation 8.1).

Finally, in the preparation of hydrogen, it is necessary to separate out the carbon dioxide, as well as any other gases that may be present due to impurities in the natural gas. This is generally accomplished by 'pressure swing adsorption' (PSA), a well-proven industrial process that yields high-purity hydrogen (up to 99.999%) with recoveries of up to 90%. Other separation methods, which employ ceramic diffusion membranes, are under development and not only promise the benefit of being able to run at higher temperatures than PSA, but also are potentially more elegant and cost-effective at the smaller scale.

Whereas it is likely that natural gas will continue to serve as the feedstock for hydrogen production, today's costs are prohibitively high for applications such as road transport while petrol is still comparatively affordable. Moreover, the fact that fossil fuels produce carbon dioxide in addition to hydrogen appears to be self-defeating on environmental grounds. Clearly, the future of steam reforming will depend on developing efficient means to capture carbon dioxide and then store it (so-called 'sequestration'), possibly in underground chambers or in the ocean.

8.4.2 Solar–thermal reforming

In Australia, the Commonwealth Scientific and Industrial Research Organisation (CSIRO) is investigating the possible use of solar energy to provide the heat required for the steam reforming of natural gas and other methane-containing gases, e.g. landfill and coal-bed methane. As the resulting syngas would contain a substantial amount of embodied solar energy (up to 25%), this alternative method of reforming offers the prospects of high thermal efficiencies and greatly reduced emissions of carbon dioxide.

To demonstrate the feasibility of solar–thermal reforming, the CSIRO erected a solar dish of 48 curved mirrors that focused the sun's rays onto a thermal receiver mounted above the dish at its focal point. Attention then turned to decentralized generation through the development of more practical 'mini' versions of the solar-tower approach coupled to a small steam reformer. Apart from the benefit of permitting the generation of hydrogen close to where it is needed, this modular technology is less expensive than a dish and is more flexible in that it allows easier integration of additional units to meet any growth in demand. The CSIRO has designed and built Australia's largest solar–thermal research facility, which consists of a solar tower and high-temperature receiver (combined height 30 m), together with a 4000-m² field of 450 heliostats; see Figure 8.2. The facility is capable of concentrating solar energy to achieve temperatures beyond 1000 °C.

FIGURE 8.2

CSIRO solar tower array

(Courtesy of the Royal Society of Chemistry from D.A.J. Rand and R.M. Dell, Hydrogen Energy: Challenges and Prospects, *2008)*

8.4.3 Coal gasification

As supplies of natural gas become depleted, gasification of coal is seen as the preferred technology of the future for electricity production, along with renewable forms of energy. This is because coal is plentiful and widely available geographically, the gasification process is more efficient than conventional coal combustion, and the carbon dioxide produced during gasification can be extracted from the syngas *before* combustion and then stored permanently underground. All these advantageous features, however, have yet to be practically realized on a commercial scale.

The syngas made from coal, like that produced by the steam reforming of natural gas, may be used to produce a variety of liquid fuels and chemicals via the above-mentioned Fischer-Tropsch process. A longer-term goal is to process the syngas further to hydrogen and carbon dioxide by the water–gas shift reaction. The ultimate aim is to develop a versatile facility that is capable of producing pure hydrogen (or organic chemicals) and electricity from coal in any desired ratio, a so-called 'poly-generation plant', as shown schematically in Figure 8.3. This scheme employs oxy-combustion of coal, which makes separation of the carbon dioxide easier than when using air with its nitrogen content. Over the past decade, the role of coal as an energy source for the future has gained renewed interest for its proven stability in supply and cost. Therefore, it is likely that coal will retain an important position in the energy mix for the foreseeable future.

8.4.4 Biomass gasification

In principle, hydrogen may be produced by the gasification of any form of biomass. The most suitable materials are (i) those with water contents in the range of 5 to 30 wt.%, which includes fast-growing energy crops such as *Miscanthus* ('silver') grass, hemp, poplars and willows, and (ii) any forestry or agricultural waste such as wood chippings, sawdust, bagasse, etc. The gasification is generally conducted at lower pressures and temperatures than for coal, and is limited to mid-size operation due to the heterogeneity of biomass, its localized production, and the relatively high costs incurred in its gathering and conveying.

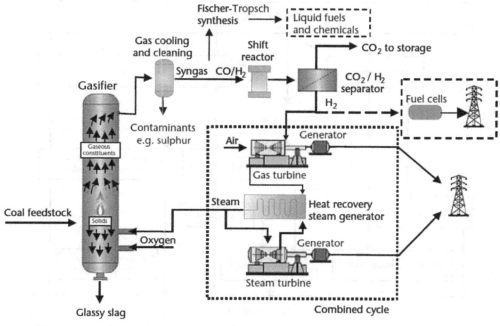

FIGURE 8.3

Concept of a poly-generation plant

(Courtesy of The Royal Society of Chemistry from D.A.J. Rand and R.M. Dell, Hydrogen Energy. Challenges and Prospects, 2008)

Wet forms of biomass, such as sewage, liquid manure and silage, may also be decomposed by anaerobic digestion at ambient temperature. The resulting 'biogas' is a mixture of predominantly methane and carbon oxides that may be converted to hydrogen by steam reforming.

8.5 Hydrogen from water

Water is a huge storeroom of hydrogen, but energy is required to split this source into its component elements. The water molecule is in fact very stable – decomposition by direct heating only starts to become significant at very high temperatures – namely, ~ 1 vol.% at 2000 °C. Moreover, it is difficult to separate the hydrogen and the oxygen at such high temperatures. Some research is in progress to produce hydrogen at lower temperatures by employing catalysts for the decomposition reaction together with ceramic membranes for efficient gas separation, but the work is very much in its early stages. Given these limitations, the splitting of water must be undertaken indirectly by processes that employ either chemical reactants or other energy forms of high thermodynamic potential (e.g. electrons or photons). The chemical reduction of water to hydrogen by means of hydrocarbons, in the form of fossil fuels, has been described above. Consequently, the following discussion is devoted principally to the generation of hydrogen from water through the use of renewable energy, i.e. via the medium of electricity or directly as solar radiation. The latter option may be accomplished in several ways, as shown schematically in Figure 8.4.

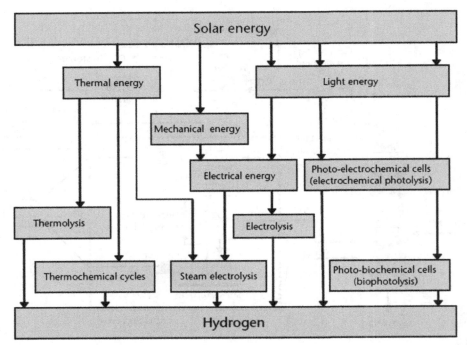

FIGURE 8.4

Solar-to-hydrogen conversion options

(Courtesy of The Royal Society of Chemistry from D.A.J. Rand and R.M. Dell, Hydrogen Energy: Challenges and Prospects, *2008)*

8.5.1 Electrolysis

The electrolysis of water to generate hydrogen or oxygen is a mature technology, but only a few percent of the total world production is obtained by this method. Electrolysis is extremely energy-intensive – the faster the generation of hydrogen, the greater is the power required per kilogram produced. Large-scale units using alkaline electrolyte typically run at 70 to 75% efficiency, while smaller systems with polymer electrolytes reach 80 to 85%. Steam electrolysers, in which some of the energy required to split water is supplied by heat, may be able to achieve very high efficiencies (over 90%) but they are not yet commercially feasible.

Conventional electrolysers are designed to operate on grid-quality power, whereas the electricity generated from renewable energy sources such as wind and solar is intermittent and of variable quality. This limitation will necessitate the development of specialized power-control and conditioning equipment. Cost is also a significant issue. Using electricity derived from renewable sources, it is estimated that hydrogen can be obtained at US$7–10 per kg, i.e. three-to-five times above that from fossil fuels. Even with the mass production of small electrolysers, should it occur, the overall economics would not change much until 'green' electricity becomes cheaper than the conventional alternative.

8.5.2 Thermochemical cycles

Because of the stability of the water molecule and the very high temperatures required to split it thermally, as discussed above, attempts have been made to accomplish the process at a more moderate upper temperature by means of an indirect route. The general idea is to decompose water by reacting it with one or more chemicals that are regenerated via a series of cyclic thermochemical reactions. In this way, the hydrogen and oxygen evolution steps are separated. Clearly, on both practical and efficiency grounds, the fewer reactions involved the better. There are several possible cycles and the most widely studied is the sulfur–iodine cycle, which consists of just three chemical reactions.

The concept of a thermochemical cycle has the attraction that the required heat could be provided by solar radiation, or by a low-carbon source such as a high-temperature nuclear reactor. Nevertheless, intensive studies of thermodynamic cycles over many years have not yet progressed beyond the pilot demonstration stage and there appears to be little prospect of developing a feasible and economic process on a commercial scale.

8.5.3 Electrochemical photolysis

Electrochemical photolysis is a prospective method for the direct production of hydrogen via the harnessing of solar radiation. Light is converted to electrical and chemical energy by using a semiconducting oxide, such as titanium dioxide (TiO_2), to absorb photons. Those photons with energies greater than that of the band gap (i.e. the energy difference between non-conductive and conductive states) generate electron (e^-)–hole (h^+) pairs that become separated by the electric field; see Figure 8.5. The

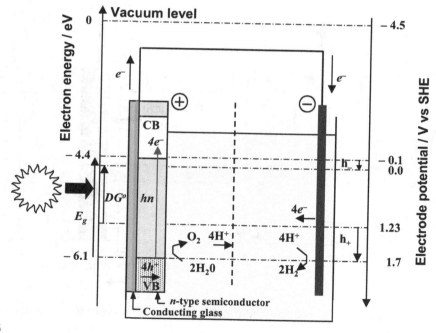

FIGURE 8.5

Operating principles of a photo-electrochemical cell for hydrogen production VB = valence band; CB = conduction band; E_g = band-gap energy; hv = energy of a photon; η_+, η_- overpotential of positive and negative electrodes, respectively; $\Delta G°$= standard change in free energy

(Courtesy of The Royal Society of Chemistry from D.A.J. Rand and R.M. Dell, Hydrogen Energy: Challenges and Prospects, 2008)

holes are driven to the surface of the TiO_2 where they oxidize water to oxygen, whilst the electrons travel round the external circuit to reduce water to hydrogen at a metal counter electrode such as platinum. By virtue of its relatively low cost, TiO_2 is most attractive as a photo-electrochemical material, but it does have a somewhat high band-gap energy (~ 3.2 eV) and therefore absorbs light energy in the ultraviolet rather than in the visible part of the spectrum. Accordingly, the efficiency of present cells is only a few percent, i.e. well below the commercial target of 10%.

Photo-electrochemical reactions may also be used to generate electricity. This is achieved in a dye-sensitized solar cell (DSSC) by means of a subterfuge, and involves separating the optical-absorption and charge-generating functions. A dye which is capable of being photo-excited is adsorbed on to the surface of porous titania (synonym for titanium dioxide) where it acts as an electron-transfer sensitizer. To date, such cells have demonstrated sunlight-to-electricity efficiencies of 11% in the laboratory, but only 5% in the field. Nevertheless, development of the DSSC opens up the possibility of improving the efficiency of hydrogen production from photo-electrochemical cells. The two technologies are operated in tandem so as to absorb complementary parts of the solar spectrum; see Figure 8.6. The front photo-electrochemical cell absorbs high-energy ultraviolet and blue light in sunlight, while longer-wavelength light in the green-to-red region of the spectrum passes to,

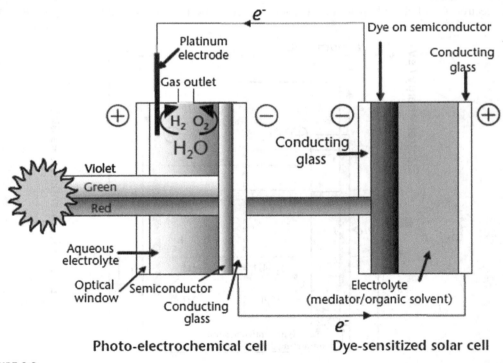

FIGURE 8.6

Operating principles of a tandem cell for enhanced hydrogen production

(Courtesy of The Royal Society of Chemistry from D.A.J. Rand and R.M. Dell, Hydrogen Energy: Challenges and Prospects, 2008)

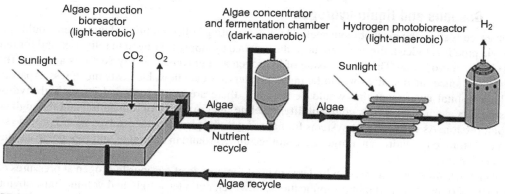

FIGURE 8.7

Schematic of developmental system for the bio-photolytic production of hydrogen

(Courtesy of The Royal Society of Chemistry from D.A.J. Rand and R.M. Dell, Hydrogen Energy: Challenges and Prospects, *2008)*

and is absorbed by, the DSSC. This boosts the energy of the electrons, which are then fed back to a hydrogen electrode in the front cell. With such an arrangement, overall photon-to-hydrogen efficiencies of up to 12% have been reported.

8.5.4 Biophotolysis

The operation of the tandem cells discussed above bears a close similarity to the processes that take place in green foliage during natural photosynthesis, where there are also two photosystems connected in series. The first step involves splitting water into oxygen and hydrogen, and then hydrogen is used to metabolize carbon dioxide into carbohydrates. There are, however, a few natural bacteria, e.g. blue-green algae (cyanobacteria), that are capable of releasing hydrogen freely into the air. This is a photo-biochemical process, which is often referred to as 'biophotolysis'. In practice, the energy-conversion efficiency of such systems is very low, often less than 1%. To improve performance, two-stage 'indirect biophotolysis' processes are being investigated – namely, photosynthetic fixation of carbon dioxide with concomitant generation of oxygen followed by dark anaerobic fermentation of the resultant algae with evolution of hydrogen. Such a system, proposed by the Hawaii National Energy Institute, is shown schematically in Figure 8.7.

8.6 Hydrogen distribution and storage

To be useful as a future fuel, hydrogen has to be conveyed to the point of use and there stored until required. The distribution and storage of hydrogen are intricately bound together and depend on both the scale of operations and the intended application. In general, the storage of hydrogen for stationary applications is less demanding than for road transport operations, where there are more severe constraints in terms of acceptable mass and volume, speed of charge/discharge and, for some storage systems, heat dissipation and supply. Finding a satisfactory solution for the onboard containment of hydrogen is one of the major challenges facing the development of fuel cell vehicles.

8.6.1 Gaseous and liquid hydrogen

In the gaseous state, the most obvious method for distributing bulk quantities of hydrogen would appear to be via pipeline. Indeed, this has long been the practice to supply hydrogen in refineries and for industrial chemical processes. The transmission of hydrogen as a universal energy vector is a more difficult proposition since the distances would be much greater and the allowable costs much less. Apart from the huge capital investment that would be required, there are many technical issues with hydrogen transmission, such as materials compatibility, pipeline integrity, leak minimization and the high costs of compression. As with delivery systems for natural gas, the pipelines themselves would provide some degree of storage for hydrogen. Large-scale storage underground in natural or anthropogenic cavities is also a possibility.

On a much smaller scale, steel cylinders are used for storage of gaseous hydrogen at pressures of up to 80 MPa. For portable and mobile applications, however, cylinder weight and volume must obviously be minimized, and to some extent this has been achieved by using all-composite vessels composed of carbon-fibre shells with aluminium liners. These containers can withstand a pressure of 55 MPa and thereby provide a hydrogen volumetric storage density of 3 to 4 MJ dm^{-3}; see Figure 8.8(a). This value approaches the 2010 target (5.4 MJ dm^{-3}) set by the US FreedomCAR Partnership for fuel cell vehicles, but is far below the 2015 target (9.7 MJ dm^{-3}). Incidentally, the corresponding gravimetric targets (7.2 and 10.8 MJ kg^{-1}, respectively) will be even more difficult to reach. In this respect, liquid hydrogen (LH$_2$) would be more attractive; its density is 850 times greater than that of the gaseous form at atmospheric pressure. There is considerable experience in the USA of the bulk liquefaction and transport of hydrogen for use as rocket fuel.

Unfortunately, the energy required to liquefy hydrogen is considerable. For large-scale production, the input is equivalent to around 30% of the energy value of the hydrogen itself, and rises to an even greater percentage for smaller operations. This inefficiency, together with the high capital cost of the cryogenic plant, is a major disincentive to conveying and storing hydrogen as a liquid. From an automotive-engineering standpoint, there are also practical difficulties in terms of designing, at acceptable cost, LH$_2$ cryostats for onboard vehicular use that are both appropriate and safe. Nevertheless, some progress has been made towards this goal by BMW and its hydrogen supplier Linde AG; see Figure 8.8(b). There are also aspects of vehicle refuelling to be addressed, especially the development of transfer lines for LH$_2$ that can be operated safely by the general public. Loss through evaporation during fuelling and boil-off on standing would also cause problems with the relatively small tanks that would typically be used in cars.

8.6.2 Metal hydrides

Certain metals and alloys can repeatedly absorb and release hydrogen under moderate pressures and temperatures via the formation of hydrides, as:

$$M + x/2H_2 \leftrightarrows MH_x + heat. \tag{8.4}$$

Heat must be removed during absorption of the hydrogen, but has to be added to effect desorption.

Hydrides have relatively poor mass storage for hydrogen; it ranges from 1 to 2 wt.% for materials such as CaNi$_5$H$_4$ that operate at near-ambient temperatures up to 3.6 wt.% for magnesium-based materials such as Mg$_2$NiH$_4$ that function above 300 °C. By contrast, hydrides have a relatively good theoretical

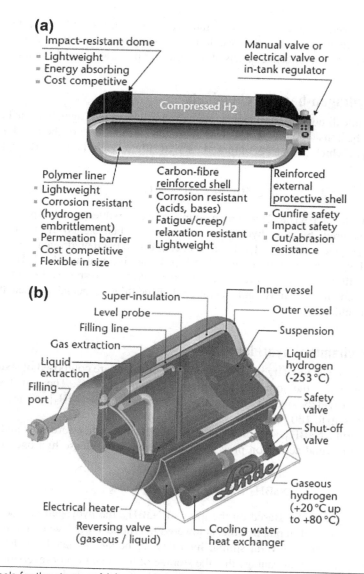

(a)

Impact-resistant dome
- Lightweight
- Energy absorbing
- Cost competitive

Manual valve or electrical valve or in-tank regulator

Compressed H₂

Polymer liner
- Lightweight
- Corrosion resistant (hydrogen embrittlement)
- Permeation barrier
- Cost competitive
- Flexible in size

Carbon-fibre reinforced shell
- Corrosion resistant (acids, bases)
- Fatigue/creep/relaxation resistant
- Lightweight

Reinforced external protective shell
- Gunfire safety
- Impact safety
- Cut/abrasion resistance

(b)

Super-insulation
Level probe
Filling line
Gas extraction
Liquid extraction
Filling port

Inner vessel
Outer vessel
Suspension
Liquid hydrogen (-253 °C)
Safety valve
Shut-off valve
Gaseous hydrogen (+20 °C up to +80 °C)

Electrical heater
Reversing valve (gaseous / liquid)
Cooling water heat exchanger

FIGURE 8.8

Schematics of vessels for the storage of (a) gaseous hydrogen; and (b) liquefied hydrogen aboard road vehicles

(Courtesy of The Royal Society of Chemistry from D.A.J. Rand and R.M. Dell, Hydrogen Energy: Challenges and Prospects, *2008)*

volumetric performance of 0.1 kg dm^{-3} hydrogen. Since lower-temperature materials would be preferred in most cases, the major research challenges are to develop new alloying techniques that increase the gravimetric density of the hydrides. In addition, operational difficulties associated with the kinetics of heat and hydrogen transfer have yet to be resolved. It has been calculated that if a typical hydride bed capable of holding 5 kg of hydrogen (to provide a driving range of ~ 500 km) were to be charged in 3 min (an acceptable time for refuelling a vehicle), it would be necessary to remove heat at a rate of ~ 500 kW. Clearly, the development of an effective means of heat transfer poses a formidable engineering challenge.

8.6.3 Simple hydrogen-bearing chemicals

Certain organic chemicals contain significant atomic proportions of hydrogen that can be recovered and therefore they may be considered as prospective hydrogen carriers. One of the best known of these is cyclohexane (C_6H_{12}), which can be decomposed catalytically to yield benzene (C_6H_6) and hydrogen, as:

$$C_6H_{12} \rightarrow C_6H_6 + 3H_2. \tag{8.5}$$

The hydrogen content is 7.1 wt.%. This is an example of a reversible (round-trip) carrier since cyclohexane is manufactured by the hydrogen reduction of benzene over a nickel catalyst at 150–200 °C. Methyl cyclohexane (C_7H_{14}) has also been proposed; it dehydrogenates to toluene (C_7H_8), with a yield of 6.1 wt.% H_2. Liquid ammonia (NH_3), hydrazine hydrate ($N_2H_4 \cdot H_2O$) and ammonia borane (NH_3BH_3) will theoretically store 17.7, 8.0 and 12 wt.% H_2, respectively. In practice, however, such materials present many safety and operational difficulties, especially when the hydrogen is intended for use in vehicles.

8.6.4 Complex chemical hydrides

Hydrogen may be stored chemically in the form of ionic salts that are composed of sodium, aluminium or boron, and hydrogen – the so-called 'complex chemical hydrides'. The alanates $Na[AlH_4]$ and $Na_3[AlH_6]$ have been widely studied. Thermal decomposition of $Na[AlH_4]$ to yield hydrogen takes place at temperatures up to 180 °C.

Sodium borohydride, $NaBH_4$, is stable up to about 400 °C and is therefore not suitable for providing hydrogen through a thermal-activation process. It does, however, release hydrogen on reaction with water, as:

$$NaBH_4 + 2H_2O \rightarrow NaBO_2 + 4H_2. \tag{8.6}$$

This is an irreversible reaction. Based on the mass of $NaBH_4$, the hydrogen released is 21 wt.% – a remarkably high output – but in practice this is lowered to around 7 wt.% when the weight of the total system is taken into account. Several similar hydrides are being evaluated for their reactivity with water, but they all have the disadvantage that the storage of hydrogen is not reversible; the spent solution has to be returned to a processing plant for regeneration of the hydride.

8.6.5 Nanostructured materials

Considerable research is being carried out on materials that have structural elements with dimensions in the nanoscale range and have high specific surface-areas ($m^2\ g^{-1}$). Amongst these materials, it has been found that carbon and boron nitride nanostructures, clathrates, and metal–organic frameworks can all store hydrogen in the molecular state via weak molecular-surface interactions.

Carbon nanotubes were some of the first nanomaterials to be investigated for hydrogen storage and many have been reported with capacities of several wt.%, which may be further enhanced by various pretreatments. There continues, however, to be considerable controversy over such findings because of the difficulty in preparing homogeneous, well-defined, pure and reproducible samples. Moreover, contrary to expectation, significant hydrogen storage in these materials usually requires either high pressure (> 10 MPa) or low temperature (at least −100 °C).

8.7 Hydrogen utilization: fuel cells

Fuel cells are a key enabling technology for a future hydrogen economy – particularly for electric vehicle propulsion – and therefore are being intensively investigated and developed in many countries. The fuel cell is an electrochemical device for the conversion of chemicals into direct-current electricity. To this extent, it resembles a primary battery. There are, however, some important differences. In a battery, all the chemicals necessary for its operation are normally confined within a sealed container. Thus, the capacity of a battery, measured in ampere-hours, is determined by the quantity of chemicals that it holds. With a fuel cell, the chemicals are supplied from external reservoirs so that the capacity of the device is limited only by the available supply of reactants.

To demonstrate the principle of the fuel cell at low power output and without regard to cost is remarkably simple and can be done in any school science laboratory. The difficulty lies in developing an engineered, high-power system at acceptable cost that will operate for long periods. The fundamental operation of a fuel cell is illustrated in Figure 8.9. Fuel is oxidized at the negative electrode with the release of electrons that pass through the external circuit and reduce oxygen at the positive electrode. The flow of electrons is balanced by a flow of charged ions in the electrolyte.

The reversible voltage, V_r, produced by the cell (i.e. when there is no current flow; also known as the 'open-circuit' voltage) is given by:

$$V_r = - \Delta G/nF, \tag{8.7}$$

where ΔG is the free energy of the cell reaction (joules per mole), n is the number of electrons involved in the reaction and F is the Faraday constant (96 485 coulombs per mole). This relationship applies also to the reverse (electrolysis) reaction. When the fuel is hydrogen, the reversible voltage under standard conditions is $V° = 1.229$ V at 25 °C, just as for an electrolyser. On drawing current from the cell, however, the voltage developed is much lower than this, typically 0.6–0.8 V, depending on the current density. Consequently, it is customary to build up the voltage to the desired level by electrically connecting cells in series to form a 'stack'.

A typical plot of cell voltage versus current is given in Figure 8.10(a). As current starts to be drawn, the voltage falls. The factors that inflict this loss in performance are reflected in different regions of the voltage–current curve. The initial fall in voltage is due to electrokinetic ('activation') limitations at the electrodes. In the central linear portion of the curve, 'ohmic' losses play a more significant role; these arise from resistance to the flow of ions in the electrolyte. Finally, mass-transport ('concentration') limitations, which result from gas consumption outstripping supply, become dominant at high current densities. There are also electrical-resistance losses in the collectors and conductors of the current – and, possibly, side-reactions and leakages of ionic current between series-connected cells – all of which may contribute to the voltage drop. Because each unit cell is prone to these effects, the working cell voltage is well under 1 V, as noted above. An example of the power performance of a cell is given in Figure 8.10

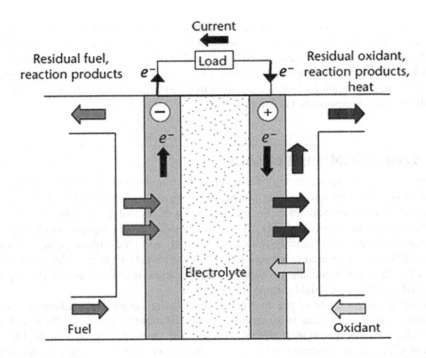

FIGURE 8.9

Schematic of fuel cell operation

(Courtesy of The Royal Society of Chemistry from D.A.J. Rand and R.M. Dell, Hydrogen Energy: Challenges and Prospects, *2008)*

(b). As the current density rises, the cell voltage declines. The power density (the product of the voltage and the current density) passes through a maximum, in this example at around 1 ampere per square cm.

The first demonstration of a basic fuel cell was by Sir William Grove in 1839, but it was not until the work of F.T. (Tom) Bacon in the 1940s that a practical system began to emerge. It was while he was at the University of Cambridge (1946–1955) that Bacon first developed the alkaline fuel cell. This was adopted by the USA in the 1960s for the Apollo space program and refined later for the Space Shuttle Orbiter. It provided all of the electricity, as well as drinking water, when the Space Shuttle was in flight.

8.7.1 Characteristics of fuel cells

There are six major types of fuel cell that can be broadly classified in terms of their temperature of operation:

- low-temperature (50–150 °C): alkaline-electrolyte (AFC), proton-exchange-membrane (PEMFC) and direct-methanol (DMFC) fuel cells;
- medium-temperature (around 200 °C): phosphoric-acid fuel cells (PAFC);
- high-temperature (600–1000 °C): molten-carbonate (MCFC) and solid-oxide (SOFC) fuel cells.

Some operational data on each type are given in Table 8.1.

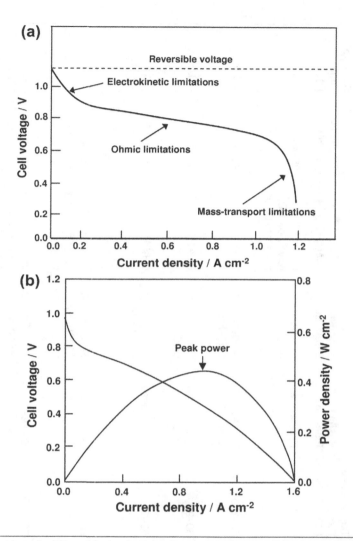

FIGURE 8.10

Typical performance characteristics of fuel cells: (a) voltage–current curve; (b) power output as a function of current density

Apart from the DMFC, which runs on methanol, hydrogen is the preferred fuel on account of its high electrochemical activity. The AFC and PEMFC are both demanding in terms of purity of the hydrogen used, which must typically contain less than 10 ppmv carbon monoxide. By contrast, the PAFC is somewhat more tolerant to carbon monoxide, while the high-temperature designs (MCFC and SOFC) will accept this impurity and a variety of hydrocarbon fuels. All the designs require the fuel to be sulfur-free.

A variety of sub-systems and components are required for a fuel-cell stack to function effectively. The exact composition of this so-called 'balance of plant' depends on the type of fuel cell, the available

Table 8.1 Principal Types of Fuel Cell

Fuel Cell Technology	Electrolyte	Temperature Range (°C)	Electrocatalyst		Fuel	Efficiency[a] (% HHV)	Start-up Time (h)
			Positive Electrode	Negative Electrode			
PAFC	H_3PO_4	150–220	Pt supported on C	Pt supported on C	H_2 (low S, low CO, tolerant to CO_2)	35–45	1–4
AFC	KOH	50–150	NiO, Ag or Au–Pt	Ni, steel or Pt–Pd	Pure H_2	45–60	< 0.1
PEMFC	Polymer	80–90	Pt supported on C	Pt supported on C	Pure H_2	40–60	< 0.1
DMFC	H_2SO_4	60–90	Pt supported on C	Pt supported on C Pt–Ru	CH_3OH	35–40	< 0.1
MCFC	Li_2CO_3	600–700	Lithiated NiO	Sintered Ni–Cr and Ni–Al alloys	H_2, variety of hydrocarbon fuels (no S)	45–60	5–10
SOFC	Oxygen-ion conductor	700–1000	Sr-doped $LaMnO_3$	Ni- or Co-doped YSZ cermet	Impure H_2, variety of hydrocarbon fuels	45–55	1–5

[a]The reported efficiency of a given type of fuel cell varies widely and often no information is provided on whether the higher heating value (HHV) or the lower heating value (LHV) of the fuel is used – refer to Box 8.2. The efficiencies that are taken here from the literature should be treated with caution as to their exact meaning and are simply included to provide an approximate comparison of the performance of the respective systems.

fuel and its purity, and the desired output of electricity and heat. Typical auxiliary sub-systems may include: (i) fuel-processing reactors, e.g. desulfurization, steam reformer, shift and partial oxidation reactors; (ii) humidifier; (iii) fuel and air delivery units; (iv) power-conditioning equipment, e.g. for inverting direct current to alternating current; (v) heat- and water-management facilities; (vi) overall system controls.

The advantages of fuel cells over other generators of electrical energy such as gas turbines, steam turbines or internal-combustion engines with alternators are as follows:

- potentially high energy-conversion efficiency over a range of sizes;
- greater efficiency when operated at part-load, in contrast to engines;
- no moving components apart from auxiliary fans and blowers;
- almost silent in operation;
- flexibility of fuel supply;
- very low exhaust emissions;
- pure water emitted when using hydrogen as the fuel;
- modular construction and ease of installation.

This impressive list of attributes has provided the incentive for much of the research that has been devoted to fuel cells in recent years.

8.7.2 Proton-exchange-membrane fuel cells

There is a consensus among automotive companies that the PEMFC is the most practical candidate for general road-transport applications. It is also acknowledged that the specifications that must be met to compete with the internal-combustion engine are stringent, particularly as regards size, performance and overall cost.

The PEMFC (also known as the 'polymer-electrolyte-membrane fuel cell' and as the 'solid-polymer-electrolyte fuel cell', SPEFC) is an acid-electrolyte fuel cell that operates on the same principle as the PAFC, although with some important differences. The electrolyte is a thin membrane of a copolymer that is highly conductive to hydrated protons (H_3O^+). The standard membrane material is that sold under the trade name Nafion®, which was produced by the DuPont Corporation in the USA, originally for application in the chlor-alkali industry. The backbone consists of polytetrafluoroethylene (PTFE, so-called 'Teflon') to which side-chains of perfluorinated vinyl polyether are bonded via oxygen atoms. Sulfonic acid groups ($-SO_3H$) at the ends of the side-chains can exchange with protons to facilitate their passage through the membrane.

Nafion® has two notable characteristics:

(i) the PTFE backbone is water-repellent and thereby prevents flooding by product water;
(ii) the sulfonic acid group attracts water molecules, which form aqueous micelles within the polymer and these, in turn, serve to conduct the protons across the membrane.

The perfluorosulfonic acid membranes, as a class, have inherent chemical and oxidative stability. They are, however, difficult to manufacture and are therefore relatively expensive to purchase. Accordingly, other companies, notably in the USA and Japan, have been searching for significantly cheaper alternatives. Efforts have focused mainly on the development of thinner membranes that, if attainable, would also offer the added performance benefits of lower electrical resistance (and hence higher power density) and improved hydration.

For PEMFCs to achieve widespread commercial success, membranes should also have greater durability, better water management and the ability to function at higher temperatures – all at reduced cost. In particular, operation at higher temperatures would yield significant energy benefits. For example, heat rejection would be easier and would warrant the use of smaller heat-exchangers in fuel cell systems. The construction of a fuel cell of this type is detailed in Box 8.1.

BOX 8.1 CONSTRUCTION OF A PROTON-EXCHANGE-MEMBRANE FUEL CELL

The central component of the PEMFC is the 'membrane–electrode assembly' (MEA), which consists of the polymer membrane sandwiched between a positive and a negative electrode, as illustrated in Figure 8.11. Typically, the MEA has a thickness of about 0.5 mm. The key objective in electrode manufacture is to produce a structure in which the electrocatalyst is finely divided and highly dispersed on a conductive substrate, so that maximum surface-area is presented to the membrane electrolyte and, in turn, the electrocatalyst loading (and hence its attendant cost) is kept to a minimum. In one method of electrode preparation, an emulsion of fine carbon powder (particle size 50–100 nm) and isopropyl alcohol is mixed with platinum particles (just a few nm in size) and

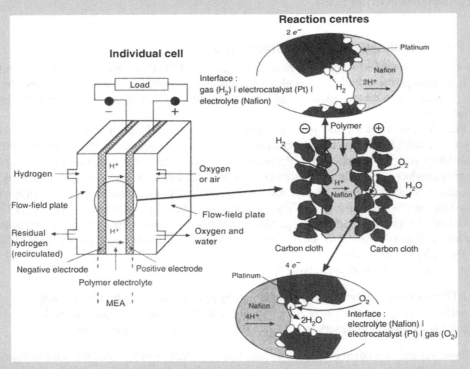

FIGURE 8.11

Schematic of a PEMFC showing pore structure and distribution of electrocatalyst at a three-phase boundary
(Courtesy of The Royal Society of Chemistry from D.A.J. Rand and R.M. Dell, Hydrogen Energy: Challenges and Prospects, *2008)*

BOX 8.1 CONSTRUCTION OF A PROTON-EXCHANGE-MEMBRANE FUEL CELL—cont'd

deionized water to form a slurry. Carbon paper (or thin carbon cloth or felt) is impregnated with an appropriate amount of the slurry and then dried. The two processes are repeated until the electrocatalyst loading reaches the desired level for the respective positive and negative electrodes. The electrolyte membrane is then placed between a pair of these electrodes and the assembly is hot-pressed. Alternatively, the slurry can be applied directly to both sides of the electrolyte membrane and the carbon paper/cloth added separately. It is customary to mix PTFE into the carbon because this polymer is hydrophobic and will assist in expelling the product water to the electrode surface, where it can evaporate. In addition to providing a mechanical support, the porous carbon substrate allows the diffusion of reactant gas to the electrocatalyst and is therefore often called the 'gas-diffusion layer'. Since the carbon paper or cloth is electrically conducting, it serves also to carry electrons to and from the electrocatalyst. It is important that product water is removed from the stack so that the electrodes do not become flooded, otherwise access of gas to the electrocatalyst sites would be impeded and hence reaction at the three-phase boundary (gas | liquid | solid) would be seriously impaired. The MEA microstructure and the electrode reaction mechanism are also shown schematically in Figure 8.11. The membrane must be kept sufficiently humidified during cell operation to permit effective conduction of protons. Accordingly, the stack may also contain a humidification unit to maintain the gases at close to their saturation levels and thereby prevent dehydration of the membrane. The distribution of reactant gas and the control of humidity are critical aspects of PEMFC design.

The MEA is held together by flow-field plates that serve to distribute the current uniformly and also act as bipoles to connect adjacent cells in series. A diagram of a single cell within a stack – separated into its components – is given in Figure 8.12. In this example, the electrocatalyst has been applied directly to the membrane. The flow-field plates are made from graphite, graphite–polymer composites or solid metal and contain channels to distribute the reactant gases so that each gas is in contact with the whole surface of its respective electrode via the porous gas-diffusion layer. In addition to this role, the plates have to distribute a cooling fluid throughout the stack, while at the same time ensuring that the reactant gases and cooling fluids do not mix.

Several versions of flow-field plate have been investigated, although major developers of fuel-cell stacks tend not to disclose their respective preferences. One of the problems commonly encountered is that of gas escape from the perimeter of the cell. This is solved by making the membrane larger than the gas-diffusion backing-plates and inserting Teflon masks in front of the graphite blocks. The entire assembly is then sealed around its perimeter. Any carbon monoxide in the fuel gas would readily poison the platinum electrocatalyst; therefore, to avoid deterioration in negative-plate performance, the concentration of this impurity should be kept to below 10 ppmv by means of a gas-cleaning process. Consequently, there is a strong incentive to develop electrocatalysts that are both cheaper and less susceptible to poisoning.

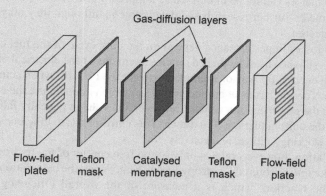

FIGURE 8.12

Schematic of a single-cell construction of a PEMFC.

Remarkable progress has been made with PEMFCs that employ platinum electrodes. In the late 1980s, cell stacks with a platinum loading of 28 mg cm^{-2} delivered a power density of about 100 W L^{-1}. Today, stacks using less than 0.2 mg cm^{-2} of platinum per electrode achieve over 1 kW L^{-1}, i.e. a ten-fold increase in power with only around 1/140th as much electrocatalyst.

Despite these advances, one of the aims of continuing research on PEMFCs is to reduce the loading of the expensive platinum electrocatalyst still further, particularly for use in electric vehicles. Modern cells can operate with a total loading (both electrodes) of 0.4 mg cm^{-2}. The target set by the US FreedomCAR Partnership is to reduce this *combined* amount (i.e. both electrodes) to 0.2 mg cm^{-2} by 2015. Nonetheless, even at the present level, the cost of platinum is no longer a major obstacle for many applications of these fuel cells. It should also be appreciated that the precious-metal catalyst can be recovered from spent fuel cells and recycled. This would be important if, in a future scenario, the majority of road vehicles were powered by PEMFCs, since the world inventory of accessible platinum might then prove to be inadequate.

8.7.3 Fuel cell efficiency

The efficiency of a fuel cell – the fraction of the energy in the fuel that is converted into useful output – is a critical issue. Much is made of the fact that fuel cells are not heat engines, so their efficiency is not limited by the Carnot cycle and therefore should be high. This reasoning has driven much of the interest and investment in the technology. The thermodynamic 'theoretical' efficiency, defined as the ratio of electrical energy output to the enthalpy of the fuel combustion reaction, can be above 80% for low-temperature fuel cells. Nevertheless, electrochemical kinetic theory says that this ratio is an upper limit that is only reached at equilibrium when the current is zero. For further explanation see Box 8.2. In practice, the efficiency is lower, particularly at high power outputs (a schematic representation of the cell voltage, and therefore the efficiency, as a function of current density has been given earlier in Figure 8.10). Just how much lower is difficult to calculate and depends on numerous kinetic and other parameters, such as the overpotentials at the electrodes, the ohmic losses in the electrolyte, the occurrence of side-reactions, fuel loss by leakage across the electrolyte, partial fuel usage and parasitic energy consumption by the auxiliary components (fans, blowers, inverters, transformers, etc.). These factors cannot be dismissed as temporary practical impediments that are simply waiting to be overcome by further research and development, although they may be reduced through advances in system design.

Since the various losses in performance depend on the duty cycle of the fuel cell and are therefore difficult to determine, reliance must be placed on measurements of the electrical output as a fraction of the input of hydrogen energy. By way of example, consider the PEMFC when acting as a power source in an electric vehicle. The stack runs at around 100 °C, at which temperature the open-circuit cell voltage is 1.19 V. During the powering of a vehicle, however, the voltage typically falls to between 0.6 and 0.8 V, as dictated by the current that is being drawn. This corresponds to efficiencies of 59 and 50% for the upper and lower working voltage, respectively.

The above calculation, however, relates only to the free energy of the hydrogen combustion process and does not take into account the entropy lost as heat. When the calculation is based on the total enthalpy (ΔH) of the reaction, which is a measure of the overall efficiency of the process, the

---

BOX 8.2 FUEL CELL EFFICIENCY

Under standard conditions (temperature 298.15 K, pressure 101.325 kPa), the thermodynamically reversible voltage (i.e. the standard voltage) for the electrolysis of water is $V° = 1.229V$. Similarly, this is the reversible voltage of a hydrogen fuel cell operating under the same conditions of temperature and pressure. On increasing the temperature, the free energy of formation of water decreases and so, therefore, does the reversible cell voltage. Thus, at 200 °C the reversible voltage of a hydrogen fuel cell is only 1.14 V. The remaining enthalpy of reaction is liberated as heat (the entropy term $T\Delta S$). From this fact, it is apparent that high temperatures favour electrolysers (since the entropy term can be supplied by heat rather than by electricity), but penalize fuel cells – unless the heat liberated by the cell reaction can be gainfully employed, as for instance in a combined heat-and-power (CHP) scheme. In practice, the output voltage of a fuel cell will be considerably less than the reversible voltage on account of losses within the cell stack. Just as in electrolysis, these losses arise from activation and concentration losses at the electrodes, ohmic losses in the electrolyte and non-uniform current distribution across the surface of the electrodes. For fuel cells, inadvertent crossover of fuel from the negative to the positive electrode and incomplete utilization of the hydrogen as it passes through the stack are further problem areas. All these factors conspire to reduce the practical cell voltage to well below the reversible value, typically to between 0.6 and 0.8 V.

Fuel cell efficiency is a topic that has given rise to much confusion in the literature. One measure of efficiency is simply the practical cell-output voltage divided by the thermodynamically reversible voltage at the stated temperature and pressure of operation. This indicates how much free energy is lost by inefficiencies in operating the fuel cell stack, without regard to how much of the enthalpy of the reaction is liberated as heat (the $T\Delta S$ term). For purposes of calculating overall energy efficiencies, it is necessary to compare the electrical output of the cell stack (in joules) with the enthalpy of the cell reaction, which for hydrogen fuel equates to the heat of formation of water. There are two values of this function: (i) a higher heating value (HHV), which corresponds to the product water being present as liquid; (ii) a lower heating value (LHV), which applies when the product water is present as uncondensed vapour or steam. The difference between the two values represents the latent heat of evaporation of water. For hydrogen, the HHV is approximately 18% greater than the LHV. In a low-temperature fuel cell, where the product is liquid water, the HHV should be employed in efficiency calculations, while for high-temperature fuel cells it may be permissible to use the LHV if the product steam is put to good use. Alternatively, the work done by the steam, for instance in driving a turbine, might be added to the electrical output and the combined figure compared with the HHV to calculate the overall efficiency. The resulting value will be 18% lower than that obtained when using the LHV for hydrogen.

In addition to the losses that originate in the cell stack, there are also external inefficiencies to be taken into account. These include electrical losses in compressing the incoming hydrogen and air, and in converting the low-voltage DC output to high-voltage AC. The total effect is a significant reduction in overall system efficiency. Finally, if the fuel cells are to be used to propel electric vehicles, for example, there are also inefficiencies in the electric motors and the drive-train to be considered.

---

corresponding efficiencies are 47 and 40%. In addition, the losses in pumps, heaters, blowers, etc., together with those in the electrical system (inverter, transformer and traction motor), must be deducted. In round numbers, the collective losses in each system can be taken as 10%. Thus, the total energy efficiency of the fuel cell system, from hydrogen to useful electrical output, lies between $0.47 \times 0.9 \times 0.9 = 38\%$ and $0.4 \times 0.9 \times 0.9 = 32\%$. This performance is still well above that of the internal-combustion engine, but nowhere near the theoretical efficiency of around 80% for a fuel cell. Moreover, no account has yet been taken of the energy consumed in transforming the primary fuel, e.g. natural gas, to hydrogen (*v.i*).

8.8 Hydrogen-fuelled road transport

8.8.1 Internal-combustion-engined vehicles

The combustion characteristics of hydrogen in internal-combustion engines are quite different from those of conventional liquid fuels and this necessitates some changes in engine design. In general, hydrogen engines tend to exhibit pre-ignition, backfire and knock, which are phenomena attributable to the low ignition energy and high flame speed of hydrogen compared with petrol. The German engineer Rudolf Erren showed in the 1930s that these malfunctions could be overcome by feeding hydrogen at a slightly increased pressure directly into the combustion chamber, rather than by introducing it with the fuel–air mixture through the carburettor.

The automotive manufacturer BMW has been working with hydrogen engines since 1978. In 2000, the company produced an experimental fleet of 15 dual-fuel BMW 7 Series, see Figure 8.13 cars, each fitted with a 140-L cryogenic storage tank that was capable of holding ~ 8 kg of liquid hydrogen (LH_2) to give the vehicle a cruising range of around 200 km. The cars also had petrol tanks that allowed a further 500 km of travel. A single switch controlled the change over from one fuel to the other;

The BMW *Hydrogen 7* is powered by a 12-cylinder internal-combustion engine that generates 194 kW. The car accelerates from 0 to 60 mph (96 km h^{-1}) in just 9.4 s and runs on LH_2 instead of the compressed hydrogen that is typical for fuel cell vehicles. The top speed is electronically limited to 143 mph (230 km h^{-1}). Because the *Hydrogen 7* has dual-fuel capabilities, it could roll out before the infrastructure for hydrogen fuelling is fully in place. There is doubt, however, whether this vehicle will ever be put into commercial production, even if hydrogen fuel technology reaches the point of economical and 'green' feasibility, and if the infrastructure required is developed. The *Hydrogen 7* uses more fuel than many trucks; it consumes 13.9 L per 100 km for petrol (gasoline) and 50 L per 100 km for hydrogen. The consumption (L per 100 km) and fuel economy (mpg) for both imperial and US gallons are listed in Table 8.2. The difference in fuel consumption is largely due to the difference in volumetric

FIGURE 8.13

Schematic of the BMW *Hydrogen 7* car. The large tank for LH_2 storage is located in the boot

(Image sourced from Wikipedia and available under the Creative Commons License)

Table 8.2 Fuel Consumption for the BMW *Hydrogen 7*

Petrol (gasoline)		Hydrogen	
L per 100 km	**mpg**	**L per 100 km**	**mpg**
13.9	20.3 imperial 16.9 US	50.0	5.6 imperial 4.7 US

energy density, with petrol yielding 34.6 MJ L^{-1} and liquid hydrogen yielding 10.1 MJ L^{-1}. Clearly, the application of hydrogen as a source of energy in an internal-combustion engine is nowhere near as efficient as in a fuel cell.

Ford, in the USA, has produced a supercharged 6.8-L V10 hydrogen engine. In 2007, 30 *E-450* hydrogen shuttle buses, each fitted with the engine, were collectively launched in the USA and Canada.

8.8.2 Fuel cell vehicles

The fuel cell vehicle (FCV) operating on hydrogen, most probably with a PEMFC, is seen by many as the ultimate solution to the increasing energy security and environmental problems that are confronting the road transport sector. The vehicles are effectively hybrids in which the internal-combustion engine is replaced by a fuel cell to provide the continuous source of energy, but this must be coupled with a secondary battery if regenerative braking energy is to be recovered, as shown schematically in Figure 8.14.

At their present stage of development, PEMFC power systems are hugely more expensive than internal-combustion engines– serious technical challenges need to be addressed to reduce materials and manufacturing costs to the point where FCVs become economically competitive. Furthermore, there are the overriding problems associated with the introduction of a hydrogen fuelling infrastructure. Nevertheless, major investment in FCVs is in place in Europe, Japan and the USA. Emphasis has been on the production of urban buses and private cars; all are employing PEMFC systems.

The hydrogen fuel could, in principle, be derived from a hydrocarbon storage tank and processed in a 'reformer' aboard the vehicle. It is has been found to be more practical, however, to supply hydrogen from an outside source to a pressurized storage tank mounted on the vehicle. Consequently, most designers of hybrid fuel cell cars have adopted this methodology; see Figure 8.15. Such vehicles can be refuelled with compressed gas at a rate of 2 kg of hydrogen per minute, but the provision of an infrastructure necessary to supply the hydrogen conveniently, i.e. from stations as widely distributed as are petrol stations today, remains a significant issue.

8.8.2.1 Efficiency considerations and carbon dioxide releases

Although fuel cells discharge no obnoxious emissions on the streets, it must be recognized that the manufacture of the hydrogen fuel on which they depend may well give rise to pollution at source. The lowest-cost means of obtaining hydrogen is from a fossil fuel precursor such as natural gas or coal, and carbon dioxide is formed in both processes. Electrolytically extracting hydrogen from water might appear to be a preferable route, but at existing electricity prices the method is more expensive in most parts of the world than that based on the processing of hydrocarbons. In addition, much electricity generation involves the combustion of fossil fuels so that the hydrogen does

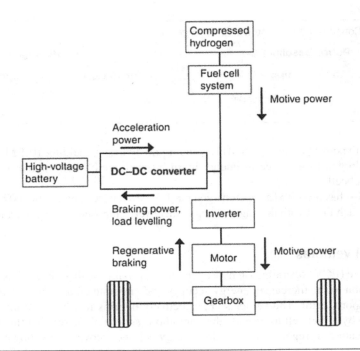

FIGURE 8.14

Example of power-train architecture for a hybrid fuel cell vehicle

(Image from C Hochgraf in Encyclopedia of Electrochemical Power Sources, *Volume 1 page 237 Elsevier copyright)*

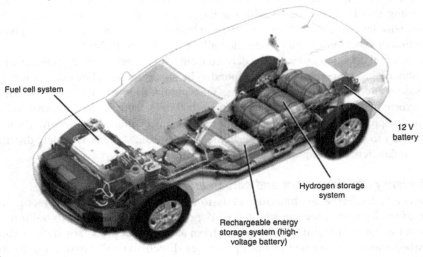

FIGURE 8.15

Schematic of the layout of the different components in a fuel cell vehicle

(Image from C Hochgraf in Encyclopedia of Electrochemical Power Sources, *Volume 1 page 238 Elsevier copyright)*

not come free of concomitant carbon dioxide emissions. Only when the hydrogen is produced using a renewable source of energy can the claim be made that the fuel cell is truly environmentally-friendly.

When comparing different transport fuels, different motive-power units and different vehicles with respect to carbon dioxide releases, it is essential to speak of 'well-to-wheels' efficiency, as described in Section 2.6, Chapter 2. This concept embraces:

(i) all of the energy consumed in extracting the fossil fuel from the ground (e.g. oil and gas from wells, coal from mines) and then refining it, conveying the refined fuel to the service station, and, finally, dispensing into the storage tank of the vehicle – the 'well-to-tank' process;

(ii) the efficiency with which, first, the fuel is combusted in the engine (or fuel cell) and, second, the resulting heat energy (or electrical energy, respectively) is converted into mechanical energy at the wheels (with due regard to the friction losses in the drive-train) – the 'tank-to-wheels' process.

It is well known that internal-combustion-engined vehicles (ICEVs) have rather low on-the-road efficiencies (at present 20 to 25%, at best), but their overall performance is in fact even lower when account is taken of the energy used in the extraction, transport and refining of the oil and then in the delivery of the petrol. Together, these extra losses typically amount to around 13% (i.e. net efficiency of 87%). On multiplying the two values, the overall 'well-to-wheels' efficiency is reduced to 17–22%.

Electrochemical generators, such as fuel cells, are certainly more efficient than internal-combustion engines, but there is considerable confusion over the numerical efficiency in real applications. For instance, use of the higher heating value (HHV) or the lower heating value (LHV) for hydrogen makes a difference of 18%; see Box 8.2. It has been shown above that the practical efficiency of a hydrogen-fuelled PEMFC (based on the HHV of hydrogen and with due consideration of electrical and mechanical losses) is likely to lie in the range of 32 to 38% (see Section 8.7.3), depending upon the current density, with the greatest efficiency at low power outputs.[1] This relationship is precisely the opposite to that of the internal-combustion engine, for which efficiency declines at low power outputs. The dependence of fuel cell performance on current delivered is especially important in vehicle applications. For fuel cells (and also electrolysers), there is a trade-off between size and capital cost on the one hand, and efficiency and running cost on the other. The larger the unit, the higher is the capital cost of construction – but the resulting low current density leads to higher efficiency of operation. For small vehicles, such as cars, where size and capital cost are at a premium, the fuel cell will probably operate at a high current density and, therefore, at a relatively low efficiency. In a hybrid system this will be mitigated to some degree by the use of a battery or electrochemical capacitor when peak power is required. In making the comparison between efficiencies of engines and fuel cells, it is also important to remember that fuel cells consume no energy when the vehicle is held up in traffic and are therefore well-suited to urban use, especially during peak hours. This advantage over an ICEV is reduced when the latter is equipped with stop–start technology; see Section 5.2.1, Chapter 5.

---

[1] Other authors have quoted somewhat higher efficiencies, usually based on the LHV of hydrogen and lower power outputs. For instance, the Ballard *Mark 902* fuel cell module was said to deliver a maximum efficiency of 48% (LHV) at partial load. That value equated to 40.6% (HHV). Thermodynamically, the HHV should be used. When electrical and mechanical losses are included, the tank-to-wheels efficiency reduces to around 33%, in line with our stated range.

Hydrogen is simply an energy carrier that has to be produced from either fossil fuels or water. Each of these processes involves the input of energy, which must be included when counting overall efficiency and carbon dioxide emissions. If the hydrogen is derived from fossil fuels there are losses in chemically reforming the feedstock (natural gas or coal); while if it is produced by electrolysis there are losses in generating and transmitting the electricity. The steam reforming of natural gas to hydrogen on a large scale is at best 75–80% efficient. Further energy is consumed and hydrogen is lost in its separation from carbon dioxide, which reduces the efficiency to, say, 70%. It is reasonable to assume a 10% energy loss in compressing the hydrogen, and another 10% in transporting it from the centralized steam reformer to the vehicle-refuelling depot. Thus, the overall efficiency from natural gas to traction effort, via hydrogen, is:

- natural gas to distributed hydrogen: $0.70 \times 0.9 \times 0.9 = 57\%$;
- hydrogen to low-voltage DC traction effort: 32–38% (say, 35%);
- 'well-to-wheels' efficiency (natural gas to traction effort): $0.57 \times 0.35 = 20\%$.

This overall figure of 20% is clearly an approximation. Nevertheless, even if the efficiency of the fuel cell were a few percentage points more than we have estimated, there would be no change to the general conclusion that the well-to-wheels efficiency of FCVs operating on reformed natural gas is not far removed from the 17–22% of present high-performance ICEVs. As discussed above, however, FCVs would offer certain other benefits. For instance, in terms of environmental pollution in urban areas, the replacement of ICEVs by FCVs would provide much better air quality and thus reduce the incidence of asthma and other chest complaints. And, of course, the average motorist is interested only in miles per gallon (or per kg hydrogen, the tank-to-wheels figure), and not with the carbon dioxide evolved in manufacturing the hydrogen. The latter is the concern of climatologists and environmentally-conscious citizens.

What if the hydrogen is produced by electrolysis rather than directly from natural gas? The practical cell voltage for the electrolysis of water is around 1.47 V. The output voltage of a PEMFC lies in the range of 0.6–0.8 V, as determined by the current density. Thus, the electrolyser and fuel-cell combination is likely to be 41–48% efficient (say, 45%). Although this is more than for a high-performance automobile, as emphasized by advocates of FCVs, the losses incurred in producing electricity from primary fuels have yet to be included. The efficiency of a conventional power station lies in the range of 30–35% (coal/nuclear-fired, say 33%) to 55% (combined-cycle gas turbine). The overall efficiency from primary fuel to traction effort is then:

- coal or nuclear plant: $0.33 \times 0.45 \times 0.9 \times 0.9 = 12\%$;
- natural gas plant: $0.55 \times 0.45 \times 0.9 \times 0.9 = 20\%$.

In these calculations, the 10% energy loss in compressing the hydrogen has been retained, while the 10% loss in distributing hydrogen has been replaced with a 10% loss in the electricity-supply system that would result from transmission/distribution, voltage reduction and rectification operations.

The above calculations are only approximate. Nevertheless, from the viewpoint of overall primary energy efficiency and carbon dioxide emissions, the analyses show that there is little incentive to replace ICEVs with FCVs, irrespective of whether the hydrogen stems directly from fossil fuels or indirectly via electricity.

Different countries use different mixes of primary energy (coal, gas, nuclear, etc.) to generate their electricity and consequently that electricity releases differing quantities of 'embedded' carbon dioxide, as shown in Figure 8.16. The highest emissions due to FCV use would be in China because most of its

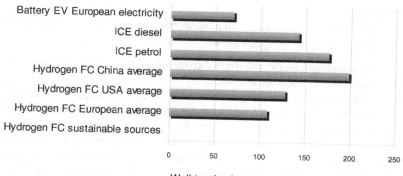

FIGURE 8.16

Regional well-to-wheels emissions of carbon dioxide for C segment (compact/small family) cars with different types of drive-train. EV = electric vehicle; ICE = internal-combustion engine; FC = fuel cell

(Figure created using data from Bernstein Black Book – Global Autos: Don't believe the Hype – Analyzing the Costs & Potential of Fuel-Efficient Technology, *2011, Bernstein Research, in cooporation with Ricardo plc)*

electricity is generated from coal; the average well-to-wheel output is 198 g per km driven, compared with 128 and 108 g per km for the USA and Europe, respectively. If the hydrogen were to be produced electrolytically with renewable sources, then the carbon dioxide liberated by FCVs would be zero. For comparison purposes, Figure 8.16 also presents the average emissions that are attributable to current diesel- and petrol-fuelled vehicles, and to a battery electric vehicle (BEV) in Europe, where, again, the production of electricity from fossil fuels degrades the advantage of an otherwise clean vehicle. Clearly, FCVs will not contribute significantly to the abatement of greenhouse gas emissions unless the carbon dioxide is sequestered centrally and disposed of permanently, or unless electricity from nuclear or renewable sources is used to generate the hydrogen. On the other hand, benefits are to be gained in heavy traffic where FCVs are more efficient than ICEVs and also non-polluting.

There are, however, considerations other than energy efficiency and greenhouse gases that may favour FCVs. Foremost among these, as discussed in Section 2.4, Chapter 2, are the world's diminishing reserves of petroleum, the concentration of the major oil resources in relatively few regions, and the huge import bills that are being incurred by many nations to keep their road transport running. A new infrastructure based on natural gas or coal, rather than on petroleum, holds many attractions, as also does electric propulsion when the range limitation of BEVs has been removed.

Whenever (and wherever) renewable electricity becomes affordable and available on a large scale, the overall energy economy of FCVs will improve. This is because the conversion of mechanical energy (e.g. wind or wave power) to electricity does not involve a Carnot cycle and the efficiency of this primary step should be 80 to 90%, rather than 30 to 55%. Solar conversion efficiencies are, for the foreseeable future, much lower. Of course, the energy losses associated with the electrolyser, the pressurization and distribution of hydrogen, and the fuel cell would remain. The extent to which such a move to renewable electricity is possible will be determined by cost considerations and political acceptability, as well as by technical aspects relating to the feeding of large amounts of renewable electricity into the grid system. So long as supplies of renewable electricity are limited, there is a strong argument for them to be utilized as such, rather than converted to hydrogen.

8.8.2.2 Fuel cell cars

When compared with hybrid electric vehicles that incorporate an internal-combustion engine, fuel cell hybrids offer three principal advantages, as follows:

(i) the fuel cell system itself produces no carbon dioxide or other harmful emissions such as nitrogen oxides, carbon monoxide or particulate matter;

(ii) the fuel cell system offers the potential for around 30% higher energy efficiency. The energy content of 1 US gallon of gasoline (2.76 kg) and 1 kg of hydrogen are similar at 120 MJ, but an FCV achieves around double the km (miles) per kg hydrogen compared with the km (miles) per gallon of gasoline achieved by a standard internal-combustion engine. On the other hand, pressurized gas cylinders occupy a far greater volume than a gasoline tank and it is this volume factor that is likely to determine how much hydrogen can be carried and, consequently, the vehicle range;

(iii) the hydrogen fuel that is consumed can be produced from a variety of renewable sources, including carbon-free methods such as the electrolysis of water.

The hybrid fuel cell vehicle also offers three advantages in comparison with BEVs, as follows:

(i) a vehicle fully charged (with hydrogen) has a greater driving range than BEVs with existing battery technology;

(ii) refuelling is very much faster and thereby enables brief maintenance stops during long journeys compared with battery recharging, which may take hours;

(iii) in cold weather, the fuel cell can be warmed up more rapidly than can a battery and thus can produce full power in a shorter period of time.

The main disadvantages of fuel cells intended for deployment in road vehicles are their present high cost, inadequate life of the fuel cell stack, poor volumetric energy density of the fuel storage system and the lack of a widespread refuelling infrastructure.

The fuel cell system in the car consists of four main subsystems, namely: the fuel supply, the air supply, a cooling system and the fuel cell stack. During operation, the fuel cell produces water and heat as well as electricity. Water is an important product and is used to humidify one, or both, of the gas streams to the cell. The electrolyte membrane must be humidified to a controlled extent to ensure its optimum functioning. Too much humidity would cause water to condense and block the reactant flows, whereas too little humidity would result in the membrane drying out and the proton conductivity becoming inadequate. Fuel cell performance is also likely to be degraded should water at critical locations within the system freeze during operation. Prevention of such an occurrence is challenging because the amount of heat lost to the ambient environment at, say, −30 °C can exceed the amount of heat that is produced when operating at low loads.

Many major automobile manufacturers have active FCV programmes but, to date, most have produced only small numbers of vehicles, principally for evaluation and demonstration purposes. Following work on prototypes that commenced in 1999, Honda became the first company to introduce a passenger car powered by a fuel cell – the *FCX-V4* – and have it certified for commercial use in Japan and the USA. The vehicle was supplied to customers in both countries in December 2002. Subsequently, Honda unveiled the *FCX Clarity* in November 2007 and production began in November 2008; see Figure 8.17(a). The vehicle was made available for lease in Southern California, in Japan and in Europe. Since then, the *FCX Clarity* has been equipped with a radically re-designed fuel cell stack that

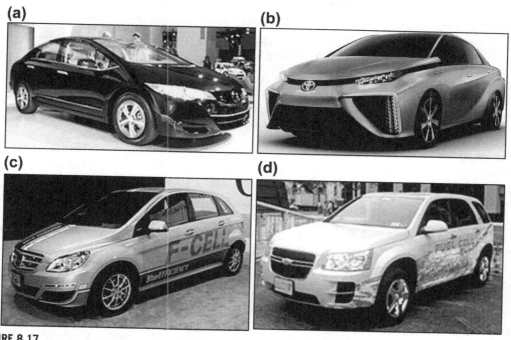

FIGURE 8.17

(a) Honda *FCX Clarity*; (b) Latest (2014) Toyota FCV; (c) Mercedes–Benz *F–Cell*; (d) General Motors fuel cell version of the *Equinox*

(Images sourced from Wikipedia and available under the Creative Commons Licence)

has allowed reductions in volume and weight together with improved stability in the generation of electric power. Key information about the performance of the vehicle is provided in Table 8.3.

The Toyota fuel cell hybrid vehicle (*FCHV*) programme had a limited commercial launch in the USA and Japan in 2002; its specifications are given in Table 8.3. The production of several prototype vehicles led to the *FCHV-adv* ('advanced'), which is mostly based on the Toyota *Highlander* sports utility vehicle (SUV). The fuel is supplied from onboard tanks of compressed hydrogen and the fuel cell is backed up by a nickel–metal-hydride battery wired in parallel. Both the battery and the fuel cell can provide power to the 100-kW driving motor, either singly or together. The mechanism is very similar to the *Hybrid Synergy Drive* in the Toyota *Prius* but with the fuel cell replacing the petrol internal-combustion engine and thereby minimizing carbon dioxide emissions at the point of use. At low speeds, all versions of the *FCHV* can run on battery alone. For high performance, such as when accelerating from rest or overtaking, the fuel cell and battery supply power in tandem. In January 2014, Toyota announced that its latest design — the *FCV Concept*, a temporary name — is due to enter the market in 2015; see Figure 8.17(b). The four-door, mid-size sedan is powered by a 100-kW (3 kW L⁻¹) fuel cell stack and has a range of 480 km (300 miles); it takes about three minutes to refill the twin tanks that store hydrogen at 70 MPa (10 150 psi).

The Mercedes–Benz *F–Cell*, shown in Figure 8.17(c), is a hydrogen fuel cell vehicle concept developed by Daimler AG. Two different versions have been produced: the first, introduced in 2002 with a top speed of 132 km h⁻¹ (82 mph), was a version of the Mercedes–Benz A-Class car; its

Table 8.3 Specifications of Some Fuel Cell Cars under Development by Major Automotive Companies

	Honda FCX Clarity	Mercedes–Benz F-Cell	Toyota FCHV	Chevrolet Equinox FCV
Vehicle class	Mid-size sedan	B-class	SUV	SUV
Motor power (kW)	100	100	90	94
Battery type	Li-ion	Li-ion (1.4 kWh)	Ni–MH (~ 1.5 kWh)	Ni–MH (1.8 kWh)
Average fuel economy (miles per kg H_2)[a]	60	53	68	35 city 45 highway
Range (miles)	280	250	400+	160–200
Top speed (mph)	100	82	100	100
Acceleration 0–60 mph (s)	10	9.8		12

[a] *1 kg of hydrogen is roughly equivalent to one US gallon of gasoline. Hydrogen price can vary over a wide range between US$5 and US$10 per kg.*

successor in 2005 was based on the Mercedes–Benz B-Class. The latter has a more powerful electric engine rated at 100 kW (134 hp) and an increased range of about 400 km (250 miles).

General Motors (GM) chose the Chevrolet *Equinox*, first introduced in 2005 and updated in 2007, as the platform for its initial production run of FCVs; see Figure 8.17(d). It provides room for four people and sufficient space to accommodate three hydrogen tanks filled at 10 000 psi (~ 70 MPa). The vehicle delivers similar performance to the petrol counterpart in terms of acceleration, braking and daily driving. The *Equinox* fuel cell model uses the fourth-generation GM hydrogen technology found in the Chevrolet *Sequel* concept, which was unveiled in 2005. The fuel cell is designed for only 80 000 km (50 000 miles) of driving, but is engineered to be functional in sub-freezing temperatures throughout its life.

The three hydrogen storage tanks give the *Equinox* a range of 320 km (200 miles). During 2007–2008, 115 of the vehicles were built and deployed by GM in several target areas – namely, New York, Washington DC and Los Angeles – as part of a comprehensive program to obtain real-world driving impressions and experiences from short-term loans of the vehicles. In the first year, 3400 drivers drove more than 800 000 km (500 000 miles) and most were generally impressed with the vehicle technology and performance.

Ford has introduced its *Focus* FCV, which is powered by a Ballard 902 fuel cell. It has a top speed of 128 km h^{-1} (80 mph) and has a driving range of 240–320 km (150–200 miles). Cars have been delivered to various cities in the USA, Germany and Canada.

Despite the fact that the performance of FCVs has reached a level that would be generally acceptable to the customer, the twin problems of high cost and very limited refuelling infrastructure remain to be resolved, although both are expected to become less serious in the future. For instance, the cost of a fuel cell car is predicted to reduce dramatically when production numbers reach commercially viable levels through the combined effect of economies of scale, reduction in the number of individual components, simplification of component structure and improvements in the production process. The infrastructure issue is considered briefly in Section 8.8.3.

Despite the current shortcomings of fuel cell cars, considerable efforts are being made within the automotive industry and within academia to realize practical systems for transport and other applications. For instance, in February 2013, Daimler, Ford and Nissan, which together have more than 60 years of cumulative experience in creating FCVs that have logged more than 10 million kilometres worldwide in test drives, signed an agreement to accelerate the commercialization of FCV technology. The objective is to design and produce a common fuel cell stack and system that can be used by each partner. By pooling their resources to optimize design commonality, and by deriving efficiencies via economies of scale, the consortium plans to launch the world's first affordable mass-market FCVs as early as 2017.

8.8.2.3 Fuel cell buses

Currently, hybrid electric buses are proving to be far more acceptable to the managers of transit authorities than are all-electric (i.e. battery) buses, for the following three reasons:

(i)   the extra cost that is associated with the larger batteries that are needed in the pure-battery bus;
(ii) the range limitation of the pure-battery design;
(iii) the lengthy period needed to recharge the battery, during which time the bus is out of commission.

Hybrid electric buses, however, are not zero-emission, even locally. Consequently, there is a perceived need for a bus that is totally clean on the streets and does not have the range restriction that limits the acceptability of the all-electric design. Fuel cell buses could ultimately meet this need, especially since one of the factors that will delay the introduction of fuel cell cars – the cost – is less of a problem for buses, which are frequently heavily subsidized.

The components of the fuel cell system for an electric bus are the same as those for a car, except that a higher-power fuel cell is now needed and the cooling operation is a greater challenge. On the other hand, the refuelling of buses is simpler because they operate from a depot where hydrogen pumps can be installed more readily than the extended network required for private cars. Furthermore, electric buses would neither emit pollutants nor waste energy when stationary in dense traffic.

Urban (transit) buses thus constitute a prime opportunity for the introduction of fuel cells. In addition to the fact that there is adequate internal space and roof area to accommodate the fuel cell stacks and the hydrogen-storage cylinders, buses have other positive features with respect to utilizing fuel cell power. The vehicles usually run on well-defined routes and schedules, are maintained by qualified technicians and only require a single hydrogen-refuelling facility to service a fleet.

A notable demonstration of fuel cell buses was conducted under an initiative known as the Clean Urban Transport for Europe (CUTE) programme that began in November 2001. A total of 27 buses were built and integrated into the public fleets of Amsterdam, Barcelona, Hamburg, London, Luxembourg, Madrid, Porto, Stockholm and Stuttgart. Each city received three buses and these were serviced through the construction of a hydrogen-refuelling station at each location. The buses were produced by DaimlerChrysler and were based on the Mercedes–Benz *Citaro* design, i.e. a full-size, low-floor vehicle that can accommodate up to 70 passengers. A similar project – the Ecological City Transport System (ECTOS) – also funded by the EU, was undertaken in Reykjavik, Iceland, between 2001 and 2005. As discussed in Section 6.4.1, Chapter 6, this country is in an unusual position in having available excess geothermal and hydroelectric power that can be used for generating hydrogen for the buses via electrolysis. Three more *Citaro* hybrid fuel cell buses were also trialled in Perth, Western Australia, from 2004 to 2007. In this demonstration, the hydrogen fuel was produced by BP in its nearby oil refinery at Kwinana. Finally, three buses were delivered to Beijing in November 2005 to enter duty on a busy 19-km route past the Summer Palace.

The CUTE and ECTOS projects were judged to be highly successful. By October 2005, the buses had carried more than four million passengers and covered almost 1.1 million km. The reliability of the buses was high, with an availability rate of over 90%. Nearly 9000 refuelling operations were carried out safely and collectively dispensed almost 200 t of hydrogen.

Subsequently, a new and improved *Citaro FuelCELL Hybrid* bus was introduced, in which the latest fuel cells are hybridized with a lithium-ion battery for energy recovery and storage on braking. High-performance electric motors that provide a continuous power of 120 kW are fitted in the wheel hubs. The overall design allows the bus to travel several kilometres on battery power alone. The fuel cells are expected to achieve an increased durability of at least five years, or 12 000 operating hours. Through improvement in the fuel cell components and hybridization with lithium-ion batteries, the new *Citaro FuelCELL Hybrid* is said to save 50% hydrogen in comparison with its predecessor. Accordingly, the number of tanks has been reduced from the nine necessary for the previously tested fuel cell buses to seven that contain a total of 35 kg of hydrogen. The range of the fuel cell buses is greater than 300 km; one of the buses is shown in Figure 8.18.

The Clean Hydrogen In European Cities (CHIC) project was launched in 2010 to integrate 26 hybrid fuel cell buses from three manufacturers into daily public transport operations and bus routes in five locations across Europe – Aargau (Switzerland), Bolzano/Bozen (Italy), London (UK), Milan (Italy) and Oslo (Norway). The initiative has been part funded by the European Union. London has had a small fleet of five fuel cell buses in commercial operation since early 2011. In July 2012, these passed a milestone of 1000 hydrogen-refuelling events at a service station located at Stratford in close proximity to the Olympic Park. With a tank of hydrogen allowing each of the buses to operate for up to

FIGURE 8.18

Citaro FuelCELL Hybrid bus in Bucharest

(Image sourced from Wikipedia and available under the Creative Commons License)

18 hours a day, the fleet has now also exceeded 100 000 miles of service. The buses offer more than a demonstration of viability and benefits of fuel-cell powered transport to thousands of Londoners; they are a permanent part of Transport for London's fleet, carrying passengers between landmarks such as Tower Bridge, the London Eye and Covent Garden.

In any hybrid system it is, of course, possible to adjust the relative size of the two power sources and the contribution that each makes. In December 2012, General Electric (GE) in the USA demonstrated a concept bus equipped with *two* different batteries and a fuel cell as range-extender. A lithium-ion battery provided power and acceleration, while a high-temperature sodium–nickel-chloride battery (the GE 'Durathon' battery; see Section 7.7.4, Chapter 7) provided better storage capability. The GE proprietary energy-management system is said to have the potential to bring down the cost of fuel cell implementation in buses by up to 50%. Such a prospect clearly emphasizes that improvements in integration and balance of plant are critical to cost reduction.

Fuel cell buses are thus emerging as a form of electric vehicle that can be totally clean on the streets and reliable in use. The combination of battery and fuel cell as the power source gives the hybrid fuel cell bus proven advantages over buses that deploy the individual technologies (battery alone, or fuel cell with a smaller battery). So long as fuel cells remain expensive, and the hydrogen infrastructure sparse, it may make sense to have battery-dominated buses with the fuel cell acting as a range-extender.

8.8.3 Infrastructure: hydrogen highways

The lack of a hydrogen infrastructure, which is signified in some futuristic plans as the 'hydrogen highway', is a major hurdle to the deployment of FCVs. For this new form of road transport to become totally acceptable, refuelling points must be made as readily available as those that service present-day automobiles. The adaptation of existing infrastructure into the plans for the distribution of hydrogen could reduce this problem to some extent. A few industrial areas within Europe – mostly where oil refineries and chemical plants are operating – are linked by a system of pipelines for hydrogen delivery. This arrangement does not, however, constitute a cohesive network for the supply of the gas as a fuel for vehicles. Although sufficient hydrogen could be produced in Europe to cater for the initial needs of FCVs, its extensive and convenient distribution to customers is likely to prove a major challenge. A handful of hydrogen stations are already in operation; most of them are in central Europe and the USA (mainly on the east coast and in California); see Figure 8.19. Plans are in hand for the number of stations to increase rapidly during the next few years. In Japan, for example, it is expected that there will be 800 stations by 2025, and 5000 by 2030.

As mentioned earlier, however, the widespread adoption of FCVs does not necessarily eliminate carbon dioxide emissions from the road transport sector. Just as the amount of carbon dioxide emission associated with the uptake of BEVs depends on the type of fuel used to generate the electricity to charge the batteries, so the emissions that arise from the use of fuel cells depend on the manner in which the hydrogen is generated. As shown in Figure 8.16, the well-to-wheels emissions of carbon dioxide from FCVs can be zero, but only if the hydrogen is generated by electrolysis using electricity derived from renewable sources. The emission of carbon dioxide that arises from the use of FCVs can be as much as 200 g per km driven if carbon-intensive hydrogen production is used.

8.9 Present status and outlook for fuel cell vehicles

The view, during the last decade of the 20th century, that hydrogen fuel cells could be expected to replace the internal-combustion engine, coupled with over-optimistic estimates from within the automotive industry about the timescale over which this could occur, led to heavy investment in fuel cell research and generated considerable political excitement. Large governmental initiatives were launched such as the 'Hydrogen Highway' in California, the Japanese Ministry of Economy, Trade and Industry

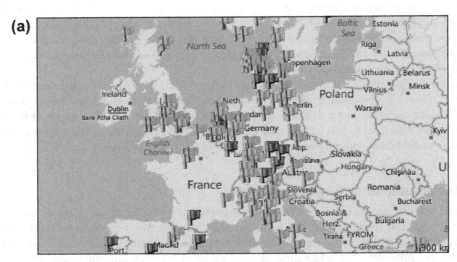

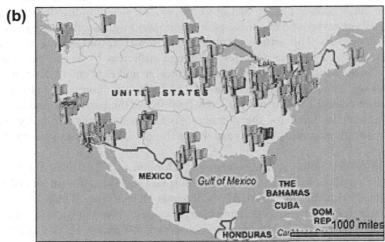

FIGURE 8.19

Hydrogen stations in (a) Europe; and (b) the USA Green flags show hydrogen stations in operation; yellow flags are stations that are planned

(Reproduced courtesy of Ludwig-Bölkow-Systemtechnik GmbH (LBST))

plan to have 50 000 fuel cell vehicles on the road by 2010, and the significant investment in hydrogen research and infrastructure within Europe.

In the event, although there has been considerable progress in the demonstration of the technical feasibility of fuel cell road vehicles, particularly buses, only small numbers of such vehicles have been produced to date and the principal obstacles to a major introduction of this revolutionary new form of propulsion are only slowly being overcome. The creation of an infrastructure for the supply of hydrogen with an adequate distribution of retail outlets will require a massive injection of funding and until it is in place there is little prospect of any legislation to mandate the use of hydrogen. Similarly, the introduction of subsidy schemes to help with the high purchase cost of fuel cell cars is unlikely until the economies brought by high-volume production can be accessed. Fuel cell buses may well become a niche opportunity as the supply of hydrogen from a central depot reduces the hydrogen infrastructure problem to a certain extent, and there is a continuing possibility of subsidies to bus fleets as governments seek to encourage the use of ultra-clean public transport. The widespread introduction of fuel cells in cars and other categories of road transport does not appear imminent, although the combined efforts of major automotive companies to achieve this end are not to be under rated.

The Shape of Things to Come

9.1 Over-arching issues

Earlier chapters are devoted principally to the various technical factors that limit the sustainability of road transport and to the search for means to overcome them. It is recognized, however, that many non-technical factors, which include environmental concerns, politics, economics and social evolution, will also play a role in shaping the way ahead. Foretelling the future with confidence is difficult but a reasonable approach is to outline the possibilities and to indicate the broader framework within which choices will have to be made. This final chapter is a consideration of the manner in which the situation may develop during the decades ahead as the push towards a sustainable system for road transport is pursued.

The invention of the internal-combustion engine had a revolutionary impact on road transport, which experienced a transition from horse-drawn conveyances and steam locomotives to motorized vehicles. The realization is now dawning that the time is approaching for a second paradigm shift that will involve massive disruptions in developed societies. Action is needed because of: (i) the non-sustainability of petroleum fuels, on account of limitations of supply for a burgeoning world population; (ii) growing urban pollution and congestion; (iii) climate change, believed to be caused in large part by greenhouse gases.

Attempts to move to an alternative system that can truly be described as sustainable must be made against a background of major social changes that are in progress around the world. Advances in communications technology are already reducing the need for business travel and this trend will grow along with teleconferencing. Similarly, computer-based face-to-face contact (such as 'Skype' and 'Facetime') will make much journeying unnecessary, since electronic communication is far less expensive and time-consuming. By contrast, leisure travel has steadily increased as society has become wealthier. The distribution of goods and mail by road will always be necessary, although advances in logistics are improving the efficiency of the process. Transitions such as those listed here will dictate both the way and the extent to which people and freight will be transported, particularly in cities.

The motive power source for road vehicles is likely to undergo a major change with the progressive electrification of the drive-train – an evolution that has already begun with the introduction of hybrid electric vehicles (HEVs) and stop–start vehicles (SSVs). The aims of reducing the consumption of petroleum and the emissions of obnoxious combustion products are further advanced by the arrival of plug-in hybrid electric vehicles (PHEVs) and by a growth in the numbers of battery electric vehicles (BEVs). Extension of these developments to large vehicle fleets should, however, be viewed in the context of the availability of electricity and the introduction of appropriate means for its generation, which will vary from country to country and from time to time. A new and important feature in electricity production is the closure, in key European countries, of large power stations that burn coal, on account of the pollutants and greenhouse gases that are emitted. In China and some other coal-rich countries, however, coal-fired utilities continue to be built. The adoption of turbines running on natural gas in plants that employ efficient combined-cycle technology will continue to offer an alternative to coal-based electricity wherever gas is available at acceptable cost. Elsewhere, there is the exploration of 'clean coal' technology, which is based on carbon capture and storage, and also expansion in the generation of electricity from natural, renewable sources. Debate continues over the future of nuclear power. Fission reactors are still being installed in certain countries but nuclear fusion may not be available until the 22nd century.

With the world population set to rise to at least 9 billion (9×10^9) – and possibly more – during the remainder of the 21st century, society is faced with a plethora of problems. Economists place their faith in increasing prosperity – brought about by technical developments and higher productivity – to ensure an acceptable standard of living for all. This outlook does not address the question as to whether there is a limit to growth imposed by resource availability or by environmental degradation. In effect, the world is faced with a 'trilemma', namely:

- how to sustain economic development;
- how to ensure an adequate supply of food, resources and energy for all;
- how to avoid severe environmental degradation, including climate change.

The resolution of this 'trilemma' is vital to the future of humankind and the challenge of providing a sustainable system of road transport represents a central component of the problem.

The global fleet of motor vehicles of all types, including two-wheelers, is now around 1.5 billion (of which 1 billion are cars) and is expected to reach 2 billion shortly after 2020. At present, there is one vehicle for every four or five people on Earth, but the spread across countries and continents is very uneven. In the USA, there is more than one car for every licensed driver, while the European Union (EU) statistic is not far behind. By contrast, the number of vehicles in China, India and the African continent is still modest on a per capita basis; see Table 9.1. In the next few years, however, new patterns of vehicle ownership are expected to evolve. A rapid increase in the number of internal-combustion-engined vehicles (ICEVs) can be anticipated in China and India, given their huge populations, improving economies and standards of living, and the understandable aspirations of the populace for personal mobility to match that of the developed nations. If, for example, the number of vehicles per 1000 people

Table 9.1 Vehicle Ownership, Population and Gross National Income (GNI) Statistics for Selected Countries

Country	Motor Vehicles per 1000 People[a]	Population (Millions)	GNI per Capita[b] (US$)
USA	812	314	48450
Brazil	259	197	10720
Russian Federation	271	143	10400
India	18	1241	1410
China	83	1346	4930

[a]Excluding motorcycles and two-wheeled vehicles.
[b]Statistics for 2011 from the World Bank.

in China and India were to reach 400 (i.e. still only half the level in the USA), it would introduce an additional billion (approximately) vehicles. Such a transformation would probably take several decades of sustained economic growth if increasing vehicle ownership was to follow the historical pattern of that in the USA (see Figure 9.1), but it is quite feasible within a 50-year horizon. In such an event, the output of carbon dioxide from cars worldwide would double and thereby nullify the effects of the abatement legislation that is currently being introduced. Consequently, far more drastic measures than are currently planned would be needed if climate change is to be halted.

The situation is quite different in the developed world, particularly in Europe. On 12 October 2012, Christine Lagarde, the managing director of the International Monetary Fund, in her speech to the organization's annual meeting, warned that, without growth, the future of the global economy would be in jeopardy. Over the previous 50 years, the advances in standards of living in many regions of the world had been spectacular, but during the past decade there has been talk of a 'crisis'. In the words of Ms Lagarde:

Perhaps the greatest roadblock [facing world economies] will be the huge legacy of public debt, which now averages almost 110 percent of GDP for the advanced economies – the highest level since World War II [see Table 9.2]. This leaves governments highly exposed to subtle shifts in confidence (in the financial markets). It also ties their hands, especially as they seek to build the infrastructure of the 21st century while respecting social promises. The needs of rapidly ageing populations add to these pressures.

Big demographic changes are taking place: vast numbers of young people in rising economic regions; greying populations in advanced and emerging economies .... By 2035, Africa will have the world's largest labour force of more than a billion people – more than India and China. But by then, there will also be more than a billion people aged over 65 in the world. Economic power is spreading from west to east .... Emerging markets and developing economies now account for half of world GDP.

The communications and technology innovations are propelling our economies and societies to ever-greater heights ... nearly three billion people [almost half the world population] are now connected to the Internet .... In short, the sands of the global economy are shifting ....

One lesson from history is clear – reducing public debt is incredibly difficult without growth. High debt, in turn, makes it harder to achieve growth.

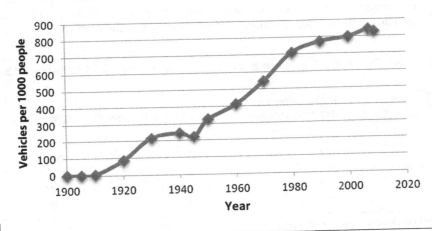

FIGURE 9.1

Increase in vehicle ownership in the USA over time

Table 9.2 List of Countries by Public Debt[a]

Country	Government Debt as % of GDP[b]
Japan	229.8
Greece	160.8
Portugal	106.8
Ireland	105.0
USA	102.9
France	86.3
UK	82.5
Germany	81.5
Spain	68.5
India	68.1
China	25.8

[a]Government gross debt in 2011 – International Monetary Fund, April 2012.
[b]GDP = gross domestic product.

Several European countries are struggling to cope with their national debts and have been faced with excruciating austerity measures in order to service them. As a result, sales of automobiles fell sharply and large car companies were obliged to lay off hundreds of their workers. There can be no more pointed reminder of the need for automotive manufacturers to adhere to their traditional stance of ensuring that their products are available at the lowest possible cost. The EU market for new cars declined each year from 2007 to 2013 and there is every sign that this trend will persist. It is expected that growth in this sector of the economy will remain elusive. Against this background, the downward pressure on car prices will be unrelenting and the budgets available to vehicle designers for the development of more fuel-efficient models will be restricted. As long as economic growth falters the affordability of sustainable road transport will be problematical.

9.2 Global climate change: extent and consequences

The terms 'climate change' and 'global warming' have been employed interchangeably in this book, as is often the case in the media. Strictly speaking this practice is incorrect as climate change can take many forms. Both terms imply gradual variations that take place over years or decades, although the consequences may be more immediate (droughts, floods, hurricanes, etc). It is often difficult to decide whether such events are natural 'weather' or a consequence of anthropogenic global warming. The challenge for energy technology is to slow down and, if possible, to reverse the deleterious effects that result from the combustion of fossil fuels. Success in this endeavour will require a revolutionary change in energy usage, which naturally includes managing the demands of road transport to optimum effect.

Attitudes to the potential consequences of climate change will exercise an influence on the direction in which road transport is developed. In 2012, the World Bank warned that the planet is on course to warm by as much as 4 °C by 2060. Such an increase in average temperature would be expected to give rise to heat waves, severe droughts and major flooding due to rises in sea level. The tropics and subtropics are among the most vulnerable regions and those areas with the world's poorest populations would be hit the hardest. Despite claims by sceptics of climate change that the World Bank's *Turn Down the Heat* report has painted an overly dire picture, the authors of the report view recent extreme meteorological events, such as the devastation caused on the coastlines of New Jersey and New York by Hurricane Sandy, as a 'new norm'. Global temperatures have risen by about 1 °C since 1860 and governments around the world take this observation sufficiently seriously to have committed to an effort that is aimed at restricting the extent of warming to 2 °C. Over the past decade, global warming does seem to have subsided somewhat, although whether this is a 'blip' attributable to other climatic factors remains uncertain.

It has been calculated that a 50% cut in total anthropogenic emissions of carbon dioxide will be required to keep the future increase in temperature to 2 °C. Such a reduction will assuredly prove a major challenge during a period when the number of cars (each of which currently produces an average 2.4 t CO_2 per annum) is predicted to double by 2020 (and possibly to treble by 2050). In fact, a wide range of scenarios and their consequences for the global atmospheric concentration of carbon dioxide have been considered. The expected outcomes for some of the principal future regimes, together with the levels of carbon dioxide at which damage is caused to marine coral reefs through acidification of the oceans, are given in Figure 9.2.

Mathematical modelling of global warming builds on the expected level of carbon dioxide in the atmosphere to predict future variations in mean global temperature. The results of some of the models are shown in Figure 9.3.

Given such forecasts of temperature, various adverse aspects of global weather patterns (rainfall, flooding, storms, etc.) can also be derived and these can lead to serious consequences for food production and price. By way of example, Figure 9.4 shows expected yields of wheat from various regions for

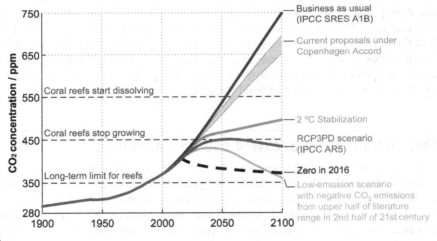

FIGURE 9.2

Atmospheric concentration of carbon dioxide (ppm) to 2100 under a variety of scenarios. IPCC SRES A1B: projection performed under scenario A1B of the Special Report on Emissions Scenarios (SRES) by The Intergovernmental Panel on Climate Change (IPCC), 2007. The Copenhagen Accord, drafted at the United Nations Framework Convention on Climate Change, 18 December 2009. RCP3PD: Representative Concentration Pathway adopted by IPCC to be representative for scenarios leading to very low emission levels, 30 November 2009. IPCC AR5: Fifth Assessment Report of the IPCC, to be finalized in 2014

(Graph used, with permission, from Sir Robert Watson, from monash.edu/research/sustainability)

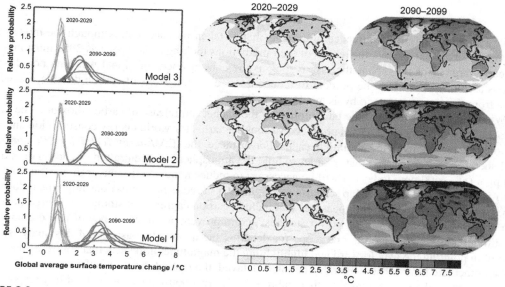

FIGURE 9.3

Results of three alternative models of global warming. Centre and right panels: projected changes in global surface temperature for the early and late 21st century relative to the period 1980–1999. Left panels: corresponding uncertainties as the relative probabilities of estimated global average warming from several different studies for the same periods

(Used, with permission, from Climate Change 2007: The Physical Science Basis. Working Group I: Contribution to the Fourth Assessment Report of the Intergovernmental Panel on Climate Change, Figure SPM.6. Cambridge University Press)

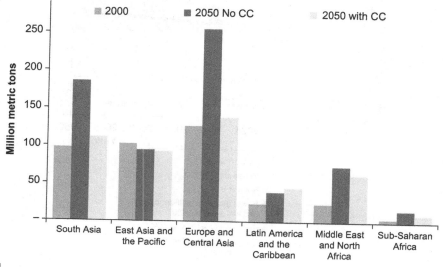

FIGURE 9.4

Impact of climate change (CC) on wheat production for various regions

(Graph used, with permission, from Sir Robert Watson, from monash.edu/research/sustainability)

2050, with and without the influence of climate change taken into account. Although the models are becoming very sophisticated, the results of their predictions looking as far ahead as 2050 must inevitably carry large error bars. Nonetheless, as long as the consequences of global warming could be as serious as forecast, it would be prudent to reduce the overall emissions of carbon dioxide, including those from the transport sector, by as much as possible.

What can be done to avoid the fate implied by the unchecked release of carbon dioxide? The International Energy Agency (IEA) has produced various scenarios for world energy consumption and carbon dioxide emissions in the coming decades. For instance, the IEA *Energy Technology Perspectives* published in 2010 concluded that in order to hold the atmospheric concentration of carbon dioxide at 450 ppm (*cf.*, ~400 ppm today), it will be necessary to reduce annual emissions from the current > 30 Gt to 5–10 Gt. The IEA Blue Map scenario see Figure 9.5 suggests how emissions might be reduced to 14 Gt through a combination of measures that include carbon capture and storage, increased penetration of renewables, more nuclear power stations, and improvements in efficiency of fuel use and fuel substitution. Although there may be scope for debating the quantitative aspects of this analysis, it is helpful in pointing the way forward and highlighting the magnitude of the task that confronts mankind in the immediate future. The longer that action is delayed, the more difficult it will be to achieve a successful outcome. Even if it transpires that anthropogenic greenhouse gases are not entirely responsible for climate change, and that natural phenomena also play a role, there will still be a need for long-term sustainability to moderate and control the consumption of fossil fuels, although the timescale might then not be quite so pressing.

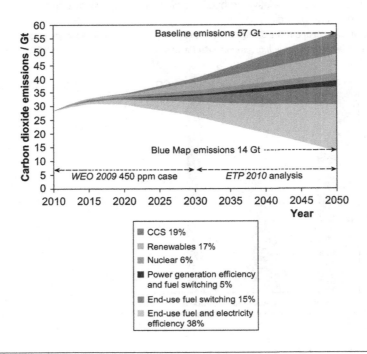

FIGURE 9.5

Key technologies for reducing carbon dioxide emissions under IEA Blue Map scenario

(Graph used, with permission, from Technology Perspectives 2010 International Energy Agency, Paris)

Recently, there has been a growing reaction against both renewable energy and nuclear power on environmental and cost grounds. In the USA, this opposition has been fuelled by the availability of a new source of indigenous and comparatively cheap energy in the form of shale gas produced by 'fracking'. In Europe, industry is concerned that its products will not be competitive with those of nations that have cheap electricity based on fossil fuels. These short-term industrial attitudes are in direct conflict with longer-term goals to abate greenhouse gas emissions.

Future trends in demographics may modify to a significant degree the energy challenge that is facing society. While world population is expected to rise from 7 billion today to some 9 billion in 2050, the urban numbers may more than double (from 3 to 6.5 billion) over the same period. The effect on the consequent evolution of road transport depends very much on the types of city that are involved. Regardless of the standard of living (i.e. gross domestic product, GDP, per capita), the private use of vehicles in cities developed according to American practice (extensive urban sprawl) is greater than that experienced by European counterparts; see Figure 9.6. Public transport is particularly favoured in Tokyo, Amsterdam and Madrid, which are labelled as 'ideal cities'. In several North American cities, 90% of journeys are made by car, while in Hong Kong the proportion is less than 20%. In general, Asian cities are more compact than those in the USA so that the uptake of private cars in the ideal cities of China and India may be less rapid than it has been in the West. Nevertheless, the increase in car ownership in each of these two countries by 2030 will be massive; see Figure 9.7. The development of road transport in cities in general is particularly important because urban travel is approximately twice as energy intense

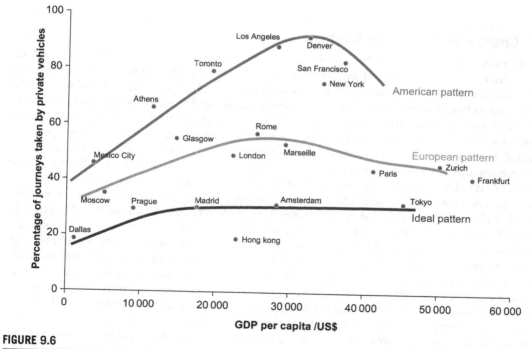

FIGURE 9.6

City design: use of private transport

(Graph used, with permission from Dame Professor J. King, Aston University)

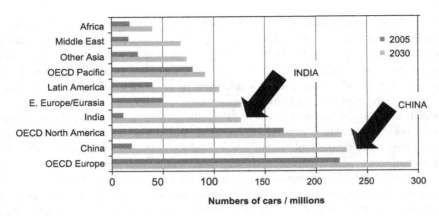

FIGURE 9.7

Numbers of cars by region in 2005 and 2030. OECD = Organisation for Economic Co-operation and Development

(Graph used, with permission from Dame Professor J. King, Aston University)

as inter-city travel. In congested urban areas, 30 to 40% of total fuel is used by cars looking for parking. The average time taken searching for a parking spot is said to be about 8 minutes.

9.3 Choice of vehicle technology

The alternatives to the internal-combustion engine as the source of power for private road vehicles, namely battery electric vehicles (BEVs) and fuel cell vehicles (FCVs), are already technically viable, albeit with serious limitations. Both technologies offer the possibility of relieving the problems associated with the use of hydrocarbon fuels, but neither is ready for wholesale uptake. The BEV is hampered by the inconvenience of restricted range and the time required for battery recharging, while the FCV is costly and has problems with the infrastructure required for the ready supply of hydrogen. Neither technology promotes sustainability unless the electricity and hydrogen, respectively, stem from non-fossil-fuel sources.

A continuation of the status quo is not acceptable, however, as is indicated by the following simple calculation. If today's global emissions of carbon dioxide are labelled as $4X$, then, as stated above, the aim is that emissions in 2050 should be $2X$ in order to meet the target of limiting global temperature rise to 2 °C. Let us assume that road transport takes its proportionate share of this burden. An optimistic view of the reduction that can be achieved by better driver behaviour and improved city design is that today's vehicle emissions could be reduced by 25%, to $3X$. By 2050, however, there could be three times as many cars globally as there are today and this takes the figure back up to $9X$. A 50% reduction due to the improvements in automobile technology still leaves the emissions at $4.5X$. In order to reach $2X$, there will need to be ~ 80% reduction in carbon dioxide emissions from road transport. An increased use of trains, buses, cycling and walking would all play a part, while road vehicles fuelled by hydrogen or electricity (battery or mains) could also help, provided that these two forms of energy are derived from non-fossil sources. An 80% reduction in overall emissions applied to all forms of industrial and domestic activity is extraordinarily demanding. Proposals as to how this all-embracing target might be achieved within the European Union are shown in Figure 9.8.

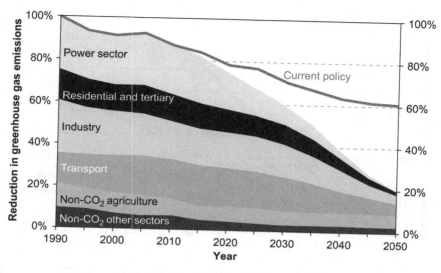

FIGURE 9.8

Pathways to an 80% 'domestic' reduction in EU greenhouse gas emissions by 2050 (100% = 1990)

(Graph used, with permission, from The European Commission, Brussels, 8.3.2011 COM(2011))

By 2020, the average emission of carbon dioxide from new cars could be as low as 95 g per km and this is close to the limit of what can be achieved by any form of internal-combustion engine of adequate power. Vehicles equipped with stop–start technology will help to achieve this goal, particularly if a 48-V system is used to facilitate regenerative braking. Such vehicles may ultimately match the performance of more costly hybrids, but neither will be capable of meeting the carbon dioxide emissions targets for 2050. By that date, it will be necessary for one or more of the fully-electric alternatives, powered by low-carbon electricity, to provide the majority of road transport services. Such a scenario seems hardly credible.

As mentioned above, the extent to which BEVs and hydrogen FCVs may reduce the production of carbon dioxide depends on the means that are used to generate electricity or to produce hydrogen, respectively. The 'effective emissions' for three categories of internal-combustion engine (ICE)-based vehicle are displayed in Figure 9.9 (the horizontal lines), where the ultimate performance achievable by an ICEV is taken to be 95 g CO_2 per km. Also shown are corresponding data for the two classes of electric vehicles (BEV and FCV – the sloping lines), which both depend on the amount of carbon dioxide that is generated during electricity production (here, hydrogen is presumed to be generated by electrolysis). For the UK, the effectiveness of the two types of electric vehicle is dramatically enhanced when the primary energy mix is improved progressively from 500 to 50 g CO_2 per kWh over the period 2010 to 2030 by retiring old coal-fired power stations and replacing them by gas-fired and nuclear utilities (as set out in plans for the electricity sector published by the UK government).

Despite the problem of 'range anxiety' associated with BEVs, and the history of several previous failed attempts to launch the technology, there are some grounds for optimism that small versions will play a role in future urban transport. There is ample evidence that limited range is not relevant for the vast majority of car journeys. The distribution of journey lengths within the UK — as reflected

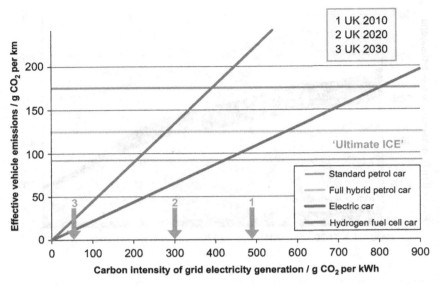

FIGURE 9.9

Emissions from conventional ICE cars, full hybrids and electric cars (BEVs and FCVs) as determined by the method of grid electricity generation

(Image sourced from The Committee on Climate change, modified by Dame Professor J. King)

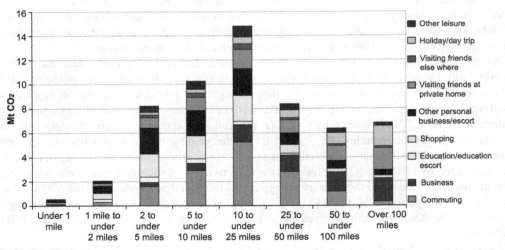

FIGURE 9.10

Carbon dioxide emissions according to journey length

(Graph from The Committee on Climate change)

by carbon dioxide releases — and their several purposes are depicted in Figure 9.10. Other data show that 99% of trips are of less than 100 miles (160 km) and thus could be accomplished in BEVs. Journeys within this range account for 88% of the carbon dioxide that is produced by ICEVs. The situation is not so very different in the USA, where 50% of journeys are of 25 miles per day or less, and 80% are less than 50 miles per day.

Most of the major car companies are prepared to offer BEVs of appropriate design for city use and if sales numbers do increase then the economies of scale may also relieve the problem of uncompetitive first costs. There is an advantage for the uptake of BEVs over FCVs in that the former can operate without the establishment of additional infrastructure to supply the fuel since the electricity distribution network already exists. The continuing rise in the price of oil-based fuels will also encourage the sale of BEVs. Of course, conventional road transport technology will not stand still over the next few decades and the efficiency of ICEVs remains far from optimized.

Several of the major car companies are committed to the development of FCVs in the belief that, on the timescale discussed here, they represent the only viable low-emission option for private travel over long distances (i.e. beyond the range of a single charge of a BEV). Toyota, in particular, anticipates that the technology will be available in 2015 and expanded after 2025. The predictions are for 100 fuelling stations in Japan in 2015 with an increase to 5000 by 2030. Nevertheless, the barriers to widespread uptake, in terms of the capital costs of (i) the vehicles and (ii) the fuelling infrastructure (100 billion Euros in the EU alone, for example), are enormous so that the lead-time may yet be much longer than expected. Deployment in buses will probably be the first application of fuel cells in the road transport sector. The specialist fuelling and servicing of fuel cell electric buses (FCEBs) can be handled by trained personnel at a central depot. In the USA, development projects that involve 25 FCEBs at eight locations are in progress. The current status and future targets for their performance parameters are shown in Table 9.3. The fuel consumption of the FCEB promises to be about half that of an average diesel bus, when making a direct comparison between the energy content of hydrogen and that of diesel fuel, although this may not be too meaningful a metric in terms of cost

Table 9.3 Performance of Fuel Cell Electric Buses in 2012 Compared with US Department of Energy/Federal Transit Administration Targets

Parameter	Units	2012 Status	2016 Target	Ultimate Target
Bus lifetime	Years/miles	5/100000	12/500000	12/500000
Power-plant lifetime	Hours	12000	18000	25000
Bus availability	%	60	85	90
Bus cost	US$	2000000	1000000	600000
Power-plant cost	US$	700000	450000	200000
Hydrogen-storage cost	US$	100000	75000	50000
Operation time	Hours per day / days per week	19/7	20/7	20/7
Maintenance cost	US$ per mile	1.20	0.75	0.40
Range	Miles	270	300	300
Fuel economy	Miles per US gallon (diesel gallon equivalent)[a]	7	8	8

[a]The amount of hydrogen it takes to equal the energy content of one US gallon of diesel.

or carbon dioxide emissions. Also, there remain issues with the lifespan and capital cost of the fuel cell itself to be addressed. Moreover, the mean time between road calls (breakdowns) appears to be worse than for diesel buses, although this is perhaps to be expected for a new technology.

In the next few years, the development of ICEVs may well focus on low-cost stop–start vehicles (SSVs) with engine downsizing supported by turbocharging and electric supercharging. There will also be continued efforts to reduce friction and drag; see Figure 9.11. In the medium-term (~2025), extreme downsizing of SSVs to two- and three-cylinder engines and 48-V systems may eventuate. There will be a move from mechanical function to electrical function for some components such as water pumps, steering and brakes. Such action will lead to significant energy savings. Later still (~2050), attempts may also be made to recoup some of the large amounts of energy that are lost as heat through the exhaust and the coolant systems.

Ultimately, travellers in private cars could even be relieved of the responsibility of driving. Driverless cars have been demonstrated with notable success in a project carried out by Google. In August 2012, the team responsible announced that they had completed over 300 000 autonomous-driving miles (480 000 km) without accident in the USA. This achievement has encouraged the states of Nevada (June 2011), Florida (April 2012) and California (September 2012) to pass laws that will allow the operation of driverless cars on their roads. The driverless car is programmed to keep within the speed limit, which is stored on its global position system (GPS) maps, and to maintain its distance from other vehicles with the aid of a laser radar (LIDAR) sensor system. The hope has been expressed that the increased accuracy of its automated driving system, in comparison with the judgement of a human driver, could help to reduce the number of traffic-related injuries and deaths while using energy and space on roads more efficiently. If, and when, autonomous driving takes over, the fact that the control of a lethal weapon had formerly been entrusted to frail humans would no doubt become a source of wonderment – as is the reaction today to the previous practices of travelling in cars without seatbelts or riding motorcycles without helmets. Many who see driving as a pleasurable recreation or as a sport will deplore the loss of personal control to automation, but if this saves lives then it will be worthwhile.

It is probably too early to estimate what the cost of the driverless system would be in mass production but, at present, the extra equipment carried by the test vehicles has a price tag of US$150 000, of

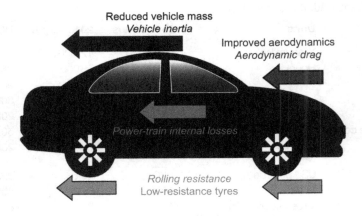

FIGURE 9.11

Energy savings are achievable by appropriate vehicle body design and tyre design

(Image courtesy of Dame Professor J. King)

which US$70 000 represents the cost of the LIDAR system. General Motors has also experimented with a small driverless vehicle named the *EN-V*. It is improbable that there will be an early break-through to enable the introduction of driving-aid technology as a complete package. More realistically, its adoption will be a stepwise process – initially, as a means for maintaining safe distances in heavy traffic to add to the cruise-control and parking-aid systems that already exist. A report from KPMG consultants foresees the first of the vehicles equipped with advanced driving-aid technology to appear in showrooms in 2019 and with self-driving supported from 2025. Nonetheless, the on-cost of such features could seriously restrict progress to the market place.

The road transport of freight, particularly food, from production centres to consumers is a key element for the support of modern society. The vehicles involved in the conveyance of goods account for up to one-third of the production of carbon dioxide in the road sector. Accordingly, much effort is currently being devoted to reducing the future carbon footprint of trucks. Whereas urban, and to a certain extent inter-city, commercial vehicles can be designed to take advantage of hybrid electric and fuel cell technologies, heavy goods vehicles (HGVs) will probably retain the ICE; other strategies will be required to render their use sustainable. In the medium-term, the focus will be on using single wide tyres with low rolling-resistance, aerodynamic trailers and bodies, and waste heat recovery. In the run-up to 2050, however, the emphasis could shift towards sustainable fuels; see Figure 9.12. Bio-diesel, for instance, is said to deliver up to 80% reduction in well-to-wheels emission of greenhouse gases, but doubts about the sustainability of bio-fuels persist. Although fuel cells are not envisaged as replacement power units for

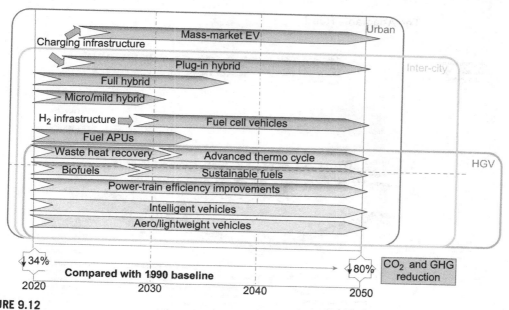

FIGURE 9.12

Long-term measures to reduce carbon dioxide emissions from commercial vehicles. Appropriate measures vary with the operating regime (EV = electric vehicle; GHG = greenhouse gas; APU = auxiliary power unit, a device that provides energy for functions other than propulsion)

(Image courtesy of the UK Automotive Council Technology Group)

ICEs in HGVs for some considerable time, it may be feasible to use them as auxiliary power sources to manage 'hotel loads' when trucks are stationary.

9.4 Roads

In the more distant future, the roads themselves could become high-technology elements. Transport researchers are already looking at the possibility of installing communication and power channels for lane control, traffic monitoring, driver information, and road-condition monitoring. Looking beyond highway monitoring and communication, the Forum of European National Highway Research Laboratories (FEHRL) has established the Forever Open Road programme. This has a vision of a new generation of roads that are adaptable, automated, and resilient to extreme weather events. The concept of the adaptable road, illustrated in Figure 9.13, is based on a prefabricated/modular system that can gradually be implemented across Europe's networks of motorways, as well as those of rural and urban roads. The road will adapt to increasing travel volumes and to changes in demand for public transport, cycling and walking.

The automated road will incorporate a fully-integrated information, monitoring and control system that will enable communication between road users, vehicles and operators. The road will also support a cooperative vehicle-highway system that will manage travel demand and traffic movements. Moreover,

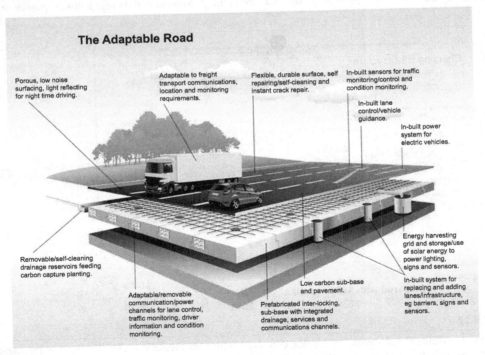

FIGURE 9.13

The adaptable road – a revolutionary concept

(Reproduced with permission. Copyright TRL (Transport Research Laboratory)

it will be capable of measuring, reporting and responding to its own condition and thereby will provide instant information on weather and incidents and travel information. The resilient road will be fully adaptable to extreme weather and conditions and the effects of climate change. It will be able to monitor flooding, snow, ice, wind and temperature change and will be able to adapt to their impacts through integrated storm drainage, and pavement heating and cooling. It will also be linked to an integrated information system for travellers and operators.

At present, the above programme of new-generation roads is little more than a vision and it remains to be seen how practical and cost-effective it will become. Nevertheless, FEHRL claims boldly that:

> *The Forever Open Road will tackle global challenges such as climate change, carbon reduction, energy saving, as well as the increasing need for journey time reliability in response to rising demand both for private car travel and the delivery of goods by road. At the same time, it will help meet European goals to provide transport infrastructure that is safe and secure; cleaner, quieter and more energy efficient; highly competitive and sustainable; provide enhanced mobility based on optimised, efficient seamless systems.*

9.5 Choice of fuel: hydrocarbon, hydrogen or electricity

Because there are alternative liquid fuels that may be used, and varied feedstocks from which they may be derived, the choice of fuels for ICEVs is a multi-faceted issue. The continuation of the use of fossil fuels in the majority of road vehicles over the next 50 years would conflict directly with attempts to restrict the increase in global temperature to 2 °C. A major switch to the use of electricity (mains or battery), and/or to hydrogen in fuel cells, as the source of energy for motive power will be under serious and continued review for the next few decades but such a consideration cannot be divorced from the much broader debate over the future for energy as a whole. Neither the generation of electricity nor the production of hydrogen will eliminate the release of carbon dioxide if those processes depend on fossil fuel feedstocks, unless carbon capture and storage is developed and practised widely and that would, of course, increase the cost of electricity significantly.

The uptake of bio-fuels would allow the continued use of ICEVs while sharply reducing their carbon footprint because it is assumed that all the crops that are consumed will be replaced in a programme of continuous replanting. There are, however, serious doubts that sufficient bio-fuels could be produced to supply the whole of the world vehicle fleet, particularly as there would be competition over agricultural land for fuel production on the one hand and for food production on the other.

The introduction of hydrogen fuel cells has progressed remarkably over recent years, particularly in the context of powering buses. But the persistent twin problems of high cost and insufficient infrastructure for hydrogen supply have to be overcome before there will be a mass market for vehicles powered by such technology.

A factor that is critical to the success of hybrid electric and battery electric vehicles is the performance of batteries. Research and development of traction batteries is still in its comparative infancy. Although the prospects for entirely new electrochemistry may be poor, the scope for developing existing battery chemistries is considerable. The lead–acid battery has been in use for 150 years yet, quite recently, there have been major advances in its design. These include the advanced valve-regulated design with added carbon, multi-tabbed electrodes and the UltraBattery™ which combines the battery with a supercapacitor in a single unit. Such advances are critical for acceptable partial-state-of-charge duty and for efficient high-rate recharge, both of which are needed for use in HEVs. There remain objections to the use of lead in batteries, based on its high mass and also on its environmental impact (recall that cadmium has been phased out of batteries on environmental grounds), but the principal attraction of lead is its low cost.

The behaviour of other battery chemistries when combined with a supercapacitor has yet to be explored and this may offer wide scope for innovation that builds on the successes achieved with the UltraBattery™. A further demonstration of the value of the UltraBattery™ was made in July 2013 when a Honda *Civic* hybrid in which the nickel–metal-hydride battery had been replaced by an Ultra-Battery™ completed 100 000 miles (160 000 km) of real-world driving on the roads of Arizona.

Lithium chemistry holds great attractions because lithium-ion cells have the highest voltage (3–4 V) of any practical secondary cell. This factor, taken together with the low atomic weight of the metal, results in batteries with high specific energy. For this reason, the lithium-ion battery has been employed in the Nissan *Leaf* and several other electric cars. Some early safety problems with lithium batteries were a result of poor voltage control on recharge and these have now largely been resolved. Lithium is comparatively expensive, but further mass production will undoubtedly lead to a reduction in the price of these batteries.

Sodium–metal-halide batteries hold promise for use in fleet vehicles, especially in extreme ambient climates where other chemistries do not perform well.

9.6 The carrot and the stick: role of governments

Governments will have a crucial role to play if the targets for the reduction of carbon dioxide emissions are to be met. Inducements can be designed to encourage both the vehicle manufacturer and the traveller to make decisions that will be compatible with the move towards a sustainable future for road transport, and/or legislation can be enacted that will impose penalties for non-compliance with 'sustainable directives'. Both instruments – the 'carrot' and the 'stick' – must be designed carefully.

Travellers will opt for public transport if a service is frequent, clean and safe. This is already the case for the world's best metro systems, for instance that in Singapore. Inter-city coach travel has been much improved by the updating of central bus stations and by the provision of onboard refreshment and lavatory facilities. Urban bus travel has been made far more tolerable by the provision of weather-proof shelters at bus stops, which are equipped with electronic boards that advise the arrival time of the next bus. Such improved facilities encourage the public to take more advantage of the services provided but often require subsidies from governments or local authorities.

Private car owners can be persuaded to purchase more eco-friendly vehicles if, by so doing, they qualify for a variety of privileges ('carrots'), which may include a subsidy towards the initial purchase cost, low (or zero) road tax, preferential parking, and/or freedom from inner-city congestion charges. Such inducements can assist the introduction of novel technology and they have been employed to support the launch of BEVs in several parts of the world. The practice cannot, however, be continued indefinitely because, once the number of vehicles being purchased becomes large, the cost to the provider of the subsidy (usually a government) becomes too large and the loss of road tax revenue (also to the government) becomes significant. Furthermore, there is a danger that, when the inducements are removed, the sales of the novel vehicles will experience a sudden slump.

The implementation of forceful legislation ('sticks') is also fraught with peril and requires the acceptance, in principle, of all the parties concerned. When the California Air Resources Board mandated that a set fraction of the vehicles sold by all the major car companies should be 'zero emission' (i.e. battery electric), it reckoned without the reluctance of the purchaser. The car companies were aware that they were going to be faced with an impossible task and fought the introduction of the mandates through several stages. Although there are still some proposals on the table, they have been watered down and thereby no longer pose a threat to the ability of the companies to sell their vehicles.

By contrast, the legislation that is appearing in various parts of the world to govern the allowable levels of carbon dioxide emissions from the whole fleet produced by each carmaker has generally found acceptance. This is because the vehicles that conform to the requirement for low emissions also offer good fuel economy to the customer at a time when fuel prices are high. There is, of course, some synergy between the various measures because the same bodies (governments) that are involved in imposing the regulations for carbon dioxide emissions are also responsible for setting taxes on vehicles and fuel.

In 2011, the World Energy Council (WEC) issued a report that outlined in some detail two distinct transport scenarios – the 'freeway' and the 'tollway' – through to 2050. The 'freeway' scenario envisaged a world where pure market forces would prevail and would create a climate for open global competition. The alternative 'tollway' scenario described a more regulated world where governments would intervene in markets to promote technology solutions and infrastructure development that would put common interests (e.g. the environment) at the forefront.

The WEC project involved detailed modelling to produce predictions of the consequences of the two scenarios. The results serve to emphasize the importance of government policies in determining the most probable pathway into the future. By 2050, the total number of cars in the world is expected to have increased from the 2010 datum by a factor of 2.2 or 2.6 according to the 'tollway' or 'freeway' scenarios, respectively. The increases will occur mainly in the developing world. At the end of the review period (2050), it is expected that conventionally-fuelled ICEVs will have a market share of 26% according to 'tollway' (see Figure 9.14(a)), and 78% according to 'freeway' (see Figure 9.14(b)). The balance of the drive-train technologies will comprise a mixture of HEVs, PHEVs and BEVs in 'tollway', and HEVs, PHEVs and gas-fuelled vehicles in 'freeway'.

The total carbon dioxide emissions from cars are expected to increase by about 35% between 2010 and 2050 according to the unfettered ('freeway') scenario. With government intervention ('tollway'), the emission of carbon dioxide from cars should fall by about 46% over the same period. Even this more optimistic scenario fails to achieve the 80% reduction that legislators around the globe are seeking.

9.7 Possible futures

Key issues for the future that will undoubtedly impact on road transport are:

- the attitude of future generations to environmental and sustainability issues and to personal car ownership;
- demographics;
- world economic development;
- further improvements in vehicle technology;
- the choice of fuels that is available;
- the role of government.

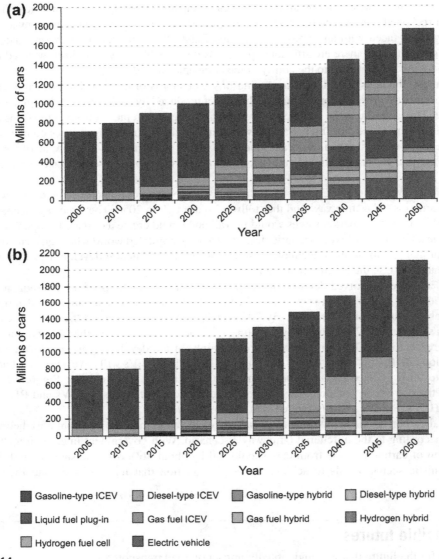

FIGURE 9.14

Technology mix for cars (millions) in the WEC (a) 'tollway'; and (b) 'freeway' scenarios

(Graphs from Global Transport Scenarios 2050, and used with permission from the World Energy Council, London, 2011)

The freedom to travel – personal mobility – is seen today as a basic human right without political restraints and is limited only by the ability to pay the cost involved. Such travel may be for business or for personal purposes (e.g. shopping, recreation, visiting friends and family, vacations), and draws on both public (trains, buses and aircraft) and private (cars) modes of travel. A mix of public and private modes has, so far, been seen as essential.

The continuing migration of populations from the countryside into cities, increased awareness of the consequences of global warming and uncertainty over the strength of national economies that are designed to depend on continuous growth may give rise to significant shifts in attitudes to personal mobility. The use of automobiles on a lease basis, rather than private ownership, is currently increasing and this feature may continue. Public transport will also expand as the numbers of city dwellers increase.

A major proportion of the residents of Manhattan, New York City, are among the few Americans who do not own cars — simply because there is nowhere to keep them. Instead, extensive use is made of taxis and the subway. Car ownership is declining in city centres around the world. Furthermore, there is an increasing reliance on smartphone 'apps' to access a mixed array of travel options, such as bicycle or car rentals, car-sharing and conventional buses. City dwellers no longer automatically associate personal mobility with the need to own a car. The move towards rental systems makes the introduction of BEVs in cities far easier, not least because the risk of purchasing an unfamiliar type of vehicle is borne by the rental company.

Car sharing companies are becoming increasingly popular. As the name suggests, this enterprise operates on the principle that cars are hired in an urban area by picking them up from selected spots and paying for their use, before returning them to designated areas. Prices are structured on an hourly or a per-day basis. Car sharing had its beginnings in Europe about 30 years ago, but has only really become commercially practical in the last decade.

Rapid transit systems may also contribute to reducing the need for private vehicle ownership. There are three categories: bus rapid transit (BRT), personal rapid transit (PRT) and group rapid transit (GRT). Bus rapid transit is used mainly in Brazil and China, where fleets operate on dedicated roads or road sections. This allows buses to run with high frequency and minimum delays. At present, however, BRT is competing with an increasing number of cars for road space. Personal rapid transit (also called the 'pod-car') features small, automated vehicles that accommodate a maximum of six persons and operate on a network of purpose-built guideways. Currently, PRT systems are operating at Morgantown (USA), Masdar City (Abu Dhabi) and London Heathrow (UK). The three facilities are all in the test phase, and plans are being made to grow them in scale if the pilot trials are successful. Group rapid transit is similar to PRT,

FIGURE 9.15

The i-ROAD, a single seat BEV from Toyota – city vehicle of the future?

(Image sourced from Wikipedia and available under the Creative Commons License)

but with higher occupancy. It can be used in areas of high density with more-or-less average peak-time travel. This type of rapid transit is seen as having a good future for urban passenger movement.

In March of 2014, Toyota introduced the *i-ROAD* (see Figure 9.15), an ultra-compact, three-wheeled, BEV intended for individuals to make short journeys in city streets. A large-scale demonstration of the vehicle is planned for Grenoble, France, at the end of 2014. The small footprint of the vehicle results in impressive maneuverability and will contribute to decreased congestion and hence reduced street-level pollution. 'Active Lean' technology automatically raises and lowers the front wheels to balance the vehicle, even when turning sharply or on uneven or sloped roads.

With a rechargeable lithium-ion battery and a highly efficient electric power-train, the *i-ROAD* has an estimated range of 42 km at a speed of 30 km h^{-1} per charge in 2014 – expected to rise to 50 km by 2016.

Despite all the various developments that are underway, the internal-combustion engine will still feature in most road vehicles for several decades, although there may be a partial transition from conventional to bio-fuels. The emission of greenhouse gases will be reduced through the adoption of smaller cars with more efficient turbo engines. Hybrid vehicles with electrical motors to supplement engines may evolve in various guises and become more affordable, but even 100% adoption of this type of road transport would be insufficient to meet the targets that are being adopted in many parts of the word for reductions in emissions.

Battery electric vehicles, and perhaps plug-in hybrids, will become more popular for urban and commuter use, particularly if their cost falls. Fuel cell vehicles pose major infrastructure issues and may not play a significant role on the world stage before 2050, except where they can be used in bus fleets that operate from central depots. It should be noted, however, that Toyota has a more optimistic outlook and is planning to introduce cars powered by fuel cells on a shorter timescale.

By 2050, most vehicles will need to have an electric component (battery or fuel cell) if the emissions targets set by governments today are to be met. True sustainability will only be ensured when all the energy for propulsion is derived from nuclear or renewable sources.

Two final words of caution:

- in 50 years' time, the technology may be available to provide pollution-free motoring, self-drive cars and adaptable roads, but unless the challenges that are now facing world economies are resolved then none of these wonders will be affordable;
- vehicle choice will remain the prerogative of the customer, who usually disregards heavily lower emissions and fuel savings in favour of reduced first cost.

Glossary of Terms

Absorbed (or absorptive) glass mat A technology used in one design of valve-regulated lead–acid battery. The electrolyte solution is absorbed in a matrix of glass fibres, which immobilizes the electrolyte solution and holds it next to the plates. The material also facilitates the transport of oxygen to the negative plate and acts as a separator. Abbreviated as 'AGM'. See **Gas recombination, Valve-regulated lead–acid battery**.

Acid stratification A phenomenon observed in lead–acid cells. During charging, high-density acid is produced in the plates and, under the action of gravity, tends to fall to the lower part of the cell. Over repeat cycles in which the electrolyte does not move, a vertical concentration gradient will therefore develop. The effect is particularly prevalent in cells that employ a flooded electrolyte. It must not be allowed to become permanent, otherwise it could irreversibly damage the battery through non-uniformity in both current distribution and active-material utilization, as well as irregular corrosion and growth of positive-plate grids. Possible countermeasures include operating the cells in a horizontal position (only appropriate for sealed, valve-regulated designs) and agitation of the electrolyte solution, e.g. by shaking the battery, 'gassing' during extended overcharge, air-bubblers or circulation with pumps.

Activated carbon A form of carbon that has been thermally or chemically treated to modify its pore structure so as to give a high surface-area and characteristic adsorption properties. Also known as 'active carbon' or as 'activated charcoal'. See **Carbon black**.

Activation overpotential The overpotential that results from the restrictions imposed by the kinetics of charge transfer at an electrode.

Active-material The material in a battery that takes part in the cell reaction during charge and discharge processes. The term usually applies to the materials contained within the electrode, hence 'positive (electrode) active-material' and 'negative (electrode) active-material'. Also known (but less commonly) as the 'active mass'. Not applicable to fuel cells, photo-electrochemical cells, or supercapacitors.

Active-material utilization The fraction, usually expressed as a percentage, of the active-material present in a positive or a negative electrode that reacts during discharge before the battery (cell) can no longer deliver the required current at a useful voltage.

Activity A thermodynamic function used in place of concentration in equilibrium constants for reactions that involve non-ideal solutions and gases.

Adiabatic process A process (e.g. expansion of a gas) that takes place without heat entering or leaving the system. In reversible adiabatic expansion: as a gas cools, its internal energy is reduced by the amount of work done by the gas on the environment.

Advanced spark Spark so timed that it occurs before the piston reaches top dead centre. Most modern engines have advanced sparking.

Advanced vehicle technologies The engineering and design processes that lead to vehicles with high energy efficiencies and low emissions; these include direct-injection, hybrid electric, fuel cell and battery-powered vehicle systems.

Aerodynamic drag The resistive force that a vehicle needs to overcome as a result of its motion through the air; the force increases in proportion to the object's frontal area, its drag coefficient and the square of its speed. See **Drag coefficient**.

Air-cooled engine Engine cooled directly by air, which is blown (usually by a fan) over the finned cylinder barrels and cylinder heads.

Alkaline battery (cell) A battery (cell) that has a strong aqueous alkaline electrolyte.

Alternating current Electric current that flows for an interval of time (half-period) in one direction and then for the same time in the opposite direction; the normal waveform is sinusoidal. Alternating current is easier to transmit over long distances than direct current, and it is the form of electricity used in most homes and business. By contrast, see **Direct current**.

Alternator Engine-driven electric generator that produces an alternating current, i.e. in contrast with the dynamo, which produces direct current. The advantage of an alternator is that it gives a higher output than a dynamo at low engine speeds. See **Alternating current, Direct current**.

Ampere-hour efficiency For a secondary battery (cell), the fraction (usually expressed as a percentage) of the electrical charge stored in the battery (cell) by charging that is recoverable during discharging. Inefficiencies arise from current inefficiencies. Also known as 'charge efficiency' or 'coulombic efficiency'.

Anaerobic Any process (usually chemical or biological) that takes place without the presence of air or oxygen.

Angular velocity Quantitative expression of the amount of rotation that a spinning object undergoes per unit time.

Anion Ion in an electrolyte that carries a negative charge and that migrates towards the anode under the influence of a potential gradient. See **Anode, Electrolyte, Ion**.

Anode An electrode at which an oxidation process, i.e. loss of electrons, is occurring. In a secondary battery (cell), the anode is the positive electrode on charge and the negative electrode on discharge. In a fuel cell, the anode is the negative electrode where hydrogen is consumed. During electrolysis, the anode is the positive electrode where oxygen is evolved. See **Battery, Electrolysis, Fuel cell**.

Anthropogenic emission Emission caused, directly or indirectly, by human activities. The emission of sulfur dioxide due to the use of fossil fuels is an example of a direct cause of emissions, and the emission of nitrogen oxides from farmland as a function of fertilizer application is an example of an indirect cause.

Anti-lock braking system A braking system that senses when any of the wheels have locked up – or are about to – and automatically reduces the braking forces to keep the wheels rolling. Commonly called 'ABS', such a system can be used to control all four wheels or only two.

Aqueous battery (cell) A battery (cell) based on an electrolyte dissolved in water.

Aquifer Underground water-bearing porous rock strata that yield economic supplies of water, sometimes heated, to wells or springs.

Aspect ratio Generally the ratio between two dimensions of an object. In tyre terminology, it applies to the unloaded sidewall height of the tyre divided by its overall width. A lower aspect ratio implies a shorter, wider tyre. When used to describe a wing, it is the span of the airfoil (the long dimension perpendicular to the airflow) divided by its chord (the dimension parallel to the airflow).

Austenitic steel Steel alloys based on austenitic iron (γ-phase iron). Austenitic stainless steel contains a maximum of 0.15 wt.% carbon, a minimum of 16 wt.% chromium, and sufficient nickel and/or manganese to retain its elasticity even at cryogenic temperature.

Balance-of-plant Additional components ('auxiliaries') integrated into a comprehensive power-system package to operate and sustain a fuel cell or certain types of battery, e.g. heat-exchangers, fans, pumps.

Barrel A measure of crude oil (petroleum), approximately 159 L.

Battery A multiple of electrochemical cells of the same chemistry, connected in series or in parallel, and housed in a single container. (Note that the term is often used to indicate a single cell, particularly in the case of primary systems.) See **Series connection, Parallel connection**.

Battery conditioning The initial application of charge–discharge cycling, carried out under carefully controlled conditions, after the manufacture of a rechargeable battery to establish full capacity.

Battery cycling Repeated charging and discharging of a secondary battery.

Battery formation During manufacture or installation of a battery, the initial charging process that converts the active-materials into the required species for proper electrochemical operation.

Battery grid The framework of a battery plate that supports the active-material and also serves as the current-collector.

Battery management   Regulation of both charging and discharging conditions, maintenance of battery components, and control of the operating temperature within the appropriate range.

Battery module   A battery unit manufactured as the basic component of a battery pack.

Battery pack   A number of batteries connected together to provide the required power and energy for a given application.

Bevel gears   A set of gears that are each shaped like slices of a cone and thereby allow the axes of the gears to be non-parallel. Bevel gears are used to transmit motion through an angle.

Biofuel   A gaseous, liquid or solid fuel that is derived from a biological source. Biofuel may be in its natural form (e.g. wood, peat) or a commercially produced form (e.g. ethanol from sugarcane residue, diesel fuel from waste vegetable oils).

Biogas   A gaseous fuel of medium energy content and composed of methane and carbon dioxide that results from the anaerobic decomposition of waste matter. Also known as 'anaerobic digester gas'.

Biomass   A collective term used to describe all biologically produced matter at the end of its life that can be used for energy production. This includes both waste matter and crops that are specially grown as a source of energy. Biomass can be solid (e.g. wood, straw), liquid (biofuels) or gaseous (biogases).

Bipolar plate   A dense electronic (but not ionic) conductor that electrically connects the positive electrode in one cell to the negative in the adjacent cell to form a 'bipolar battery'. The cells are series-connected and so allow the voltage to be built up. Bipolar plates are also employed in fuel cells to distribute fuel or air to the electrodes, to remove reaction products, and for heat transfer. Depending on the type of electrochemical cell, the plate may be made out of carbon, metal or a conductive polymer (which may be a carbon-filled composite). Also (but less commonly) known as a 'duplex' electrode, especially in Leclanché batteries.

Bituminous coal   A dense coal (carbon content 45–85 wt.%) that is black, but sometimes dark brown, often with well-defined bands of bright and dull material. It serves primarily as fuel in electricity generation, with substantial quantities also used for heat and power applications in manufacturing and to make coke or coking coal, an essential ingredient in making steel. By contrast, see **Lignite**.

Bottom dead centre   Lowest position a piston reaches in its cylinder bore, at the bottom of its stroke. See **Stroke**.

Breakdown voltage   The voltage at which a discharge occurs across the dielectric in a capacitor. See **Dielectric, Capacitor**.

Busbar   An electrical conductor that makes a common connection between several circuits. Also used to describe: (i) the rigid metallic conductor that connects plates of the same polarity within a battery; (ii) the conductors for an electrical system to which the battery terminals are respectively attached. Often designated simply as a 'bus'.

Calendar-life   Period of time within which a battery can be used before its performance decreases to an unacceptable value.

Cam   Eccentric projection on a shaft for moving another component as the shaft revolves, e.g. on the engine camshaft that opens the inlet and exhaust valves. See **Camshaft**.

Camshaft   A shaft fitted with several cams, whose lobes push on valve lifters to convert rotary motion into linear motion. The opening and closing of the valves in all piston engines are regulated by one or more camshafts.

Capacitance   The ability of a capacitor to store electric charge, measured in farads. See **Capacitor**.

Capacitive current (density)   The current (or current density) flowing through an electrochemical cell that is charging or discharging the electrical double-layer capacitance. This current does not involve any chemical reactions (charge transfer), it only causes accumulation (or removal) of electrical charges on the electrode and in the electrolyte solution near the electrode. There is always some capacitive current flowing when the potential of an electrode is changing, and the capacitive current is generally zero when the potential is constant. Also known as 'non-Faradaic current' or 'double-layer' current. By contrast, see **Faradaic current**.

Capacitor   An electrical device that serves to store electricity or electrical energy. It has three essential parts: two electrical conductors, which are usually metal plates, separated and insulated by the third part called the dielectric. The plates are charged with equal amounts of positive and negative electrical charges, respectively. This is a 'physical' storage of electricity as compared with the 'chemical' storage in a battery. See **Electrolytic capacitor**, **Electrochemical capacitor**.

Capacity   The amount of charge, usually expressed in ampere-hours, that can be withdrawn from a primary or secondary battery under specified conditions; see **Rated capacity**. Alternatively, the charge on a capacitor (measured in farads).

Carbon black   An amorphous form of carbon produced commercially by thermal or oxidative decomposition of hydrocarbons. It has a high surface-area to volume ratio – although this ratio is low compared with activated carbon – and is therefore often used to support electrocatalysts in some types of battery and fuel cell. See **Activated carbon**.

Carbon capture and storage   See **Sequestration**.

Carbon fibre   Threadlike strands of pure carbon that are extremely strong in tension (that is, when pulled) and are reasonably flexible. Carbon fibre can be bound in a matrix of plastic resin by heat, vacuum or pressure to form a composite that is strong and light – and very expensive.

Carbon footprint   A measure of the impact of human activities on the environment in terms of the amount of greenhouse gases produced; expressed as tonnes of carbon dioxide or carbon emitted, usually on a yearly basis.

Carburettor   A device in the fuel system that mixes petrol with air and thereby atomizes the petrol in varying proportions to suit engine operating conditions.

Carnot cycle   The most efficient ('ideal') cycle of operation for a reversible heat engine. It consists of four successive reversible operations, as in the four-stroke internal-combustion engine, namely: isothermal expansion and heat transfer to the system from a high-temperature reservoir, adiabatic expansion, isothermal compression and heat transfer from the system to a low-temperature reservoir, and adiabatic compression that restores the system to its original state. See **Adiabatic process**, **Stroke**.

Carnot efficiency   The maximum efficiency with which thermodynamic work can be produced from thermal energy flowing across a temperature difference.

Carnot theorem   The principle that no heat engine can be more efficient than a reversible engine operating between the same temperatures. It follows that the efficiency of a reversible engine is independent of the working substance and depends only on the temperatures between which is it operating.

Cast-on-strap   A term used to designate that step in the manufacturing process of lead–acid batteries where the plate lugs of several electrodes of the same polarity (parallel connection to increase the capacity) are connected via a strap made of lead alloy. The cast-on-strap joint is formed by introducing the lugs in molten alloy held in a mould and then solidifying the assembly by air cooling.

Catalyst   A substance that increases the rate of a chemical reaction, but that is not itself permanently changed.

Catalytic converter   Often simply called a 'catalyst', this is a stainless-steel canister fitted to a car's exhaust system that contains a thin layer of catalytic material spread over a large area of inert supports. The material used is some combination of platinum, rhodium and palladium; it induces chemical reactions that convert an engine's exhaust omissions into less harmful products. So-called 'three-way catalysts' are particularly efficient; their operation, however, demands very precise combustion control, which can be produced only by a system for feedback control of the fuel–air ratio. See **Feedback fuel–air ratio control**.

Cathode   An electrode at which a reduction process, i.e. gain of electrons, is occurring. In a secondary battery, the cathode is the negative electrode on charge and the positive electrode on discharge. In a fuel cell, the cathode is the positive electrode where oxygen is consumed. During electrolysis, the cathode is the negative electrode where hydrogen is evolved. See **Battery**, **Electrolysis**, **Fuel cell**.

Cation   Ion in an electrolyte that carries a positive charge and that migrates towards the cathode under the influence of a potential gradient. See **Cathode**, **Electrolyte**, **Ion**.

Cell reversal   Inversion of the polarity of the electrodes of the weakest cell (or cells) in a battery, usually as a result of overdischarge, when differences in the capacity of individual cells result in one or more cells reaching complete discharge before the others.

Charge   The supply of electrical energy to a secondary battery for storage as chemical energy.

Charge-acceptance   The ability of a battery in a defined state-of-charge to convert active-material during charge into a form that can be subsequently discharged; it is quantified as the ratio, expressed as a percentage, of the charge usefully accepted at a specified temperature and charge voltage during a small increment of time to the total charge supplied during that time.

Charge efficiency   Same as **Ampere-hour efficiency**.

Charge factor   The inverse of **Ampere-hour efficiency**.

Charge profile   The sequence of current and voltage used to charge a battery.

Charge rate   The current applied to charge a battery to restore its available capacity. See *C*-rate.

Charge retention   The ability of a battery to retain its charge under zero-current conditions.

Chassis   A general term that refers to all of the mechanical parts of a car attached to a structural frame. In cars with unitized construction, the chassis comprises everything but the body of the car.

Clean coal technology   An umbrella term used to describe technologies that aim to reduce the environmental impact of the generation of electrical and mechanical energy from coal. These technologies include: chemical washing of minerals and impurities from the coal; gasification for use with combined-cycle gas and steam turbines; treatment of flue gases with steam to remove sulfur dioxide; carbon dioxide capture and storage; and dewatering of lower-rank coals (brown coals) to improve the calorific quality, and thus the efficiency of the conversion into electricity. See **Combined cycle, Integrated gasification combined-cycle, Sequestration**.

Climate change   A statistically significant change of climate that is attributed directly or indirectly to human activity, which alters the composition of the global atmosphere and is in addition to natural climate variability observed over comparable time periods. Note that climate is usually defined as the 'average weather', which, in turn, means using statistics to describe weather (temperature, precipitation and wind) in terms of the mean and variability over a period of time. The World Meteorological Organization uses periods of 30 years, but periods can be as short as months or as long as tens of thousands of years. See **Global warming**.

Clutch   A mechanical means of connecting and disconnecting the drive from the engine to the gearbox (and hence the road wheels). Most clutches are of the friction type, with the gripping load provided by a spring, or springs, or by centrifugal force.

Coal gas   A fuel gas, which is usually rich in methane, carbon monoxide and hydrogen, produced when coal is heated in the absence of air (so-called destructive distillation or pyrolysis). It is a by-product in the preparation of coke and coal tar. Coal gas was a major source of energy in the late 19th and early 20th centuries and was also known as 'town gas'. The use of this gas declined with the increasing availability of natural gas. See **Pyrolysis**.

Cogeneration   See **Combined heat and power system**.

Coil spring   A bar of resilient metal wound into a spiral that may be compressed or extended without permanent deformation. Coil springs have many automotive applications but are particularly important as suspension springs.

Cold-cranking amps   A performance rating for an automotive battery. Usually defined as the current that the battery can deliver for 30 s while maintaining a terminal voltage greater than or equal to 1.2 V per cell at $-18\,°C$, when the battery is new and fully charged.

Cold start   Start-up of a fuel cell after a shut-down while the system is at ambient temperature.

Combined cycle   A technology to improve the thermal efficiency of a power station that uses natural gas as fuel. The gas is first burnt in a gas turbine and the waste heat contained in the exhaust gases is then recovered and used to raise steam to drive a steam turbine. See **Integrated gasification combined-cycle, Thermal efficiency**.

Combined heat and power system An installation where there is simultaneous generation of power (either electrical or mechanical) and useful heat (e.g. process steam) in a single process. Also known as 'cogeneration'.

Combustion chamber The space within the cylinder when the piston is at the top of its travel – it is formed by the top of the piston and a cavity in the cylinder head. Since most of the combustion of the air–fuel mixture takes place in this space, the design and shape of the chamber can greatly affect the power, fuel efficiency and emissions of the engine.

Composite Any material that consists of two or more components – typically one or more of high strength and one that is an adhesive binder. The most common composite is fibreglass, which consists of thin glass fibres bonded together in a plastic matrix. The structural properties of composites can be altered by controlling the orientation and configuration of the high-strength components.

Compression ratio The ratio between the combined volume of a cylinder and a combustion chamber when the piston is at the bottom of its stroke, and the volume when the piston is at the top of its stroke. The higher the compression ratio, the more mechanical energy an engine can squeeze from its air–fuel mixture. Higher compression ratios, however, also make detonation more likely. See **Stroke**.

Concentration overpotential The potential difference caused by differences in the concentration of the charge-carriers between the bulk solution and the electrode surface. It occurs when the electrochemical reaction is sufficiently rapid to lower the surface concentration of the charge-carriers below that of the bulk solution. The rate of reaction is then dependent on the ability ('mass transfer') of the charge-carriers to reach the electrode surface. Also known as 'mass-transport overpotential', and less commonly as 'diffusion overpotential'.

Connecting rod The metal rod that connects a piston to the crankshaft. See **Crankshaft**.

Coulombic efficiency Same as **Ampere-hour efficiency**.

Coupé A closed car with two side doors and less than 33 cubic feet (0.93 m^3) of rear interior volume, according to measurements based on SAE standard J1100. A two-door car is therefore not necessarily a coupé.

Crankshaft A shaft with one or more cranks, or 'throws', that are coupled by connecting rods to the engine's pistons. Together, the crankshaft and the connecting rods transform the reciprocating (up-and-down) motion of the piston into rotary motion.

C-rate The discharge rate or charge rate, in amperes, that is numerically equal to the rated capacity of a battery in ampere-hours. The charging and discharging current of a cell is often expressed as a multiple of C. (Example: the 0.1 C and 5 C currents for a battery with a rated capacity of 20 Ah are 2 A and 100 A, respectively.) See **Discharge rate**, **Rated capacity**.

Current collector A material included in the electrodes of an electrochemical power source to conduct electrons to or from the active-material. See **Active-material**.

Current density In an electrochemical cell, the current flowing per unit electrode area.

Current efficiency The fraction, usually expressed as a percentage, of the current passing through an electrolytic cell (or an electrode) that accomplishes the desired chemical reaction. Inefficiencies may arise from reactions other than the intended reaction taking place at the electrodes, or side reactions consuming the product. The expected production can be theoretically calculated and compared with the actual production. See **Electrolytic cell**.

Cut-off voltage The selected voltage at which charge or discharge is terminated.

Cycle A single charge–discharge of a battery.

Cycle-life The number of cycles that can be obtained from a battery before it fails to meet selected performance criteria.

Cylinder The round, straight-sided cavity in which the pistons move up and down. Typically made of cast iron and formed as a part of the block.

Cylinder head The aluminium or iron casting that houses the combustion chambers, the intake and exhaust ports, and much or all of the valve-train. The head (or heads, if an engine has more than one bank of cylinders) is always directly above the cylinders.

Cylindrical cell   A cell in which the positive and negative plates are rolled up and placed into a cylindrical container (as opposed to stacking the plates in a prismatic cell design).

Deep discharge   A qualitative term that indicates discharge of a large proportion of the available capacity of a battery.

Depth-of-discharge   The ratio, usually expressed as a percentage, of the ampere-hours discharged from a battery at a given rate to the available capacity under the same specified conditions. Abbreviated as 'DoD'.

Dielectric   A substance (solid, liquid or gas) that is a non-conductor of electricity (i.e. an insulator). An electric field in a dielectric does not give rise to a net flow of electricity. Rather, an applied field causes electrons within the substance to be displaced and, thereby, creates an electric charge on the surface of the substance. The phenomenon is used in capacitors to store charge. See **Capacitor**.

Differential   A special gearbox designed so that the torque it receives is split and delivered to two outputs that can turn at different speeds. This allows the road wheels to be driven at different speeds when the vehicle is proceeding on a curved path. On a corner, the outside wheels cover a greater distance than the inside wheels, and have to travel faster. Differentials within axles are designed to split torque evenly; but when used between the front and rear axles in four-wheel-drive systems (a 'centre differential'), they can be designed to apportion torque unevenly.

Direct current   Electric current that flows in one direction only, although it may have appreciable pulsations in its magnitude. It is the form of electricity produced by electrochemical cells. By contrast, see **Alternating current**.

Disc brakes   Properly called 'calliper disc brakes', a type of brake that consists of a disc that rotates at wheel speed, straddled by a calliper that can squeeze the surfaces of the disc near its periphery. Disc brakes provide a more linear response and operate more efficiently at high temperatures and in wet conditions than drum brakes. By contrast, see **Drum brakes**.

Discharge profile   The current–time sequence used in the discharge of the battery.

Discharge rate   The current at which a battery is discharged. The current is usually expressed in terms of the rated capacity of the battery, C. See **C-rate**. Note that the charge delivered by lead–acid batteries is highly rate dependent and therefore the discharged rate should be expressed as C_X/t, where X is the hour rate and t is the specified discharge time, usually in hours.

Distributed energy   Local power generation, storage and metering/control systems that allow power to be used and managed in a distributed and small-scale manner, thereby placing generation close to load in order to minimize electricity transmission and maximize waste-heat utilization. Also known as 'distributed power generation' or 'embedded generation'.

Double-layer capacity   The charge required to establish the double layer at the electrode | electrolyte solution interface. See **Electrical double layer**.

Drag coefficient   A dimensionless measure of the air resistance (or 'aerodynamic sleekness') of an object. A sleek car has a drag coefficient, or 'C_d', of about 0.30; the value for a square flat plate is 1.98. See **Aerodynamic drag**.

Driveability   The general qualitative evaluation of operating qualities of a power-train that include idle smoothness, cold and hot starting, throttle response, power delivery, and tolerance for altitude changes. See **Power-train**.

Driveshaft   The shaft that transmits power from the transmission to the differential. See **Transmission, Differential**.

Drive-train   The elements of the propulsion system (including engine, transmission, driveshaft and differential) that deliver mechanical energy from the power source to drive the wheels of a given vehicle.

Drum brakes   A type of brake, with an iron casting shaped like a shallow drum, which rotates with the wheel. Curved brake shoes are forced into contact with the inner periphery of the drum to provide braking.

Duty cycle   The operating regime of a battery in terms of rate and time of both charge and discharge, and time in standby mode.

Electric vehicle   In terms of road transport: (i) a vehicle powered solely by an electrochemical power source, such as a battery or a fuel cell. Power assistance may also be provided by a supercapacitor; (ii) a vehicle powered by mains electricity, such as a trolleybus or a tramcar (tram).

Electrical double layer   A region existing at the boundary of two phases and assumed to consist of two oppositely charged layers. One phase carries an excess of negative charge, which is balanced by a positive excess of equal magnitude on the other phase. (For example, a layer of negative ions adsorbed on colloidal particles that attracts a layer of positive ions in the surrounding electrolytic solution.) Also known simply as 'double layer'. See **Double-layer capacity**.

Electrocatalyst   A substance that accelerates the rate of an electrochemical (electrode) reaction, but that is not itself permanently changed.

Electrochemical capacitor   A capacitor that stores charge in the form of ions (rather than electrons), adsorbed on materials of high surface-area. The ions undergo redox reactions during charge and discharge. The device is also known as an 'electrochemical double-layer capacitor', a 'supercapacitor', or an 'ultracapacitor'. See **Capacitor, Electrolytic capacitor, Supercapacitor**.

Electrochemical cell   A device that converts chemical energy into electrical energy by passing a current (reverse flow of electrons) between a positive and a negative electrode, through an ionically-conducting electrolyte phase. During discharge of a battery (or operation of a fuel cell), an oxidation reaction take place at the negative electrode (anode), which yields electrons to the external circuit, and a reduction reaction takes place at the positive electrode (cathode), which takes up electrons from the external circuit. Note that during charging of a secondary battery (or the operation of an electrolytic cell), a reduction reaction takes place at the negative electrode, which is now a cathode, and an oxidation reaction takes at the positive electrode, which is now an anode. The positive and the negative electrodes in the secondary battery keep the same polarity during both charge and discharge. Also described by the archaic terms '**Galvanic cell**' or 'voltaic cell'. See **Battery, Electrolyte, Electrolytic cell, Fuel cell, Negative electrode, Positive electrode**.

Electrode   An electronic conductor that acts as a source or sink of electrons that are involved in electrochemical reactions.

Electrode potential   The voltage developed by a single electrode, either positive or negative; usually related to the standard potential of the hydrogen electrode, which is arbitrarily set at zero volts. The algebraic difference in voltage of any pair of electrodes of opposite polarity equals the cell voltage. See **Standard electrode potential**.

Electrolysis   The production of a chemical reaction by passing a direct electric current through an electrolyte medium.

Electrolyte   A chemical compound that ionizes when dissolved or molten to produce an electrically conductive medium; also solid materials that are conductive due to the movement of ions through voids, or empty crystallographic positions, in their crystal lattice structure, e.g. yttria-stabilized zirconia, used primarily in solid oxide fuel cells. Note that in the case of dissolved materials, it is fundamentally incorrect to refer to the 'electrolyte solution' as the 'electrolyte'. Nevertheless, the former terminology has become common practice.

Electrolytic capacitor   A storage device similar to any other type of electrical capacitor, except that only one of its conducting phases is a metallic plate; the other conducting phase is an electrolyte solution. The dielectric is a (passive) oxide film on the surface of the metal (typically aluminium or tantalum); the latter constitutes one conducting phase of the capacitor. There is also another metal electrode immersed in the solution that serves only as the electrical contact to the solution. Electrolytic capacitors typically have much larger capacitance than classical counterparts because the dielectric is very thin (of the order 10^{-6} cm). See **Dielectric, Electrochemical capacitor**.

Electrolytic cell   An electrochemical cell that consists of a positive and a negative electrode and an electrolyte, through which an externally generated electric current is passed in order to produce an electrochemical reaction. Note that, during charging, a secondary battery essentially acts as an electrolytic cell. Also known as an 'electrolysis cell'. See **Positive electrode, Negative electrode, Electrolyte**.

Electrolyser   An electrochemical plant designed to effect the process of electrolysis.

Electromotive force   An archaic and loose term for the voltage produced by a battery or generator in an electrical circuit or, more precisely, the energy (expressed per unit of charge, i.e. volts) supplied by a source of electric power in driving a unit charge around the circuit. It should be noted that the term 'force' is a misnomer. Nevertheless, the term is so well established that its use continues even though it is incorrect. Abbreviated as 'emf'.

Electron   An elementary particle with negative electric charge of 1.602×10^{-19} coulombs and a mass of 9.109×10^{-31} kg.

Electronegative   A term to describe elements that tend to gain electrons and form negative ions. The halogens (fluorine, chlorine, bromine, iodine, astatine) are typical electronegative elements.

Electropositive   A term to describe elements that tend to lose electrons and form positive ions. The alkali metals (lithium, sodium, potassium, rubidium, caesium, francium) are typical electropositive elements.

Embodied energy   A term often used to describe the amount of energy used to create a product, i.e. the energy required to extract raw materials, manufacture the product and supply it to the point of use. A product that requires large amounts of energy to obtain and process the necessary raw materials, or a product that is transported long distances during processing or to market, will have a high embodied energy level. Also known as 'embedded energy'.

End of life   The stage at which a battery can no longer achieve the required performance criteria.

Endothermic reaction   A chemical reaction during which heat is absorbed. By contrast, see **Exothermic reaction**.

End voltage   Same as **Cut-off voltage**.

Energy   The ability to do work or produce heat (measured in joules, J, and often expressed in Watt-hours, Wh).

Energy crops   Trees and grasses grown specially for use as a fuel or for extracting plant oils or alcohols that may be used as fuels in internal-combustion engines.

Energy density   Stored energy per unit volume, usually expressed in Wh L^{-1}.

Energy efficiency   The ratio of the energy output from a device to the energy input; usually expressed as a percentage. Also known as a 'round-trip efficiency' or as Watt-hour efficiency.

Engine control unit   A computer system that regulates the operation of an engine by monitoring certain engine characteristics (rpm, coolant temperature, intake airflow, etc.) through a network of sensors and then controlling key variables (fuel metering, spark timing, exhaust-gas recirculation, etc.) according to pre-programmed schedules. Abbreviated as 'ECU'.

Enthalpy   A thermodynamic quantity (H) equal to the total energy content of a system when it is at constant pressure. The gain or loss of energy of a system when it reacts at constant pressure is expressed by the change in enthalpy, symbolized by ΔH. When all of the energy change appears as heat (Q), the change in enthalpy is equal to the heat of reaction at constant pressure, i.e. $\Delta H = Q$. The values of ΔH and Q are negative for exothermic reactions (heat evolved from system) and positive for endothermic reactions (heat absorbed by system).

Entropy   A thermodynamic quantity that represents the amount of energy in a system that is no longer available to do useful work. When a closed system undergoes a reversible change, the entropy change (ΔS) equals the energy lost from, or transferred to, the system by heat (Q) divided by the absolute temperature (T) at which this occurs, i.e. $\Delta S = Q/T$. At constant pressure, the amount of heat (Q) is equal to the change in enthalpy (ΔH).

EPA fuel economy   Laboratory tests of fuel economy, administered by the United States Environmental Protection Agency (EPA), that use simulated weight and aerodynamic drag to re-create real driving conditions.

Equalizing charge   A charge regime designed to restore undercharged cells in a battery to the fully charged state without damaging those already fully charged. Also known more simply as 'equalization', or as 'reconditioning'.

Equilibrium potential   See **Reversible potential**.

Equilibrium voltage   See **Reversible voltage**.

Exhaust-gas recirculation   A method of reducing exhaust emissions of NO_x (nitrogen oxides) from an engine by recirculating some of the exhaust gas into the intake manifold. The exhaust gas serves as an inert filler that absorbs heat during the combustion process and thereby reduces the peak temperature reached during combustion. Abbreviated as 'EGR'. See **Intake manifold**.

Exhaust manifold   The network of passages that gathers the exhaust gases from the various exhaust ports of engine and routes them toward the catalysts and silencers (mufflers) of the exhaust system. A manifold with free-flowing passages of a carefully designed configuration, called a 'header', can improve engine breathing.

Exhaust port   The passageway in the cylinder head that leads from the exhaust valves to the **exhaust manifold**.

Exothermic reaction   A chemical reaction during which heat is evolved. By contrast, see **Endothermic reaction**.

Failure mode   A process that results in a battery failing to meet the required performance criteria. There is no recognized standard method of rating or defining exactly when a battery has reached the end of its life.

Faradaic current   The current that is flowing through an electrochemical cell and gives rise to (or is caused by) chemical reactions (charge transfer) at the electrode surfaces. By contrast, see **Capacitive current**.

Faradaic reaction   A heterogeneous charge-transfer reaction occurring at the surface of an **electrode**.

Fast charging   A charging procedure that generally involves the rapid return of energy to a battery at the C-rate or greater. See C-rate.

Feedback fuel–air ratio control   A feature of a computer-controlled fuel system. By using a sensor to measure the oxygen content of the exhaust from an engine, the system maintains the fuel–air ratio very close to the value for chemically-perfect combustion. Such tight control of the fuel–air ratio is mandatory for the proper operation of three-way catalysts. See **Catalytic converter**.

Fischer–Tropsch process   A catalysed chemical reaction in which synthesis gas, a mixture of carbon monoxide and hydrogen, is converted into liquid hydrocarbons of various forms. Also known as 'Fischer–Tropsch synthesis'. See **Synthesis gas**.

Float charging   A constant-voltage charge regime applied over extended periods to maintain a battery in the fully charged state. When the float voltage is applied to a battery, a current known as the 'float current' flows into the battery and exactly cancels the battery's own internal self-discharge current. The float voltage is the ideal maintenance voltage for the battery that maximizes battery life. See **Self-discharge**.

Flooded lead–acid battery   Term used to denote lead–acid batteries with a free, i.e. not immobilized, electrolyte solution. By contrast, see **Valve-regulated lead–acid battery**.

Floorpan   The largest and most important stamped metal part of the car body. Usually assembled from several smaller stampings, the floorpan forms the floor and fixes the dimensions for most of the external and structural panels of the car. It is also the foundation for many of the mechanical parts.

Flywheel   A heavy disc attached to the crankshaft on an engine to increase its rotary inertia, thereby smoothing its power flow. See **Crankshaft**.

Fossil fuels   Carbonaceous deposits (solid, liquid or gaseous) that derive from the decay of vegetable matter over geological time spans.

Fuel cell   An electrochemical device that directly converts the chemical energy of a fuel (often hydrogen) and an oxidant (usually air/oxygen) to low-voltage direct-current electricity, together with the release of heat. Fuel cells differ from most types of battery in that the reactants are not contained within the cell, but are supplied from outside to keep pace with their consumption.

Fuel cell vehicle   An electric-drive vehicle that derives the power for its motor(s) from a fuel cell system; hybrid fuel cell vehicles also obtain propulsion power from a supplementary battery and/or supercapacitor.

Fuel injection   Any system that meters fuel to an engine by measuring its needs and then regulating the fuel flow, by electronic or mechanical means, through a pump and injectors. See **Port fuel injection**, **Throttle-body fuel injection**.

Galvanic cell   An archaic term for an electrochemical cell in which stored chemical energy is converted into electrical energy on demand. See **Electrochemical cell**.

Gasification   A special type of pyrolysis where thermal decomposition takes place in the presence of a small amount of air or oxygen. See **Coal gas**, **Pyrolysis**.

Gasoline   Term used for petrol in the United States of America.

Gas recombination   For lead–acid batteries, this term refers to the reaction of oxygen evolved at the positive electrode on charge with the active-material of the negative electrode, thereby partially discharging this electrode and suppressing hydrogen evolution (note that a hydrogen-recombination cycle is kinetically impossible). Since the negative electrode is simultaneously on charge, the discharge product (lead sulfate) is immediately reduced electrochemically to lead and the chemical balance of the cell is restored. Thus the practice of water maintenance ('topping up') experienced with conventional lead–acid batteries is no longer necessary and the unit may be operated as a sealed unit. Also known as the 'internal oxygen cycle' or 'oxygen recombination'. See **Valve-regulated lead–acid battery**.

Gassing   The evolution of a gaseous product at one or both electrode polarities. The action usually results from local self-discharge or from the electrolysis of water present in the electrolyte solution during charging. See **Self-discharge**.

Gearbox   Device for overcoming the inability of the engine to produce sufficient torque or turning effort at low running speeds. It comprises an assembly of gear sets that serves to multiply the engine torque by various amounts with proportional reductions in the output speed of the output shaft. If, for example, the shaft is made to revolve at half engine speed, its torque will be double that of the engine.

Gel battery   A battery in which the liquid electrolyte is immobilized through the formation of a gel. For alkaline batteries, the gelled medium generally consists of an alkaline aqueous solution and gelling agents such as starch, cellulose and synthetic organic polymers. For lead–acid batteries, the sulfuric acid electrolyte solution is immobilized by the addition of finely-divided silica powder.

Gibbs free energy   The energy liberated or absorbed in a reversible process at constant pressure and constant temperature. Put another way, if the change in energy (ΔG) is positive, it is the minimum thermodynamic work (at constant pressure) needed to drive a chemical reaction (or, if ΔG is negative, it is the maximum work that can be done by the reaction). Thus, the Gibbs free energy is a thermodynamic quantity that can be used to determine if a reaction is spontaneous or not. The change in free energy, ΔG, in a chemical reaction is given by $\Delta G = \Delta H - T\Delta S$, where ΔH is the change in enthalpy and ΔS is the change in entropy. This is known as the 'Gibbs equation'. See **Enthalpy**, **Entropy**.

Global warming   The observed and projected increases in the average temperature of the earth's atmosphere and oceans. See **Climate change**.

Gravimetric efficiency   The ratio of the mass of stored fuel to the mass of the storage device.

Gravimetric energy density   Same as **Specific energy**.

Gravimetric power density   Same as **Specific power**.

Greenhouse effect   The trapping of heat by greenhouse gases which allow incoming solar radiation to pass through the earth's atmosphere, but prevent the escape to outer space of a portion of the outgoing infrared radiation from the surface and lower atmosphere. This process has kept the earth's atmosphere about 33 °C warmer than it would be otherwise; it occurs naturally but may also be enhanced by certain human activities, e.g. the burning of fossil fuels. See **Greenhouse gases**.

Greenhouse gases   Any of the gaseous constituents of the atmosphere, both natural and anthropogenic, that absorb and re-emit radiation at specific wavelengths within the spectrum of infrared radiation emitted by the earth's surface, the atmosphere and clouds. Greenhouse gases include water vapour, carbon dioxide, methane, nitrous oxide, halogenated fluorocarbons, ozone, perfluorinated carbons and hydrofluorocarbons. See **Greenhouse effect**.

Half-shaft   Axle shaft that conveys the drive from the differential gear to one of the road wheels. See **Transmission**.

Handling   A general term used to cover all the aspects of the behaviour of a car that are related to directional control of the vehicle.

Heat-exchanger A device in which heat is transferred from one fluid stream to another without mixing. Heat-exchanger operations are most efficient when the temperature differentials are large.

High-rate partial-state-of-charge operation The discharge and charge of a battery at high currents conducted within a certain state-of-charge window, e.g. 30 to 70%. Abbreviated as 'HRPSoC'. See **State-of-charge**.

Hour rate See **Discharge rate**.

Hybrid electric vehicle A vehicle that derives part of its propulsion power from an internal-combustion engine and part of its propulsion power from an electric motor, or that uses an internal-combustion engine to power a generator to charge a battery that in turn powers one or more electric-drive motors. See **Parallel hybrid electric vehicle**, **Series hybrid electric vehicle**.

Hydrogen economy The concept of an energy system based primarily on the use of hydrogen as an energy carrier and fuel, especially for transport and for distributed power generation. See **Distributed energy**.

IEA Blue Map scenario A scenario produced by the International Energy Agency (IEA) that sets the goal of halving global energy-related carbon dioxide emissions by 2050 (compared with 2005 levels) and examines the least-cost means of achieving that goal through the deployment of existing and new low-carbon technologies.

Insolation A measure of the solar energy incident on a given area over a specified period of time. Usually expressed in $kWh\ m^{-2}$ per day.

Intake manifold The network of passages that directs air or an air–fuel mixture from the throttle-body to the intake ports in the cylinder head. The flow typically proceeds from the throttle-body into the plenum chamber which, in turn, feeds individual tubes, called 'runners', that lead to each intake port. Engine breathing is enhanced if the intake manifold is configured to optimize the pressure pulses in the intake system. See **Plenum chamber**, **Throttle-body**.

Intake port The passageway in a cylinder head that leads from the intake manifold to the intake valve(s). See **Intake manifold**.

Integrated gasification combined-cycle A technology employed in some coal-fired power stations. Instead of feeding powdered coal directly to the boilers to raise steam, it is first converted to gas which is a mixture of carbon monoxide, hydrogen and nitrogen. The gas is then used as fuel for a combined-cycle power plant (gas turbine/steam turbine). See **Clean coal technology**, **Combined cycle**.

Intercalation The reversible insertion of atoms or molecules into a layered compound. For example, lithium can be intercalated between the structural layers of graphite (as occurs during charging of a lithium-ion cell).

Intercooler A heat-exchanger that cools the air (or, in some installations, the intake charge) that has been heated by compression in any type of supercharger. An intercooler resembles a radiator; it houses large passages for the intake flow, and uses either outside air or water directed over it to lower the temperature of the intake flow inside. See **Supercharger**.

Internal resistance The opposition to current flow that results from the various electronic and ionic resistances within an electrochemical cell. See **Electrochemical cell**.

Internal short circuit Same as **Short circuit**.

Inverter An electronic device that converts direct current to alternating current.

Ion An atom that has lost or gained one or more orbiting electrons, and thus becomes electrically charged.

Ion-exchange membrane A plastic film formed from ion-exchange resin. The utility of such membranes is based on the fact that they are permeable preferentially only to either positive ions (cation-exchange membrane) or negative ions (anion-exchange membrane).

IR drop The decrease in battery voltage that results from current flow through the internal resistance. Also known as 'ohmic loss', 'ohmic overpotential', or 'resistance overpotential'.

Joule heating Heat solely arising from current flow in a conductor. Also known loosely as 'I^2R heating', where I represents the magnitude of the current (in amperes) and R is the electrical resistance of the conductor (in ohms), but note that the amount of heat liberated is I^2Rt joules, where t is time in seconds.

Knock   A condition in which, after the spark plug fires, some of the unburned air–fuel mixture in the combustion chamber explodes spontaneously ahead of the flame front, set off only by the heat and pressure of the air–fuel mixture that has already been ignited. The resulting high-frequency pressure waves force parts of the engine to vibrate, which, in turn, produces an audible knock. The abnormal combustion (also known as 'detonation', 'pinging', or 'pinking') greatly increases the mechanical and thermal stresses on the engine.

Leaf spring   A long, flat, thin and flexible piece of spring steel or composite material that deflects by bending when forces act upon it. Leaf springs are used primarily in suspensions.

LiDAR (light detection and ranging)   A remote sensing technology that measures distance by illuminating a target with a laser and analysing the reflected light.

Life-cycle analysis   A method for evaluating 'the whole life of a product'. That is, all the stages involved, such as raw materials acquisition; manufacturing; distribution and retail; use and reuse; and maintenance, recycling and waste management, in order to create products that are less environmentally harmful. The process consists of three parts: (i) inventory analysis (selecting items for evaluation and quantitative analysis); (ii) impact analysis (evaluation of impacts on the ecosystem); (iii) improvement analysis (evaluation of measures to reduce environmental loads).

Lignite   The lowest rank of coal (carbon content 25–35 wt.%), used almost exclusively as fuel for electric power generation; also referred to as brown coal. By contrast, see **Bituminous coal**.

Lithium-ion battery   A rechargeable battery in which lithium ions move through an organic-based electrolyte solution from the negative electrode (anode) to the positive electrode (cathode) during discharge, and in reverse, from the positive electrode (anode) to the negative electrode (cathode), during charge. Abbreviated as 'Li-ion battery'.

Load-levelling   A method for reducing the large fluctuations that occur in electricity demand, for example by storing excess electricity during periods of low demand (usually overnight) for use during periods of high demand (usually during the day).

Membrane–electrode assembly   A core component of a fuel-cell structure that consists of a polymer electrolyte membrane coated with catalyst–carbon–binder layers ('electrodes'). The so-called membrane–electrode assembly (MEA) is placed between bipolar plates to form the basic unit of a fuel-cell stack. Electrochemical reactions commence when a fuel (e.g. hydrogen) and an oxidant (e.g. oxygen) are applied, respectively, to the negative electrode (anode) and the positive electrode (cathode) sides of the assembly.

Memory effect   A phenomenon that occurs when a rechargeable battery is repeatedly cycled to less than a full depth-of-discharge and the rest of the capacity becomes inaccessible at normal voltage levels; usually applies to nickel–cadmium cells, but also found with nickel–metal-hydride cells. The memory effect is reversible.

Mid-engine   A chassis layout that positions the engine behind the passenger compartment but ahead of the rear axle.

Monobloc   Same as **Battery module**.

Monocoque   A type of body structure that derives its strength and rigidity from the use of thin, carefully shaped and joined panels, rather than from a framework of thick members. Also called 'unit' or 'unitized' construction.

Monopolar   The conventional method of battery construction in which the component cells are discrete and are externally connected to each other.

Nameplate capacity   Same as **Rated capacity**.

Negative electrode   The electrode in an electrochemical cell that has the lower potential.

Nominal capacity   Same as **Rated capacity**.

Nominal voltage   The voltage of a battery calculated from equilibrium conditions. In practice, this parameter cannot be readily measured, but the open-circuit voltage is a good approximation to the nominal battery voltage for battery systems in which the non-ideal effects are low. See **Open-circuit voltage**.

Non-renewable energy.   Energy that is derived from sources (usually hydrocarbons) that were formed from vegetation by processes that have taken place on geological time scales. By contrast, see **Renewable energy**.

Ohmic loss   Same as **IR drop**. Also known as 'ohmic overpotential', 'resistive loss', 'resistance overpotential'.

Oil   The terms oil and petroleum are used synonymously for crude (unrefined) oil.

Oil sands   A mixture of sand, clay, water and bitumen from which oil may be recovered and refined.

Oil shale   Rocks rich in organic material (kerogen) from which petroleum may be recovered by dry distillation.

Open-circuit voltage   The voltage of a power source, such as a battery, fuel cell or photovoltaic cell, when there is no net current flow.

Otto cycle   Named after its inventor, Nikolaus August Otto (1832–1891), the cycle involves four engine strokes with combustion occurring instantaneously at constant cylinder volume (i.e. at the top of the stroke); the spark-ignition petrol engine operates on an approximation to this cycle. See **Stroke**.

Overcharge   The supply of charge to a battery in excess of that required to return all the active-materials to the fully charged state.

Overdischarge   The discharge of a battery beyond the level specified for correct operation.

Overhead cam   A type of valve-train arrangement in which the camshaft(s) is(are) in the cylinder head. When the camshaft(s) is(are) placed close to the valves, the valve-train components can be stiffer and lighter, thus allowing the valves to open and close more rapidly and the engine to run at higher rpm. In a single-overhead-cam layout, one camshaft actuates all of the valves in a cylinder head. In a double-overhead-camshaft layout, one camshaft actuates the intake valves, and one camshaft operates the exhaust valves.

Overpotential   The shift in the potential of an electrode from its equilibrium value as a result of current flow.

Overvoltage   A voltage that is higher than the normal or predetermined limiting value for a system.

Parallel connection   The connection of like terminals of cells or batteries to form a system of greater capacity, but with the same voltage. By contrast, see **Series connection**.

Parallel hybrid electric vehicle   A type of hybrid electric vehicle in which the alternative power unit is capable of producing motive force and is mechanically linked to the power-train. See **Power-train**. By contrast, see **Series hybrid electric vehicle**.

Parasitic load   The natural self-discharge of a battery; the constant electrical load that is present even when no power demand is being made on the battery by the user. See **Self-discharge**.

Peak power   The sustained pulsed power that is obtainable from a battery under specified conditions; usually measured in watts over a period of 30 seconds.

Petrol   Term used in the United Kingdom for a light hydrocarbon liquid fuel for spark-ignition internal-combustion engines. Other terms for such fuel are 'gasoline', 'gas', and 'motor spirit'.

Petroleum   See **Oil**.

pH   A measure of the acidity/alkalinity (basicity) of a solution. The pH scale extends from 0 to 14 (in aqueous solutions at room temperature). A pH value of 7 indicates a neutral solution. A pH value of less than 7 indicates an acidic solution; the acidity increases with decreasing pH value. A pH value of more than 7 indicates an alkaline solution; the basicity or alkalinity increases with increasing pH value.

Photovoltaic   Relating to or designating devices that absorb solar radiation and transform it directly into electricity. Commonly called a 'solar cell'. See **Solar array**, **Solar cell module**.

Photovoltaic cell   A semiconductor device for converting light energy into low-voltage direct-current electricity.

Plate   A common term for battery electrode.

Plate group   Within a cell, a set of electrodes which are connected in parallel. See **Parallel connection**.

Plenum chamber   A chamber, located between the throttle-body and the runners of an intake manifold, used to distribute the intake charge evenly and to enhance engine breathing. See **Throttle-body**.

Polarity   Denotes the electrode that is positive and the electrode that is negative.

Port fuel injection   A form of fuel injection with at least one injector mounted in the intake port(s) of each cylinder. Usually the injector is positioned on the air intake manifold close to the port. Port fuel injection improves fuel distribution and allows greater flexibility in intake-manifold design, which can contribute to improved engine breathing. By contrast, see **Throttle-body fuel injection**.

Positive electrode   The electrode in an electrochemical cell that has the higher potential.

Power density   The power output of an energy device per unit volume; usually expressed in $W\ L^{-1}$ and often quoted at 80% depth-of-discharge.

Power-train   The elements of a vehicle propulsion system that include all drive-train components plus an electrical power inverter and/or controller, but not the battery or fuel-cell system. See **Drive-train**.

Primary battery (cell)   An electrochemical battery (or cell) that contains a fixed amount of stored energy when manufactured, and that cannot be recharged after that energy is withdrawn. By contrast, see **Secondary battery (cell)**.

Prismatic cell   A cell in which the positive and negative plates are stacked rather than rolled as in a cylindrical cell.

Proton-exchange membrane   The component in a proton-exchange-membrane fuel cell that acts as an electrolyte through which protons, but not electrons, can pass (to move along the electrode and generate a current), as well as a barrier film to separate the hydrogen-rich feed in the negative-electrode compartment of the cell from the oxygen-rich positive-electrode side. Proton-exchange membranes are also employed in certain designs of electrolysis cells. See **Fuel cell**.

Pyrolysis   Thermal decomposition of a substance at elevated temperatures in the absence of air or oxygen.

Rack-and-pinion   A steering mechanism that consists of a gear in mesh with a toothed bar, called a 'rack'. The ends of the rack are linked to the steered wheels with tie rods. When the gear is rotated by the steering shaft, it moves the rack from side to side and thus turns the wheels.

Ragone plot   A graphical illustration of the specific energy of a battery (Wh kg^{-1}) as a function of its specific power (W kg^{-1}).

Rated capacity   The capacity of a battery as specified by the particular manufacturer. It is the minimum expected capacity when a new, but fully formed, battery is discharged under the prescribed conditions.

Rated power   The power capability of a battery as specified by the particular manufacturer.

Rechargeable battery   See **Secondary battery (cell)**.

Rectifier   An electrical device for converting alternating current to direct current.

Regenerative braking   The recovery of some fraction of the energy normally dissipated during braking of a vehicle and its return to a battery or some other energy-storage device. The process of slowing a vehicle involves using the kinetic energy of the vehicle to drive the motor/generator, thereby supplying drag to the vehicle and charge to the battery. Most hybrid electric vehicles employ regenerative braking.

Renewable energy   Forms of energy (such as hydropower, sunlight, tidal energy, wave power and wind) that flow through the earth's biosphere and are available for human use indefinitely, provided that the physical basis of their flow is not destroyed. Also known as 'renewables'. By contrast, see **Non-renewable energy**.

Reserve capacity   The number of minutes that a lead-acid battery can maintain a useful voltage (over 1.75 V per cell) at a constant 25-A discharge rate and at a specified temperature (e.g. 5 °C).

Resistance overpotential   Same as **IR drop**; also known as 'ohmic loss', 'ohmic overpotential', or 'resistive loss'.

Reversible potential   The potential of an electrode when there is no net current flowing through the cell.

Reversible voltage   The difference in the reversible potentials of the two electrodes that make up the cell.

Round-trip efficiency   Same as **Energy efficiency**.

Society of Automotive Engineers   The professional association of transportation-industry engineers that sets most automotive-industry standards for the testing, measuring and designing of automobiles and their components. Abbreviated as 'SAE'.

Secondary battery (cell)   A battery (or cell) that is capable of repeated charging and discharging. Also known as a 'rechargeable battery (or cell)'. By contrast, see **Primary battery (cell)**.

Self-discharge   The loss of capacity of a battery under open-circuit conditions as a result of internal chemical reactions and/or short circuits. The rate of self-discharge is dependent on the type of battery and the ambient temperature. Self-discharge decreases the shelf life of batteries and causes them to have less charge than expected when actually put to use. See **Short circuit**.

Separator   An electronically non-conductive, but ion-permeable, material that prevents electrodes of opposite polarity from making contact.

Sequestration   The capture of carbon dioxide from streams of mixed gases and its subsequent indefinite storage. Also known as 'carbon capture and storage'.

Series connection   The combination of unlike terminals of cells or batteries to form a battery of greater voltage, but with the same capacity. See **String**. By contrast, see **Parallel connection**.

Series hybrid electric vehicle   A type of hybrid electric vehicle that runs on battery power like a pure electric vehicle until the batteries discharge to a set level, at which point an alternative power unit is engaged to recharge the battery. By contrast, see **Parallel hybrid electric vehicle**.

Service-life   The duration over which a battery continues to meet the required performance criteria for a particular application.

Shelf-life   The period over which a battery can be stored and still meet specified performance criteria.

Shock absorber   A device that converts motion into heat, usually by forcing oil through small internal passages in a tubular housing. Used primarily to dampen suspension oscillations, shock absorbers respond to motion. Their effects, therefore, are most obvious in transient manoeuvres.

Short circuit   The direct connection of positive and negative electrodes either internal or external to the battery.

Solar array   An assembly of solar (photovoltaic) cell modules (also called panels) electrically connected together. Also known as a 'photovoltaic array'. See **Solar cell module**.

Solar cell module   Groups of encapsulated solar (photovoltaic) cells framed in glass or plastic units; usually the smallest unit of solar electric equipment available to the customer.

Solid | electrolyte interface (or interphase)   A passivating layer that is formed on a lithium negative electrode when it is in contact with either an organic electrolyte solution liquid or a polymer electrolyte. This layer makes practical lithium batteries possible because lithium metal is too reactive to be stable in contact with the electrolyte media. The layer acts as an interface between the metal and the solution and has the properties of a solid electrolyte. Abbreviated as 'SEI'.

Specific capacity   The capacity output of a battery per unit weight; usually expressed in Ah kg^{-1}.

Specific energy   Stored energy per unit mass, expressed in Wh kg^{-1}.

Specific heat   The quantity of heat that unit mass of a substance requires to raise its temperature by one degree. It is expressed in units of J kg^{-1} K^{-1}.

Specific power   The power output of a battery or a fuel cell per unit weight; usually expressed in W kg^{-1}.

Spiral wound   An electrode structure of high surface-area that is created by winding the electrodes and separator into a spiral 'jellyroll' configuration.

Spoiler   An aerodynamic device that changes the direction of airflow in order to reduce lift or aerodynamic drag and/or to improve engine cooling.

Standard cell voltage   The reversible voltage of an electrochemical cell with all active-materials in their standard states.

Standard electrode potential   The reversible potential of an electrode with all the active-materials in their standard states. Usual standard states specify gases at a pressure of 101.325 kPa and unit activity for elements, solids and 1 mol dm^{-3} solutions — all at a temperature of 298.15 K.

Starting–lighting–ignition battery   A battery in an automobile that is used almost exclusively for the tasks of starting the engine, powering the lights and providing the energy used for fuel ignition. Commonly known as an 'SLI battery'.

State-of-charge   The fraction, usually expressed as a percentage, of the full capacity of a battery that is still available for further discharge, i.e. state-of-charge = [100 − (% depth-of-discharge)]%. Abbreviated as 'SoC'.

State-of-health   The state-of-health of a battery is an arbitrary measure of its condition, compared with its ideal conditions. There is no consensus in the industry on how the state-of-health should be determined. Any of the following parameters (singly or in combination) may be used to derive an arbitrary value: internal resistance, impedance, conductance, capacity, voltage, self-discharge, ability to accept a charge, number of charge–discharge cycles. Abbreviated as 'SoH'.

Steam reforming   The reaction of fossil fuels with steam at high temperature to generate a mixture of hydrogen and carbon monoxide. It is the dominant method of commercial hydrogen production and is based on reacting methane (natural gas) with water; carbon dioxide is formed as a by-product.

Storage-life   Same as **Shelf-life**.

String   A number of cells or batteries connected in series.

Stroke   The distance between the extremes of travel by a piston in a cylinder.

Strut   A suspension element in which a reinforced shock absorber is used as one of the locating members of a wheel; typically by solidly bolting the wheel hub to the bottom end of the strut.

Supercapacitor   Supercapacitors – sometimes called electrochemical capacitors, electrochemical double-layer capacitors (EDLC) or ultracapacitors – do not employ a conventional solid dielectric. They bridge the gap between conventional capacitors and rechargeable batteries. They have the highest available capacitance values per unit volume and the greatest energy density of all capacitors. See **Capacitor, Electrochemical capacitor, Energy density**.

Supercharger   An air compressor used to force more air into an engine than it can inhale on its own. The term is frequently applied only to mechanically driven compressors, but it actually encompasses all varieties of compressors – including turbochargers. See **Turbocharger**.

Sustainability   The ability of energy technologies to meet humanity's needs on an indefinite basis without producing irreversible environmental effects. (Note that various definitions exist in the literature, but they all convey the same principal message.)

Synthesis gas   A mixture of carbon monoxide and hydrogen made by reacting natural gas with steam and air or oxygen. Also known as 'syngas'.

Synthetic natural gas   Methane produced by the catalytic reaction of carbon monoxide with hydrogen or from coal by reaction with hydrogen.

Tapered battery charging   A simple, economical method of battery charging that initially delivers high currents to discharged batteries, then gradually reduces the current as the battery voltage increases and the fully charged state is approached.

Theoretical capacity   The charge output of a battery assuming 100% utilization of the active-materials.

Theoretical specific energy   The energy output of a battery referred to the weight of only the active-materials and a 100% utilization of these materials. See **Active-material**.

Thermal efficiency   For a heat engine, the ratio of the work done by the engine to the mechanical equivalent of the heat supplied in the steam or fuel.

Thermal runaway   A condition under which an uncontrolled increase in temperature occurs and destroys the battery; usually caused by the passage of increasing current (e.g. during constant-voltage charging or short-circuit discharge) as the temperature rises.

Thermolysis   The dissociation or decomposition of a molecule by heat.

Throttle-body   A housing that contains a valve used to regulate the airflow through the intake manifold. The throttle-body is usually located between the air cleaner and the intake plenum. See **Plenum chamber**.

Throttle-body fuel injection   A form of fuel injection in which the injectors are located at the throttle-body of the engine, and thereby fuel is fed to more than one cylinder. Such an arrangement saves money by using fewer injectors; but because it routes both fuel and air through the intake manifold, it eliminates some of the tuning possibilities offered by port fuel injection.

Throughput   The total energy output provided by a battery during its life, i.e. the sum of the energy output during all discharges, usually quoted in terms of the number of 'capacity turnovers' achieved.

Top dead centre   Position of a piston at the top of its stroke in the cylinder. See **Stroke**.

Torque   The rotational equivalent of force, measured in Newton metres (N m) or foot pounds force (ft lbf).

Torque converter   A particular type of fluid coupling with a third element added to the usual input and output turbines. Called the 'stator', this additional element redirects the churning fluid against the output turbine and thereby increases the torque, but at the expense of rpm and efficiency.

Town gas   See **Coal gas**.

Traction battery   A battery designed to provide motive power.

Trafficators   Earlier type of turn indicators with hinged illuminated arms instead of flashing lights.

Transformer   An electrical device that steps up voltage and steps down current (or vice versa). Transformers work with alternating current only.

Transmission   In North America, the gearbox is often referred to as the 'transmission', whereas it is actually only one component of the transmission, which additionally includes the driveshaft, differential and half-shafts.

Three-phase boundary   The three-phase boundary of a fuel cell is the area of contact between the phases necessary for electrochemical reactions at the electrode, namely: ion-conducting phase, electron-conducting phase and gas phase. Also known as 'triple-phase boundary'.

Turbocharger   A supercharger powered by an exhaust-driven turbine. Turbochargers always use centrifugal-flow compressors, which operate efficiently at the high rotational speeds produced by the exhaust turbine.

Turbo lag   Within the operating range of a turbocharger, the lag is the delay between the instant the accelerator in a car is depressed and the time the turbocharged engine develops a large fraction of the power available at that point in its power curve.

Ultracapacitor   See **Supercapacitor**.

Uninterruptible power supply   A unit (e.g. a storage battery) employed to maintain a consistent power supply to an electronic device (e.g. a computer) in the event of an unwanted loss or sag (or a surge) in mains power; used for critical applications such as medical, scientific or military facilities; network servers; and telecommunication systems.

Valence   A number that indicates the combining power of one atom with others. For most elements, the number of bonds can vary. The IUPAC definition limits valence to the maximum number of univalent atoms that may combine with the atom, that is, the maximum number of valence bonds that is possible for the given element.

Valve-regulated lead–acid battery   A sealed design of lead–acid battery that depends on pressure valves opening only under extreme conditions, instead of simple vent caps on the cells, to let gas escape. The technology is based on an internal oxygen cycle and the electrolyte solution is immobilized. See **Absorbed (or absorptive) glass mat**, **Gas recombination**, **Gel battery**.

Voltage   The difference in potential between the two electrodes of a cell, or the two terminals of a battery.

Voltaic efficiency   The ratio, usually expressed as a percentage, of the average voltage during discharge to the average voltage during charge.

Volumetric energy density   Same as **Energy density**.

Volumetric power density   Same as **Power density**.

Watt-hour efficiency   Same as **Energy efficiency**.

Yaw   The rotation about a vertical axis that passes through the centre of gravity of a car.

Index

Printed in the United States
By Bookmasters